Diana10.1
土木工程有限元分析

柴舜 编著

微信扫一扫获取
本书的.py文件

南京大学出版社

图书在版编目(CIP)数据

Diana10.1 土木工程有限元分析 / 柴舜编著. — 南京：南京大学出版社，2018.7
ISBN 978-7-305-20282-7

Ⅰ. ①D… Ⅱ. ①柴… Ⅲ. ①土木工程－有限元分析－应用软件 Ⅳ. ①TU-39

中国版本图书馆 CIP 数据核字(2018)第 111586 号

出版发行	南京大学出版社
社　　址	南京市汉口路 22 号　邮编 210093
出 版 人	金鑫荣
书　　名	Diana10.1 土木工程有限元分析
编　　著	柴　舜
责任编辑	陈兰兰　王南雁　　编辑热线 025-83593923
照　　排	南京理工大学资产经营有限公司
印　　刷	南京大众新科技印刷有限公司
开　　本	787×1092　1/16　印张 25.25　字数 611 千
版　　次	2018 年 7 月第 1 版　2018 年 7 月第 1 次印刷
ISBN	978-7-305-20282-7
定　　价	78.00 元

网　　址：http://www.njupco.com
官方微博：http://weibo.com/njupco
微信服务号：njuyuexue
销售咨询热线：(025)83594756

* 版权所有，侵权必究
* 凡购买南大版图书，如有印装质量问题，请与所购
  图书销售部门联系调换

目 录

前言 ·· I

第一章 Diana 概述 ··· 1

  1.1 背景介绍 ·· 1

  1.2 Diana 软件的主要功能和安装运行 ·· 3

  1.3 Diana 软件常用单元类型 ·· 6

  1.4 Diana10.1 文件系统 ··· 33

  1.5 Diana10.1 图形用户界面 ··· 35

  1.6 Diana 有限元分析流程 ·· 40

  1.7 DianaIE Python 语言 ·· 42

  1.8 Diana 单位 ··· 44

第二章 Diana 材料模型和规范简介 ·· 46

  2.1 Diana 各类材料模型简介 ··· 46

  2.2 混凝土结构的裂缝模型 ·· 56

  2.3 钢筋本构模型 ·· 65

  2.4 Diana 长期性能问题的本构模型介绍 ·· 69

  2.5 Diana 规范简介 ··· 72

第三章 Diana 土木工程非线性建模案例分析 ··· 81

  3.1 案例一:平面预应力框架非线性分析 ··· 81

  3.2 案例二:箱梁体外粘贴钢板加固案例 ·· 98

  3.3 案例三:预应力混凝土连续梁长期性能分析 ································· 138

  3.4 案例四:平面单元钢筋混凝土梁的裂缝分析 ································· 160

  3.5 案例五:门式框架 ·· 175

  3.6 案例六:剪力墙滞回分析 ··· 197

  3.7 案例七:预应力筋粘结滑移模型 ·· 225

  3.8 案例八:钢筋混凝土结构非线性动力分析 ···································· 238

第四章　水化热反应···259
4.1　案例一：管廊节段瞬态水化热分析案例······································259
4.2　案例二：大体积混凝土水化热开裂指数·······································281

第五章　预制节段拼装构件非线性数值分析···316
5.1　案例一：预制拼装混凝土块键齿受剪破坏·····································316
5.2　案例二：预制节段拼装梁阶段性分析··332
5.3　案例三：预制节段拼装梁随机场分析案例·····································368

第六章　Diana不足之处和对后续版本建议··393

附录　Diana10.1快捷操作及默认术语··395

参考文献···396

前 言

　　Diana(Displacement ANAlyzer，又称作 DIANA，为统一形式，本书以 Diana 统称。)是一款同时适用于工程设计单位和科研单位的土木工程有限元非线性分析软件，于上世纪 70 年代诞生于荷兰计算力学部门。目前 Diana 跨越了 8.1 版到 10.2 版，其中 Diana10.1 版同时兼具 DianaIE 和 iDiana 的两种 GUI 界面操作和不同语言类型的命令流操作。截至目前为止，除了软件自带的英文手册之外，国内尚无相关书籍系统介绍该软件的操作。因此本书以 Diana10.1 软件作为平台，基于作者熟悉的土木工程研究领域和多年来使用多个不同版本 Diana 经验，并侧重于当前流行的结构非线性分析，介绍了该软件的单元类型、材料本构类型、规范类型在 GUI 界面操作中的设置以及 Python 语言的编辑，以期能为初学者尽快入门这款软件提供一定帮助。为方便读者学习，在第三章至第五章配有一定数量操作例题和相关 Python 语言的命令流附件。

　　本书作者长期使用 Diana 软件进行有限元非线性分析，对该软件的操作有很多学习心得和体会，这对解决初学者在使用和操作上的疑惑相信会有一定的帮助。正如前文所言，该软件是一款同时适用于工程设计单位和科研院校的软件，因此本书不仅适用于工程设计单位的设计人员和科研人员，同时还适用于从事土木工程、岩土工程和市政工程的相关科研人员，亦可作为高校教师、博士生、研究生和本科生的软件学习教程。

　　本书介绍的操作方式均为 Diana10.1 下的 GUI 可视化操作和 Python 命令流语句操作。为方便读者学习，书中所有附带有命令流语句例题的 Python 语句命令流文件均会在相应的官网上贴出来。其实 Diana 软件是一种适用于诸多领域和诸多方向的软件，具有非常广阔的应用前景，绝不仅仅局限于土木工程结构方向的应用。由于作者在热力学分析、水工结构、市政结构和流体结构、岩土和地下结构等其他诸多领域的研究较少，因此本文的重点在于土木工程的结构方向，同时偏向于以迭代方法为主的非线性分析计算。也希望相关领域的前辈和专家能够早日撰写出更优质的书籍。

　　为方便使用 Python 命令流语句完成建模、非线性分析和后处理等一系列操作，读者可在扉页微信扫一扫获取第三章到第五章部分算例的 Python 语句命令流.py 文件,请广大读者自行下载学习。

  由于编者水平有限,成书时间仓促,书中难免存在错误与缺陷,恳请广大读者批评指正,也欢迎各位 Diana 领域同行共同交流研讨。在此特别感谢上海敦楪代理公司的张辰经理升级这款优质的有限元分析软件,感谢敦楪公司华东区 Diana 负责人贾明杰总工程师和荷兰阿纳姆 Diana 总公司的马燊总工程师,感谢他们在 Diana10.1 软件交流会中对我的指点,正是他们不厌其烦的讲解,以及在使用软件过程中给予的指导,在掌握 Diana10.1 操作过程中给了我许多帮助,使我能够在最短的时间内从 9.4 版 Diana 旧式命令流编辑操作过渡到新版 10.1 版本的 GUI 用户可视化界面操作和 Python 语句命令流编辑操作。**同时本书也起着抛砖引玉的效果,希望更多的专家和大师今后能够在 Diana 领域写出理论程度更深、质量水平更佳的著作。**

第一章 Diana 概述

1.1 背景介绍

  Diana(Displacement ANAlyzer,又称作 DIANA,为统一格式,本书统称为 Diana)于二十世纪七十年代诞生于荷兰计算力学部门。是一款由 TNO Diana 公司研发,适用于土木工程各个领域,性能优异的有限元分析软件。近年来尤其在结构、桥梁、岩土、地下以及桩基等工程领域有着广泛运用。近二十年来,Diana 软件在界面操作、命令流语句的简化、单元库类型与材料本构的丰富等方面经历着不断的发展和完善。相对于国外其他有限元软件而言,Diana 软件因其在钢筋混凝土结构非线性有限元分析中对混凝土结构开裂的数值模拟、水化热模拟、砂土液化、随机场预测以及建筑结构时变性能和抗震等各方面具有良好的模拟效果从而受到了全世界科研人员和设计人员的高度重视,应用越来越广泛。而迄今为止,国内尚无相关书籍系统介绍此类软件的操作。由于语言关系,对软件系统自带的英文手册的查阅和学习也成为一部分用户理解和学习此类软件的瓶颈。鉴于此,作者希望通过本书的编写来填补空缺,通过系统介绍这款性能优异的土木工程有限元软件,帮助国内高校科研人员和工程设计人员在使用本书并配合手册学习这款软件时能够有的放矢,快速掌握该软件的基本操作。

  本书的一大特色就是通俗易懂。相对于介绍全面,理论分析精湛的用户手册而言,本书基于 Diana10.1 的软件平台,根据多年使用新旧版本 Diana 的经验基础上,淡化了大量内容晦涩、难度较深和理解起来较为吃力的有限元基本理论知识部分,尽量用浅显易懂的语言和简单易懂的阐述方式来说明该软件的学习重点,这将有助于初学者快速掌握这款软件在建模中的基本操作技能。同时本书对该软件在土木工程相关领域的适用特点给予阐述,并根据作者多年经验,结合 Diana9.4 版本的界面操作、命令流操作范例,将在施工现场所见的一些工程结构进行抽象和简化后建立的模型和近年来国内相关研究热点结合,编写了一定数量的例题,对 Diana 软件本构参数的设置和具体操作流程进行了实例展示,希望能帮助读者快速入门。与此同时,为使初学者易于理解,本书在后面提供的部分例题中还附上了对应的 Python 语言的命令流,相信这些例题和命令流会对读者的学习产生一定的帮助。本书对 DianaIE 软件模块的界面操作进行拆分,以逐一介绍常用的主要功能作为编写思路,突出详略和重点之处,不仅对该软件的总体进行了宏观把握,还对该软件应该介绍的重点部分进行了较为详细的介绍。本书的第二大特色就是融合作者本人的数值模拟、试验和工程经验编制了相关例题。例题分别在三、四、五三个章节,其中第三章是采用 Diana 软件对传统土木工程结构建立非线性计算数值模型;第四章主要是介绍 Diana 的一大特色名片——"水化热"操作;第五章为国内新兴的预制节段拼装结构,这种结构由于受力特性和结构性状均较为复杂,对几何模型的建立、界面单元和非线性计算各个方面都有着较高的要求,因此该部分将是本书着重介绍的部分,同时也是本书的闪光点之一。本书另外一大闪光点是每一章节的例题除大量采用图形和表格方式介绍 DianaIE 的模块使用及操作以外,大部分例题还附上了 Python 语言命令流,供读者对比学习两种殊途同归的建模方式,以期达到适应不同后缀名称的文件的打开、运行、编辑和保存方式,具备建模的入门水平。本书的第三大特色就是在后面三章的章节例题编写中加入了学习要义,即将每一节模型中具有代表性的四五个经典操作有针对性地提炼出来,放置在每节开头,学习要义本着从易到难、从简到繁的方针,使读者在阅读中能够保持循序渐进、有的放矢,确保头脑清醒,学习时不迷茫。与此同时,由于 10.1 版软件界面采用英文风格,考虑到有些初学者可能对英文 GUI 界面版本的操作理解困难,因此编者在一开始的内容介绍和后续例题中加入了对应操作界面中文术语翻译,并在后面附上括号,输入软件原版的英文术语,力争使读者能够快速进入对 Diana 英文版界面的熟悉状态。这里需要指出的是,例题编写过程中采用的数据均为作者臆想和假定的数据,没有经过专业性土木工程试验的检验和论证,所以难免会和实际情况有所偏差,读者在阅读本书例题时理应具有批判性眼光,一方面应着眼于重点学习软件的操作要义,对书中的计算分析结果做到批判性吸收,对作者在书中提到的建模操作方法的学习能够做到举一反三、触类旁通;另一方面在学习这些具体操作的同时还应着眼于书中有哪些值得改进之处,或者站在读者自身角度思考是否有更好的本构模型抑或是更为快捷的建模思路与建模方式,怎样可以做得更好,模拟得更精确。

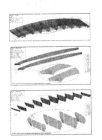

(a)玻璃移动桥(芬兰)　　(b)英国伦敦滑铁卢和城市线人行连接隧道
(Waterloo and City Line pedestrian link tunnel)

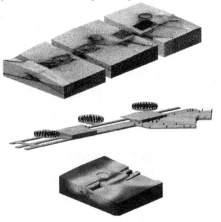

(c)芬兰赫尔辛基西延线凯拉涅米站(Helsinki subway Western extension, Keilaniemi station)

图 1.1-1　Diana10.1 数值模拟案例

本书的内容分为以下几个章节展开：第一章主要介绍 Diana 的相关特性，从软件的主要功能，安装运行，常用单元类型，文件系统，用户图形 GUI 操作界面以及常用的 Python 语言命令流等几个方面进行介绍，且重点介绍 DianaIE 的用户可视化界面操作 GUI 的主要功能，以方便初学者能比较快速入门。同时本书作者也考虑到对那些 Diana 操作水平较高，从老版的 iDiana 软件系统刚刚升级到 Diana10.1 软件的老用户而言，在书中也添加了一些老版的元素与新版相同内容进行对比验证和对比学习。第二章主要介绍 Diana 软件中的混凝土和钢材本构类型、混凝土裂缝类型和全世界各混凝土规范和钢材规范的本构特性。第三章编制了大量例题介绍 Diana10.1 最基本的前处理建模和非线性计算的操作，为使得使用 Diana9.6 之前版本的老版用户在升级后能够缩短熟练使用 Diana10.1 版的学习时间，其中个别例题来自于 Conrete Tutorial 9.4 版本教程中非常具有代表性的例题，本书重新采用 Diana10.1 建模，使得这部分读者可以用最快的速度融会贯通，掌握新版软件的基本操作技能。除此之外，对于土木工程研究热点中的低周循环加载下的滞回分析和动力学抗风抗震领域时程分析等热点问题也有涉及。第四章主要涉及施工阶段中存在的水化热暂态反应，并根据日本、欧洲和美国不同规范，采用 Diana 软件对大体积混凝土进行热分析。第五章围绕预制节段拼装梁的受剪破坏、长期性能和随机场预测编写了三节的例题。第六章基于作者建模经验，对该软件目前存在的一些不足和可以改进之处提出了一些建议。

1.2 Diana 软件的主要功能和安装运行

Diana 是一款具有丰富材料本构和强大非线性分析功能的结构有限元分析软件，也是在结构、桥梁、岩土、隧道、地下、水利、市政消防等各项工程领域有着广泛应用的高端非线性分析有限元软件。所谓的非线性，就是在荷载作用下结构产生较大位移和变形，而这种变形对平衡有着较大的影响，因此结构的变形协调方程建立于变形后的状态。对于非线性计算往往采用迭代计算方法。在 Diana 中也可以精确模拟钢筋混凝土结构从初始状态到产生裂缝，不断扩展，继而最终倒塌的整个过程。这种模拟精确考虑了结构的实际几何形状、混凝土的材料本构模型特点，在嵌入式钢筋杆单元和嵌入式钢筋网片与混凝土单元耦合方式以及钢筋与混凝土相互之间有粘结（Fully-bonded）、无粘结（Non-bonded）、粘结滑移（Bond-slip）等力学特性方面有着较为精确的模拟。在抗震方面，Diana 不仅提供可应用于结构设计的线性动力分析功能，还提供考虑地震反复荷载作用历程的非线性动力分析功能。Diana 在岩土开挖、大坝分析等方面也具有优异功能。例如，施工阶段分析、土体与结构的协同分析、流固耦合分析、用户自定义本构模型、多种界面单元、大变形及大应变分析、材料非线性分析、考虑时间影响及环境影响效应的分析、非线性动力分析等。同时，Diana 在隧道工程领域也有很多应用。常规的隧道分析和设计主要针对隧道的开挖或者衬砌的受力进行分析。Diana 的热力学还被广泛用于结构在火灾中的分析，大体积混凝土水化热反应等温度影响下的结构性能分析。

Diana10.1 版软件主要包含两个模块，一个是由荷兰 Diana 研发团队最新开发 GUI 界面操作的 DianaIE 模块，一个是保留了 Diana9.6 版软件前处理功能的 iDiana 模块。新老用户在使用过程中，根据自己的熟练程度，既可以在 DianaIE 交互环境下直接使用 GUI 操作界面进行建模计算,也可以在传统的 iDiana 模块中导入编辑.bat 二进制模型程序文件生成模型,此外,使用 DianaIE 文件还可以导入 IGES,STEP IGES,STEP 等 CAD 模型,并且在导入的模型的基础上进行后续的建模操作。

接下来作者对该软件在土木工程分析中较为成熟的应用作一些简要的介绍。

其中 Diana10.1 运用的主要领域有:

(1) 钢筋混凝土结构的开裂分析。
(2) 大体积混凝土的水化热分析。
(3) 混凝土结构收缩徐变等长期时变性能下的非线性分析。
(4) 结构的阶段性分析。
(5) 砌体结构的抗震分析。
(6) 运用随机场理论预测裂缝宽度。
(7) 低周试验及循环荷载试验下的钢筋混凝土结构滞回分析。

Diana10.1 版本和 9.4 版本类似,用户需要先购买软件并经总公司注册许可之后才能使用。与同时适用于 XP 系统、32 位 Windows 系统和 64 位 Windows 系统的 9.4 版不同的是,新版的 Diana10.1 软件只支持 Windows 64 位计算机系统。与此同时,Diana10.1 版安装环境必须在计算机联网状态下才能进行。为保证后续安装程序顺利运行,用户在安装前应关闭所有杀毒软件或者将 Diana 中的相关安装文件(Setup.exe 和 Dianahasprus.exe)添加进信任区。Diana10.1 版软件采用软件密钥文件加密和计算机物理绑定方式。用户在购买后首先应将 c2v 文件发送到代理公司进行许可证状态的激活,激活后用户可登陆 Diana 官网(http://tnodiana.com/Diana-downloads)下载相关的 Diana 10.1 安装模块。

Diana9.4 版软件包主要由三个部分组成,即启动程序文件 Setup.exe 文件安装、硬件密钥 HASPUser Setup.exe 文件安装以及 MIDAS 模块的安装,而 Diana10.1 软件的安装主要涉及到三个方面:

(1) DianaIE 的安装。下载 Setup.exe 文件并进行安装。安装成功后的 Diana 软件主要包括 bin 启动文件夹、binseg 文件夹、执行注册功能的 lib 文件夹,以及用于编辑 Python 语言的 python 文件夹,包括目前暂以 9.6 版 Diana 手册内容为主的 PDF 文件夹,如图 1.2-1 所示。

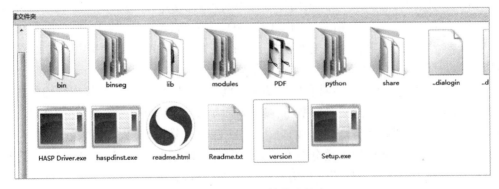

图 1.2-1　Diana 软件文件夹

(2) Dianahasprus.exe 文件密钥的安装。在安装 Dianahasprus.exe 之前,用户必须购买该软件并向总公司或就近向本地区所在的代理服务公司提出 c2v 文件密钥激活和许可证状态激活申请,待文件在公司总部激活成功并更新许可证信息后方可安装。安装注册中密钥号信息,激活 c2v 文件及更新许可证信息程序资料所有权归公司所有,用户一律不得擅自披露。

(3) iDiana 模块的安装。由于 iDiana 模块保留了之前 Diana9.6 版本的前处理特色,因此该部分安装方式与老版本相同。

待完成 Dianahasprus.exe 的信息注册与升级后,点击 Setup.exe,弹出如图 1.2-2 所示界面:

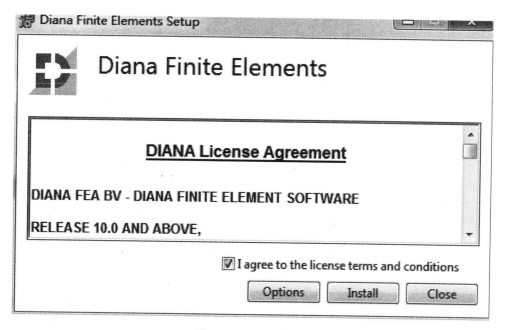

图 1.2-2　Setup 界面

点击 Install,出现初始化程序界面,进入程序安装流程,如图 1.2-3 所示。

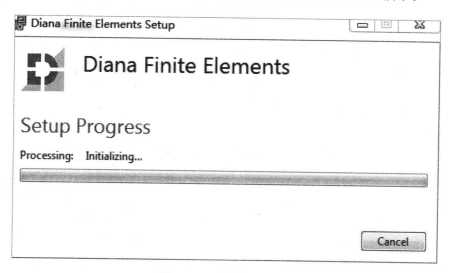

图 1.2-3　程序安装流程

安装完成后,点击OK,弹出DianaIE界面,表示DianaIE安装成功。如图1.2-4所示。

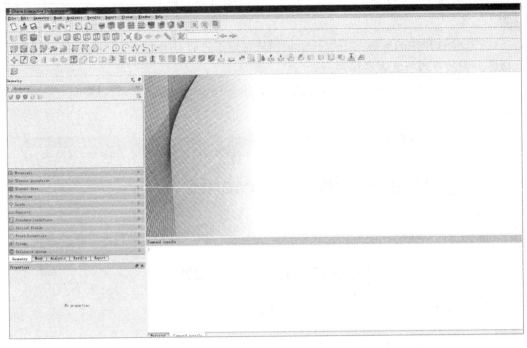

图1.2-4 DianaIE界面

Diana文件具有以下两种特性:

(1)向下兼容性

在Diana软件中,高版本的软件配置兼容低版本的配置,即用户在使用升级版高版本的软件时,可以打开任意一种低版本的文件。

(2)同级别通用性

版本数中小数点前整数位相同的Diana软件运行文件可以相互打开。(例:Diana9.4可以打开Diana9.3版本生成的二进制模型文件,模型材料文件可以相互打开。Diana10.1可以打开Diana10.0版生成的二进制dpf文件、python语言文件。但是10.1版本的打不开9.4版本的文件。)

1.3 Diana软件常用单元类型

作为土木工程领域一款性能优异的软件,Diana的单元类型众多,适用于各种结构分析。在Diana10.1版本中,单元类型选择为自动选择,即用户并不能直接输入或指定单元类型和名称,而需要首先指定模型维数、划分类型和划分单元的次数。当选择二次单元时,用户还需要设置每个单元中间位置处节点位置的确定方式(Mid-side node location)是采用线性插值(Linear interpolation)或者是在单元形状上(On shape)即可。在后续网格划分步骤中,用户还需根据需要划分的对象,再次设置划分类型及单元中间位置处节点位置的确定方式,软件根据用户设置的这些参数,待网格智能划分成功后在GUI可视化界面properties属性栏名目下的划分类型meshing栏中会自动显示网格的单元类型。Diana结构中单元分类方式众多。按照单元的维数划分可以分为 1D 单元、2D 单元和 3D 单元。根据数值积分类型的不同,又可以分为按面域积分(Area Integration)和高斯数值积分(Gauss Integration)两种方式。按照形状分类,可以分为线单元(Line Elements)、面单元(Face Elements)、体单元(Solid Elements),其中面单元内部又可以分为三角形单元(Triangle Elements)、四边形单元(Quadrilateral Elements);体单元内部又可分为四面体单元(Pyramid Elements)、三棱柱单元(Wedge Elements)和六面体单元(Brick Elements);按照单元坐标变换和位移函数所用形函数阶次是否相等可以分为等参单元和非等参单元;按照力学特征不同总体上还可分为桁架单元(Truss Elements)、梁单元(Beam Elements)、平面应力单元(Plane Stress Elements)、平面应变单元(Plane Strain Elements)、弯曲板单元(Plate Bending Elements)、轴对称单元(Axisymmetric Elements)、平面壳单元(Flat Shell Elements)、曲面壳单元(Curved Shell Elements)、实体单元(Solid Elements)、界面单元(Interface Elements)、接触单元(Contact Elements)、弹簧单元(Spring Elements)、组合单元及其他可用于结构分析的单元。本书重点介绍土木工程领域中常见的各类单元。对于与流体或热分析等其他领域有关的单元,这里不予赘述。

1.3.1 桁架单元(Truss Elements)

桁架单元可分为 2D 桁架单元以及 3D 桁架单元,根据位移变量的不同又可以分为普通桁架单元(Regular Element)和高级桁架单元(Enhanced Element)。桁架单元的特点是垂直于单元长度方向的直径相对于单元的杆件长度而言可以忽略不计。桁架的变形只有沿着杆长方向的轴向伸缩变形,无剪切或弯曲变形。按照维数、节点数以及自由度数的不同可以分成 L2TRU、L4TRU 以及 L6TRU 等单元。

1. 普通桁架单元(Regular Element)

L2TRU 单元属于普通桁架单元。该单元由两个节点构成。可模拟桁架、弹簧或者预应力筋受力特征。其单元名称中 L 表示单元形状为直线(Line),数字 2 表示该单元的自由度数,TRU 表示该单元是桁架类型(Truss)单元。每个节点只有一个沿着轴向的伸缩位移,可以沿着节点 X 或 Y 或 Z 单一坐标轴平动方向移动,如图 1.3-1 所示。单元的位移插值函数为线性方程。该单元只能承受拉压,不能承受弯矩作用。在单元的局部坐标系 x 方向上存在着质量分布,但在整体坐标系 X,Y,Z 轴方向上质量无法分布,因此在动力分析中不可使用该单元。普通桁架单元的特性见表 1-1。

图 1.3-1　L2TRU 桁架单元位移自由度示意图

表 1-1　普通桁架单元特性表

单元特性	参　　数
单元名称	L2TRU
节点个数	2

(续表)

单元特性	参 数
维数	2
单个节点自由度	1
单元总自由度数	2
位移插值函数	$u_x(\xi)=a_0+a_1\xi$
位移	U_X
几何特征参数	截面面积
适用范围	沿着单轴伸缩的2维桁架、拉索、弹簧、体外预应力筋等

2. 高级桁架单元(Enhanced Element)

L4TRU 是一个两节点的桁架单元,每个节点上有沿着 X 和 Y 两个方向的 2 个自由度,单元形状为直线,单元名称中数字 4 代表了该单元自由度总数。每个节点具有沿着节点坐标系 X 方向和 Y 方向的平动位移,可以沿着坐标轴方向移动。该单元只能承受拉压,不能承受弯矩作用。单元的参数特征见表 1-2。

表 1-2 L4TRU 单元的参数特征

单元特性	参 数
单元名称	L4TRU
节点个数	2
维数	2
单个节点自由度	2
单元总自由度数	4
位移插值函数	$u_i(\xi)=a_0+a_1\xi$
位移	U_X, U_Y
几何特征参数	截面面积
适用范围	2维桁架、拉索、弹簧

L6TRU 桁架杆单元同样由两个节点构成,单元形状为直线,单元名称中数字 6 代表了该单元自由度总数。每个节点具有沿着节点 X,Y,Z 三个方向的平动位移,可以沿着 X 或 Y 或 Z 方向平动。该单元同样也只能承受拉压,不能承受弯矩作用。单元的参数特征见表 1-3。

表 1-3 L6TRU 单元的参数特征

单元特性	参 数
单元名称	L6TRU
节点个数	2
维数	3

(续表)

单元特性	参 数
单个节点自由度	3
单元总自由度数	6
位移插值函数	$u_i(\xi)=a_0+a_1\xi$
位移基本变量	U_X, U_Y, U_Z
几何特征参数	截面面积
适用范围	3维桁架、拉索、弹簧

3. 索单元(Cable Elements)

索单元形状皆为曲线,并且每一个单元相较于桁架单元有更多自由度,因此这些单元适用于悬索结构的大变形的几何非线性问题的分析,同时也适用于单根曲线预应力钢束和普通钢筋混凝土结构中的曲线钢筋模拟。根据维数和单元自由度的不同,索单元可以分为CL6TR(2维单元,6个自由度)、CL8TR(2维单元,8个自由度)、CL10T(2维单元,10个自由度)、CL9TR(3维单元,9个自由度)、CL12T(3维单元,12个自由度)、CL15T(3维单元,15个自由度)。

1.3.2 梁单元

Diana梁单元具有广泛的工程应用背景,可在实际结构中模拟钢筋混凝土梁、预应力混凝土大跨结构。根据维数以及单元自由度的不同,梁单元可分为 L6BEN,L7BEN,CL19BE,CL12B,CL15B(其中开头的字母C是Curved缩写,表示该梁的形状是曲线梁的意思),根据梁单元形状的不同又可以将梁单元分为直线梁单元(以字母L开头的单元)和曲线梁单元(以C字母开头的单元),而根据梁单元适用理论和受力特性的不同,梁单元还可以分为第一类梁单元、第二类梁单元和第三类梁单元。

1. 第一类梁单元(CLASS-I Beams)

第一类梁单元主要是直线型梁单元,此类单元基于平截面假定和欧拉-贝努利梁理论(Euler-Bernoulli beam theory),在分析中不考虑剪切变形,建模中还可以根据用户的需要将Diana材料定义后的截面几何特性赋予不同形状的截面(例如矩形、T形、I形、箱形等)和所需的截面尺寸。第一类梁单元主要应变类型有纵向拉伸应变、弯曲应变,以及对3维梁单元而言具有的平面外的扭转变形。应力由法向应力和弯矩组成。在第一类梁单元中,位移和转角被认为是非独立变量,故通常用单元y方向位移的二阶导数来表示曲率。具有"剪力锁死"特性的铁木辛柯梁也属于这类单元。此类梁单元不仅适用于分析采用梁单元建模的混凝土结构,同时也适用于解决预应力筋单元中采用离散式钢筋单元模拟单根具有无粘结、有粘结或者粘结滑移特性的预应力筋与混凝土结构的耦合问题。通过形状系数的设置可实现该类梁单元与考虑剪切变形和转动惯性,具有"剪力锁死"特征的铁木辛柯梁单元之间的转换。

2. 第二类梁单元(CLASS-II Beams)

与第一类梁单元相比,第二类梁单元同样基于平截面假定和贝努利理论,单元的位移与转角同样被认为是非独立的变量。在此类梁单元中,同样不考虑剪切变形的影响。与第一类不同的是,第二类梁单元考虑梁单元沿轴向相对变形。由于此类单元在截面上沿单元坐标轴方向采用插值类型数值积分,因而这类梁单元适用于几何非线性分析和物理非线性分析。常用的 L7BEN 即属于这种类型的梁单元。

3. 第三类梁单元(CLASS-III Beams)

与前两类梁单元相比,第三类梁单元主要为曲线梁单元,与第二类梁单元类似,此类单元在截面上沿单元坐标轴方向采用插值类型数值积分,在 Diana 有限元分析中该类型梁单元考虑剪切变形的影响,且位移和转角为相互独立的变量,即由节点上的法向位移和转角各自独立插值。由于单元上节点较多,因此该类单元往往为曲线梁单元(**Diana9.6 版之前的第三类梁单元均为曲线梁**),位移函数也往往具有较高的阶数,因此在与其他单元相连时具有较好的位移协调性和单元之间的边界适应性。

Diana 10.1 版中中三种类型的梁单元对应的特征见表 1-4。

表 1-4 三类梁单元

单元类别	单元名称	维 数	节点个数	总自由度数	受力特征	位移基本变量
第一类梁单元	L6BEN	2	2	6	不计剪切变形和轴向相对变形	U_x, U_y, ϕ_z
第一类梁单元	L12BE	3	2	12	不计剪切变形和轴向相对变形	$U_x, U_y, U_z, \phi_x, \phi_y, \phi_z$
第二类梁单元	L7BEN	2	2	7	考虑轴向相对变形,不计剪切变形	$U_x, U_y, \phi_z, \Delta u_x$
第二类梁单元	L13BE	3	2	13	考虑轴向相对变形,不计剪切变形	$U_x, U_y, U_z, \phi_x, \phi_y, \phi_z, \Delta u_x$
第三类梁单元	CL9BE	2	3	9	考虑剪切变形	U_x, U_y, ϕ_z
第三类梁单元	CL12B	2	4	12	考虑剪切变形	U_x, U_y, ϕ_z
第三类梁单元	CL15B	2	5	15	考虑剪切变形	U_x, U_y, ϕ_z
第三类梁单元	CL18B	3	3	18	考虑剪切变形	$U_x, U_y, U_z, \phi_x, \phi_y, \phi_z$
第三类梁单元	CL24B	3	4	24	考虑剪切变形	$U_x, U_y, U_z, \phi_x, \phi_y, \phi_z$
第三类梁单元	CL30B	3	5	30	考虑剪切变形	$U_x, U_y, U_z, \phi_x, \phi_y, \phi_z$

注:在计算初应变和初应力的时候,第二类梁单元上作用的自重和分布荷载在计算中不予考虑。

1.3.3 平面单元

平面单元是一种以 2D 为主的单元,在结构的中轴线位置建立模型,确定单元类型后,再通过几何属性模块赋予单元厚度即生成平面单元。3D 平面单元在诸多方面与壳单元具有很强的相似性。在 Diana 中,根据单元受力形式的不同,平面单元又可分为平面应力单元(Plane Stress Elements)、平面应变单元(Plane Strain Elemnts)。

1. 平面应力单元(Plane Stress Elements)

平面应力单元是由平板形状的单元组成,故又称为平板单元。一个单元上的所有节点都必须位于同一个平面之内,平面应力单元的两个重要的特征是:① 厚度方向的尺寸相对于单元的长度和宽度可以忽略不计;② 垂直于板面方向的局部应力为 0,即垂直于平面沿厚度方向的应力 $\sigma_{zz}=0$。Diana10.1 软件根据用户实际建模的需要,可提供 2D 和 3D 的平面应力单元,而 3D 平面应力单元在横向与具有刚度的其他单元相连时,可以在非平面类的几何单元内存在。通常情况下,将 3D 平面应力单元称为膜单元(3D Membrane Elements)

2D 平面应力单元每个节点有 2 个自由度,分别为节点 x,y 两个方向的平动自由度。3D 平面应力单元的位移变量主要是沿着坐标轴方向的平动位移变量 U_x,U_y 和 U_z。应变变量主要由 x,y,z 三个方向的正应变 ε_{xx},ε_{yy},ε_{zz} 和剪切应变 γ_{xy} 组成。对应的应力为正应力 σ_{xx},σ_{yy},σ_{zz} 和切应力 τ_{xy},其中 Z 方向应变对应的正应力 σ_{zz} 为 0。通过厚度积分,Diana10.1 可以自动计算法向和切向力,即 $\{n_{xx},n_{yy},n_{xy},n_{yx}\}^T$,其中切向力满足互等定理 $n_{xy}=n_{yx}$。而根据基本位移变量的不同,在 Diana 中主要分为常规的 2D 和 3D 平面应力单元和面内转动的平面应力单元(Elements with Drilling Rotations)。后者除了具有与常规平面应力单元在整体坐标系下沿各坐标轴方向相同的平动位移之外,还有一个绕坐标轴 Z 轴方向的转动变量 φ_z。

平面应力单元的厚度在 Diana 中的定义方式比较特别,对于各向同性的平面应力单元,厚度的赋予与截面的方向无关,而对于各向正交异性的平面应力单元在定义厚度时仅适用于某几类比较特殊的平面应力单元。Diana 软件对于单元厚度的定义通常分为单元厚度等值赋予和不等值赋予两种方式。对于厚度等值赋予情况而言,当在几何属性定义模块中确定单元截面的几何形状之后,只需赋予一个厚度值即可,该厚度即代表所有单元各节点的厚度相同且为此厚度值。而对于厚度不同的单元而言,在确定单元截面的几何形状之后则需要逐个输入各点处厚度值才能形成需要的单元。平面应力单元厚度的等值赋予和不等值赋予简图如图 1.3-2 和图 1.3-3 所示。在厚度净值赋予的情况下,通常平面单元需要赋予厚度的个数与所选单元的类型和节点个数有关,例如 4 节点矩形单元要赋予 4 个不同厚度值,或者 8 节点等参单元需要赋予 8 个不同的厚度值,根据单元类型和单元的插值阶数,三角形单元节点个数需要输入厚度值可以是 3 个不同厚度值,也可能是 6 个。

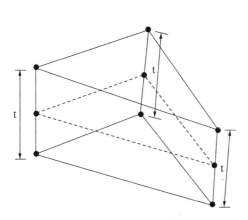 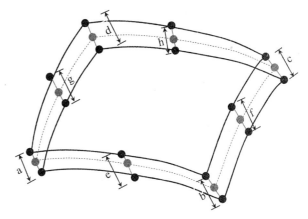

图 1.3-2 平面应力单元厚度等值赋予图示　　图 1.3-3 平面应力单元厚度不等值赋予图示

各类平面应力单元的特性见下表 1-5。

表1-5 各类平面应力单元单元特性

单元名称	单元形状	维数	总自由度数	插值次数及积分方式	特征
T6MEM	3节点三角单元	2D	6	线性插值 1点面域积分	适用于2D模型中，位移具有沿着X,Y两个坐标轴方向的平动自由度Ux,Uy
Q8MEM	4节点四边形单元		8	线性插值 2×2高斯积分	
CT12M	6节点三角单元		12	二次插值 3点面域积分	
CQ16M	8节点四边形单元		16	二次插值 2×2高斯积分	
CQ18M	9节点四边形单元		18	二次插值 3×3高斯积分	
T9GME	3节点三角单元	3D	9	线性插值 3点面域积分	适用于3D模型中，具有沿X,Y,Z的3个坐标轴方向平动自由度Ux,Uy,Uz
Q12GME	4节点四边形单元		12	线性插值 2×2或3×3高斯积分	
CT18GM	6节点三角单元		18	二次插值 3点减缩面域积分	
CQ24GM	8节点四边形单元		24	二次插值 2×2或3×3高斯积分	
T9MEM	3节点三角单元	2D	9	线性插值 3点面域积分	面内转动单元。除了沿X,Y坐标轴方向的平动自由度Ux,Uy之外，还具有绕垂直于单元面的法向Z轴的转动自由度φ_z
Q12ME	4节点四边形单元		12	线性插值 2×2面域积分	
T6OME	3节点三角单元	2D	6	线性插值 1点面域积分	正交各向异性厚度单元。各方向厚度不同，要分开赋予。位移插值方程为线性形式，位移变量有Ux和Uy
Q8OME	4节点四边形单元		8	线性插值 2×2高斯积分	
CT12O	6节点三角单元		12	二次插值 3点面域积分	
CT16O	8节点四边形单元		16	二次插值 2×2高斯积分	
T9MWE	3节点三角单元	3D	9	线性插值 1点面域积分	适用于翘曲分析，位移变量仅有沿着三个坐标方向的平动位移，但应力向量包括沿坐标轴三个方向正应力和切应力

常用的平面应力单元为 CQ16M 单元，这种单元是一种 8 节点高斯积分的二次插值等参矩形单元。如图 1.3-4 所示，单元上的每个节点一共有沿着 x 和 y 两个方向的自由度，常用于模拟 2D 平面内梁以及楼板等混凝土模型。除混凝土结构外，该单元也可单独赋予钢材本构特性用于模拟开孔钢板结构以及钢材疲劳损伤等特性。该单元非常适用于混凝土中弥散开裂(Smeared Cracking)模型，可以与作为嵌入其中的模拟纵筋和预应力筋的 Bar 单元以及模拟钢筋网片的 Grid 单元有良好的耦合，对于非线性计算也有着较好的协调性和收敛性。

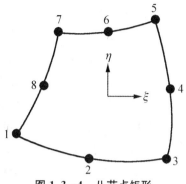

图 1.3-4　八节点矩形单元 CQ16M

2. 平面应变单元(Plane Strain Elements)

平面应变单元节点位于 XOY 平面内，单元的节点在整体坐标系下的 Z 方向坐标值为零。垂直于单元面方向的法向 Z 方向应变值为零，荷载必须作用在单元面内。对于平面应变单元而言，单元的应变主要由正应变和剪切应变组成。与平面应力单元表述刚好相反，平面应变单元 Z 方向正应变为 $\varepsilon_{zz}=0$。平面应变单元基本位移量为整体坐标系下的 X 方向和 Y 方向的平动位移 U_X 和 U_Y。与平面应力单元类似，平面应变单元中也含有面内转动(Elements with Drilling Rotations)的平面应变单元。

与平面应力单元类似，平面应变单元应变变量为 x，y，z 三个方向的正应变 ε_{xx}，ε_{yy}，ε_{zz} 和切应变 γ_{xy}，其中 z 方向的正应变值为 0。与此同时，对应的应力值为 x，y，z 三个方向的正应力 σ_{xx}，σ_{yy}，σ_{zz} 和切应力 τ_{xy}。

平面应变单元可承受线荷载和单元均布荷载，还可以承受温度作用，所有的面荷载和线荷载的施加方式均表现为在节点上施加节点荷载，与平面应力单元赋予厚度的方式类似，对平面应变单元荷载值的赋予主要是通过确定每一个节点位置上的节点力大小和方向，如果不加说明，仅仅赋予一个荷载值，则 Diana 默认为沿着平面外方向所有节点具有相同的荷载值，即每个节点上承受的荷载值大小相同，且指向均为 Z 方向。如果要赋予不同方向、不同大小的荷载值，Diana 中必须对单元上的每个节点定义荷载大小和方向，方向可沿着坐标轴方向(X，Y，Z 方向)，还可以通过法向(NORMAL)和切向(SHEAR)的方式来定义平面应变单元上的各节点荷载值。以图 1.3-5 为例，该图表示 8 节点平面应变单元 1，2，3 三个节点承受法向荷载，荷载大小 $F_1=300$ N，$F_2=400$ N，$F_3=500$ N，则在几何模型.dat 文本文件中相应表示为：

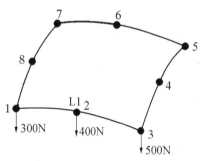

图 1.3-5　平面应变单元不等值力赋予

'LOADS'

CASE 1

ELEMEN

1 EDGE L1

```
FORCE 300 400 500
DIRELM NORMAL
```

根据单元实际建模要求的不同,用户还可以改变和调整单元上各节点受力的大小、方向。

除了常规平面应变单元之外,Diana 中还有一种基于平面应变假设的分析 3D 问题的完全平面应变单元。同常规的平面应变单元相比,这种单元消除了对平面外位移、应变和应力分量的限制,即增加了沿着 Z 轴方向的平动位移 U_z,而在应变变量上增加了沿着平面外 Z 轴方向的正应变 ε_{zz} 和剪切应变 γ_{yz},γ_{xz}。与此同时应力变量在对应的平面外的剪切方向相应增加了切应力 σ_{yz} 和 σ_{xz}。除此之外,一种形状类似于梁单元的有限壳单元的单元类型也于近年应运而生。这类单元严格意义上来说属于介于平面应力单元和壳单元之间的一类特殊的平面应变单元,单元上的每个节点仅仅具有沿着 x 轴和 y 轴两个方向的平动位移 u_x,u_y 和绕 Z 轴的转动变量 φ_z 这三个自由度。这类单元对"剪切锁死"的问题比较敏感。当单元几何形状为直杆时,沿着局部坐标系 x 方向正应变 ε 为常数;当单元形状为曲线时,沿着局部坐标系 x 方向正应变 ε 线性变化。厚度 ζ 方向积分点和积分方式与 x 方向不同,当结构进行物理线性分析时,厚度 ζ 方向为 2 点高斯积分;当结构需要进行物理非线性分析时则宜采用辛普森积分,且积分点的个数与物理非线性的程度息息相关。而在 Diana 软件的平面应变单元库中独有一种特殊橡胶类型的平面应变单元,这类单元适用于在非线性条件下模拟具有橡胶受力特性的超弹性结构和构件,结合一些弹簧单元和阻尼器(Dashpot)本构特性还可以专门用来模拟隔震结构或隔震支座,如对橡胶隔震垫和阻尼隔震层的模拟。

各平面应变单元类型见表 1-6。

表 1-6 平面应变单元

单元名称	单元形状	维数	总自由度数	插值次数及积分特征	特征
T6EPS	3 节点三角单元	2D	6	线性插值 1 个面域积分	单元内的每个节点位移变量只具有沿着 X 方向和 Y 方向的平动自由度 U_x,U_y
Q8EPS	4 节点四边形单元		8	线性插值 2×2 高斯积分	
CT12E	6 节点三角单元		12	二次插值 3 点面域积分	
CQ16E	8 节点四边形单元		16	二次插值 2×2 高斯积分	
CT30E	15 节点三角单元		30	四次差值 12 点面域积分	除具有上述普通平面应变单元特性外,特别适用于岩土结构的非线性分析领域
CQ20E	8 节点四边形单元	2D	20	二次插值 3×3 高斯积分	均为四边形二次单元,均采用高斯积分,单元中的每个节点只有 X 和 Y 轴两个方向平动位移 U_x,U_y
CQ22E	9 节点四边形单元		22	二次插值 3×3 高斯积分	

(续表)

单元名称	单元形状	维数	总自由度数	插值次数及积分特征	特征
CT18GE	6节点三角单元	3D	18	二次插值 3点面域积分	单元上每个节点具有沿X,Y,Z轴三个方向平动自由度Ux,Uy,Uz。沿平面外指向的坐标轴,应力和应变值与坐标值相互独立。每种单元内的积分点个数除了默认设置之外还有可供选择的其他合适积分点数。但积分点个数一旦超过上限,该单元便不可使用
CQ24GE	8节点四边形单元		24	二次插值 面内2×2高斯积分	
CT27GE	9节点三角单元		27	三次插值 7点面域积分	
CQ36GE	12节点四边形单元		36	三次插值 面内3×3高斯积分	
L6PE	2节点直线形		6	位移插值函数为线性 单元内1×2数值积分	单元形状为线形,局部坐标系下的x方向为单元形状的切线方向,单元上每个节点仅具有沿着局部坐标系x和y轴方向的平动自由度和绕着z轴的转动自由度。Y方向的正应力为零,对剪切锁死比较敏感,物理线性分析时,厚度ζ方向为2点高斯积分;当物理非线性时宜采用辛普森积分
CL9PE	3节点曲线形		9	位移插值函数 最高阶次为二次 单元内2×2数值积分	

1.3.4 弯曲板单元(Plate Bending Elements)

就几何尺寸而言,弯曲板单元与平面应力单元类似,弯曲板单元的单元节点坐标必须位于一个平板单元之内,其次弯曲板单元的单元厚度相对于宽度尺寸可以忽略不计。然而从受力特性而言,如果作用于单元上的荷载仅为平行于单元面的纵向荷载,则这类单元称为平面应力单元。如果作用于单元上的荷载为垂直于单元面的横向荷载,则这类单元称为弯曲板单元。弯曲板单元上作用力必须垂直于单元面,且垂直于平面沿厚度方向的应力 $\sigma_{zz}=0$。而与平面应力单元不同的是,除了作用力之外,弯曲板单元上还可以在平面内作用力矩,力矩的方向为绕着某个局部坐标轴。板单元需要同时满足变形协调条件和平衡条件。弯曲板单元在变形前后满足平截面假定,所能承受的荷载包括集中点荷载(Point Load)、线荷载(Edge Load)、面荷载(Face load)、温度作用(Temperature)、集中荷载(Concentration

Load),以及初始应力(Initial Stress)等不同形式的荷载。

弯曲板单元的位移变量与平面应力单元相比有较大不同。首先,由于平面内弯矩的存在,平面板单元的位移变量存在绕着 x 轴和 y 轴的两个平面内转动变量 φ_x,φ_y。与此同时,弯曲板单元还具有沿着 Z 方向的平动位移 U_Z。与平面应力单元不同的是,由于弯曲板单元承受面内弯矩,因此弯曲板单元的应变变量只有五个,且没有正应变,只有 x 方向、y 方向、xOy 面三个方向的曲率应变 κ_{xx},κ_{yy},κ_{xy} 和扭曲率 Ψ_{yz},Ψ_{zx}。从整个单元类型的力学特性来看,弯曲板单元的力学特性可近似看做一种介于平面应力单元与曲壳单元之间的过渡类型单元。弯曲板单元的应力形式比较复杂,在 Diana 中,可以输出两种类型的应力形式:一种是以弯曲应力和荷载集度方式输出的弯曲板单元应力形式,另一种是以柯西(Cauchy)应力形式输出的弯曲板单元应力。前者由弯曲应力 m_{xx},m_{yy},m_{xy} 和切向集度应力 q_{yz},q_{xz} 组成,而后者的柯西应力主要是由三个方向的正应力 σ_{xx},σ_{yy},σ_{zz} 和切应力 τ_{xy},τ_{yx},τ_{xz},τ_{zx},τ_{yz},τ_{zy} 组成,其中沿着 Z 方向的正应力 $\sigma_{zz}=0$。

弯曲板单元的厚度赋予方式与平面应力单元类似,同样有等厚度值赋予和非等厚度值赋予两种方式,关于厚度赋值的方式和类型在前文平面应力单元中已详细介绍,这里不再赘述。

弯曲板单元主要有两种类型,一种是基于离散型克希霍夫直线法,在单元内部或者是在单元边界上的点保持克希霍夫直线法假设的离散克希霍夫弯曲板单元;另一类是基于明德林平板原理的弯曲板单元,即这种单元中垂直于板中性面的直线在变形后仍然保持直线,但由于横向剪切变形的影响,不一定垂直于变形后的中面。

表 1-7

单元名称	单元形状	总自由度数	积分方式	特 征
T9PLA	3 节点三角形弯曲板单元	9	线性插值 转角方程为二次	基于离散型克希霍夫理论,考虑剪切变形影响
Q12PL	4 节点四边形弯曲板单元	12	线性插值 Z 向位移和绕 x,y 轴转角方程均为线性	
CT18P	6 节点三角形单元	18	二次插值 面域积分 Z 向位移和绕 x,y 轴转角方程均为完全二次	基于明德林平板原理的弯曲板单元,垂直于板中性面的直线在变形后仍然保持直线
CQ24P	8 节点四边形弯曲板单元	24	二次插值 Z 向位移和绕 x,y 轴转角方程均为线性	

1.3.5 轴对称单元(Axisymmetric Elements)

在采用 Diana 软件对实际结构进行模拟时,实际结构往往为对称结构,即绕着某一根或某几根轴成轴对称图形。而实际结构数值模型又往往非常庞大,在非线性计算中会非常耗费时间和资源,为缩减计算时间,提高计算效率,往往需要对模型取半结构或是 1/4 结构,这时候用户既可以选择直接对原结构取半结构或 1/4 结构模型,并且在正确施加对应的约束后进行非线性计算,同时又可以在某些特殊情况下采用轴对称单元进行计算。

Diana 中的轴对称单元主要分为两类:一类是坐标系下仅具有 U_x 和 U_y 基本位移量,包含三角形或四边形单元截面的实体环状单元,根据单元特性又可以更进一步分为普通实体环状单元(Regmar Slid Rings)和橡胶实体环状单元(Ruber Solid Rings)。这类单元形状规则简单,计算流程简便,计算效率较高,因而是普遍使用的一类轴对称单元;另一类为改进的壳单元,这类单元位移变量为沿着 X 轴与 Y 轴的平动自由度和绕着 Z 轴转动的转动自由度。与普通的平壳单元和曲壳单元类似,需要赋予一个厚度值,也具有等厚度值赋予和不等厚度值赋予两种方式,并且厚度的尺寸相对于壳单元的长度而言可以忽略不计。各轴对称单元特性见表 1-8。

表 1-8

单元名称	单元形状	自由度总数	插值次数及积分方式	特　性
T6AXI	3 节点三角轴对称单元	6	线性插值 1 点面域积分	对称实体环状单元,其截面主要由一系列三角平面单元和四边形平面单元组成,且这些单元每个节点仅具有沿着 X 方向和 Y 方向的平动位移变量 U_x,U_y
Q8AXI	4 节点四边形轴对称单元	8	线性插值 单元内 2×2 高斯积分	
CT12A	6 节点三角轴对称单元	12	二次插值 4 点面域积分	
CQ16A	8 节点四边形轴对称单元	16	二次插值 2×2 高斯积分	
CT30A	15 节点三角轴对称单元	30	四次插值 12 点面域积分	
CQ20A	8 节点四边形橡胶轴对称单元	20	截面上位移为二次插值,压力为线性插值 3×3 高斯积分	均为实体环状的对称四边形单元,插值方式均采用拉格朗日插值。力和位移变量的插值阶数不同,具有不协调性,适用橡胶结构中超弹性非线性计算
CQ22A	9 节点四边形橡胶轴对称单元	22	截面上位移为二次拉格朗日插值,压力为拉格朗日线性插值 3×3 高斯积分	
L6AXI	2 节点直杆单元	6	位移插值方程为线性 厚度方向 ζ 采用 1×2 点高斯积分	单元形状类似于 3D 梁单元,力学特性与平壳和曲壳单元类似,均由可退化的实体单元组成,通过赋予厚度值生成单元。厚度相对于单元长度可忽略不计,单元内每个节点具有沿着 x 轴、y 轴方向的平动位移 U_x,U_y 和绕着 z 轴转动的转角 φ_z。物理线性分析时厚度方向 ζ 采用 2 点高斯积分,但在物理非线性分析时厚度方向应采用辛普森积分
CL9AX	3 节点曲线杆单元	9	位移插值方程为线性 厚度方向 ζ 采用 2×2 点高斯积分	

1.3.6 壳单元

1. 平壳单元(Flat Shell Elements)

平壳单元的荷载作用方式众多。荷载既可以垂直作用于单元面上,又可以作用在单元内,单元上作用的力矩必须位于单元之内。平壳单元主要有两种形式。一种为常规的平壳单元,另一种为面内转动的平壳单元。平壳单元本质上是平面应力单元和弯曲类平面单元组合。厚度方向尺寸相对于宽度方向可忽略不计,平壳单元满足明德林平板理论假定,即在变形前垂直于单元面的中线仍然保持直线。平壳单元主要分为两种类型,一类是常规的平壳单元(Regular Elements),这类单元每个节点的基本位移量为沿着单元局部坐标系下 x, y, z 三个坐标轴方向的平动自由度 U_x, U_y, U_z 和绕着 x 轴和 y 轴的转动自由度 φ_x 和 φ_y;另一类则是面内转动单元(Elements with Drilling Rotations)。相对于常规平壳单元而言,这类单元在原有的位移变量基础上增加了一个绕着 z 轴转动的变量 φ_z。这类壳单元又被称为第二类平壳单元,它可以避免整体刚度矩阵病态。在 3D 单元建模中,采用平壳单元往往比采用 3D 的平面应力单元和弯曲板单元具有更好的模拟效果。

平壳单元的应力也分为两大类:柯西应力(Cauchy Stress)、弯矩应力和荷载集度应力(Moments and Forces)。前者的应力变量与弯曲板单元完全一致,即由正应力 σ_{xx}, σ_{yy}, σ_{zz} 和切应力 τ_{xy}, τ_{yx}, τ_{xz}, τ_{zx}, τ_{yz}, τ_{zy} 组成,而在弯矩应力和荷载集度应力类型中除了弯矩应力变量与弯曲板单元保持一致,在荷载集度上增加了沿着 x 轴、y 轴方向的法向集度 n_{xx}, n_{yy},垂直于 x 轴、平行于 y 轴的集度 n_{xy},以及垂直于 y 轴、平行于 x 轴的集度 n_{yx}。

在平板单元中,还有一种特殊的样条单元(Spline Elements),这种单元形状为分段的矩形,力学特性与普通的平壳单元相似,且厚度方向的尺寸相对于长宽方向均可忽略不计。与常规平板单元不同的是,样条单元是被分为数段的矩形单元,每一段矩形单元的宽度和厚度均相同,但是在局部坐标系下单元的纵向长度方向可以划分为不同长度的矩形单元。与其他弯曲板单元相比,样条单元本身还考虑了横向的剪切变形。单元形状如图 1.3-6 所示。

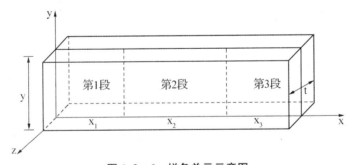

图 1.3-6 样条单元示意图

表 1-9

单元名称	单元形状	自由度总数	插值次数及积分方式	特 性
T15SF	3 节点三角单元	15	几何和位移双线性插值 1 点数值积分	
Q20SF	4 节点四边形单元	20	位移转角方程双线性插值 2×2 数值积分	基于明德林平板理论
CT30F	6 节点三角单元	30	二次插值 3 点数值积分	

(续表)

单元名称	单元形状	自由度总数	插值次数及积分方式	特性
CQ40F	8节点四边形单元	40	几何和位移函数双二次插值 2×2数值积分	
T18SF	3节点三角单元	18	几何函数线性插值 位移函数二次插值 3点数值积分	
T18FSH	3节点三角单元	18	适用于后屈曲分析和非线性振动问题分析,单元分析形式为解析综合,只能显示单元中点和各节点处的输出结果	同常规平壳单元相比,这类单元增加了一个绕着Z轴转动的自由度。部分单元的几何函数和位移插值函数最高阶数不相同,显示出位移与几何的非协调性。
Q24SF	4节点四边形单元	24	几何函数线性插值 表征X,Y方向的平动位移函数二次插值 Z方向的平动和转动位移值函数均为线性 2×2数值积分	
CT36F	6节点三角形单元	36	几何函数和位移函数均为二次插值 只允许3点数值积分	
CQ48F	8节点四边单元	48	几何函数和位移函数均为二次插值 2×2数值积分	
Q48SPL	8节点三段矩形单元	48	X方向样条插值 Y方向双线性插值 2×2×2高斯积分	样条单元的单元形状均为分段矩形单元,插值方式为样条插值,每一部分的宽度和厚度均相同,但纵向长度可赋予不同值。
Q56SPL	10节点四段矩形单元	56	X方向样条插值 Y方向双线性插值 2×2×2高斯积分	

2. 曲壳单元(Curved Shell Elements)

Diana中的曲壳单元是基于各向材料同性的可退化实体单元,曲壳单元具有与平壳单元完全类似的受力特性。根据节点类型和自由度可以分为 T15SH,Q20SH,CT30S,CQ40S,CT45S 以及 CQ60S 等单元。根据单元的形状特性,可以分为三角形单元和矩形单元。根据单元ζ方向厚度赋予特性,又可以分为普通曲壳单元和分层曲壳单元。采用曲壳单元建模的结构首先应当在结构的中心线或者中轴线处建立结构有限元模型,在确定单元类型之后再对真实结构赋予厚度。曲壳单元由于其形状多为二次和三次曲线且节点个数众多,因而具有良好的边界适应性、较好的单元协调性以及较高的计算收敛性,广泛应用于3D薄壁结构中,尤其在薄壁箱梁桥结构的非线性分析中往往是优先考虑的选项。Diana9.4和Diana10.1版中的曲壳单元见表1-10。

表 1-10

单元名称	单元形状	自由度总数	插值次数及积分方式	特　性
T15SH	3 节点三角单元	15	线性插值 单元内采用 3 点面域积分 厚度方向 ζ 默认为 3 点辛普森积分	每个节点均具有在整体坐标下 U_X, U_Y, U_Z 三个平动自由度和 Rotx, Roty 两个转动自由度，沿着厚度方向材料属性和厚度不分层，一旦确定了某个节点位置处的材料属性或厚度，那么沿着 ζ 方向该节点材料或厚度完全一样。厚度方向 ζ 默认为 3 点辛普森积分，用户也可采用 2 点高斯积分。当厚度方向的积分点多于 3 个时仅仅在非线性计算时适用。
Q20SH	4 节点四边形单元	20	线性插值 单元内采用 2×2 高斯积分 厚度方向 ζ 默认为 3 点辛普森积分	
CT30S	6 节点三角单元	30	二次插值 单元内采用 3 点减缩面域积分 厚度方向 ζ 默认为 3 点辛普森积分	
CQ40S	8 节点四边形单元	40	二次插值 单元内只能采用 2×2 高斯积分 厚度方向 ζ 默认为 3 点辛普森积分	
CT45S	9 节点三角单元	45	三次插值 单元内 7 节点面域积分 厚度方向 ζ 默认为 3 点辛普森积分	
CQ60S	12 节点四边形单元	60	三次插值 单元内 3×3 高斯积分 厚度方向 ζ 默认为 3 点辛普森积分	
CT30L	6 节点三角单元	30	二次插值， 各层 ξ 和 η 面内方向均采用 3 点减缩面域积分 沿着分层的厚度方向 ζ 默认为逐层 2 点积分	每个节点均具有 U_X, U_Y, U_Z 三个平动自由度和 Rotx, Roty 两个转动自由度分层单元，每一层单元可以单独赋予其材料属性、厚度，并且可以在各自的层区内独立进行数值积分。厚度方向 ζ 处采用默认逐层 2 点积分之外还可以用 3 点积分。但高于 3 个积分点情况仅适用于非线性分析
CQ40L	8 节点四边形单元	40	二次插值 各层 ξ 和 η 面内方向均采用 2×2 减缩高斯积分， 沿着分层的厚度方向 ζ 默认为逐层 2 点积分	

曲壳单元每个节点具有沿着整体坐标系下的平动位移 U_X, U_Y, U_Z 以及绕着局部坐标轴 x 轴和 y 轴切平面方向转动位移 Rotx, Roty，应变变量不仅包括沿着局部坐标系 x, y, z 三个方向的正应变 $\varepsilon_{xx}, \varepsilon_{yy}, \varepsilon_{zz}$，也包括三个切应变 $\gamma_{xy}, \gamma_{xz}, \gamma_{yz}$。而曲壳单元中同时存在局部应变和整体应变，通过坐标转换矩阵实现局部应变矩阵和整体应变矩阵之间的转化。与平面应力单元类

似的是,根据平板假定和克希霍夫定律,单元局部坐标系 z 方向的应力值为零。

曲壳单元和平壳单元的应力方式类似,也同时存在着两种类型的应力:柯西应力(Cauchy Stress)和整体弯矩应力(Generalized Moments and Forces)。

在这两种情况中,Diana10.1 的曲壳单元的厚度赋予与平面应力单元类似,在模型的中轴线处建立模型,待建立好结构的几何模型,确定截面形状后再赋予壳单元结构的几何厚度即完成对厚度的赋予。曲壳单元的厚度赋予方式与平面应力单元基本相同,即存在厚度等值赋予和不等值赋予两种情况。如果在曲壳单元中仅仅定义一个厚度值,则 Diana 默认该单元的每个节点均具有相同的厚度,且沿着 ζ 方向。若定义的单元上各个节点厚度值不同,则用户需要在几何属性模块中确定单元截面形状后,对各个节点的厚度值一一给予指定。

通常情况下的曲壳单元可以看做是平面应力单元和弯曲板单元的叠加。有一种说法是:很多时候往往认为壳单元是一种"伪 3D 单元",即壳体单元是介于平面类单元特性和实体单元特性的一种特殊的过渡性 3D 单元,更像是解决三维坐标系下的二维问题。原因是壳单元虽然使用了三维坐标系在结构的中心线或中轴线上建模,但曲壳单元和这两种平面单元赋予厚度的方式完全相同,同时通过在软件截面几何属性部分赋予几何属性特性形成单元。因此本质上说,曲壳单元可以看成一种使用三维坐标系建模的平面应力单元与弯曲板单元的组合。因此 3D 下的三维坐标建立几何模型成了壳单元和平面应力单元唯一的不同点。

下面重点介绍常用的曲壳单元——矩形八节点可退化曲壳单元(CQ40S)。

CQ40S 是八节点组合退化曲壳单元,是作者在后续例题中常用的单元。该单元是一种典型的曲壳类单元,由上下两个曲面以及母线方向的壳体厚度形成的曲面围成。其中 C 表示单元形状为曲边(Curved),Q 表示单元形状为二次的四边形单元(Quadrilateral),40 指的是单元的总自由度个数,S 表示单元类型为壳单元。该单元在厚度方向为单一的厚度层。这种曲壳单元在土木工程有限元建模分析中适用于模拟薄壁钢筋混凝土结构或薄壁箱梁结构,同时对嵌入其中的分布钢筋网片和线性预应力筋 bar 单元也有比较良好的模拟效果。该单元每个节点包含 u_x,u_y,u_z 3 个平动自由度以及 φ_x 和 φ_y 两个转动方向的自由度。通常为了避免单元刚度过大造成刚度矩阵畸变或者是剪力锁死,该壳面单元面内积分采用 2×2 的积分方式,厚度方向选用 3 点辛普森积分,如图 1.3-7 所示。

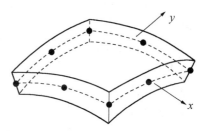

图 1.3-7 CQ40S 曲壳单元示意图

CQ40S 曲壳单元的单元特性见表 1-11。

表 1-11 CQ40S 曲壳单元的单元特性

单元特性	参 数
单元名称	CQ40S
节点个数	8
维数	3D
单个节点自由度	5

(续表)

单元特性	参　数
单元总自由度数	40
插值特性	面内采用2×2高斯积分 厚度方向选用3点辛普森积分
单元内每个节点自由度	$U_X, U_Y, U_Z, Rotx, Roty$
几何参数	壳单元厚度,单元方向对应的整体坐标系方向
适用范围	薄壁箱梁结构、薄壁钢筋混凝土结构

1.3.7 实体单元(Solid Elements)

实体单元又称为固体单元。与之前各类单元不同,实体单元是由四边形与四边形、四边形与三角形以及三角形与三角形相互组合成的各种不同形状的几何体——六面体单元(Hexahedron/Brick Elements)、棱柱体单元(Wedge Elements)以及四面体单元(Pyramid Elements),因此根据实体单元的立体形状的各个面是由相同形状平面还是两种不同形状平面组成,实体单元的积分方式分为两种。对于四面体和六面体这种各个面形状均为三角形或四边形的单元而言,积分形式具有如下规律:单元的节点数越多,插值函数的阶数越高,单元体内的积分点个数往往越多。而对于三棱柱这种顶面和侧面形状不同的单元而言,需要分别考虑 $\xi\eta$ 方向上积分和 ζ 方向上积分,即在四面体或三棱柱体这种单元顶面和底面的三角形或四边形域内的面内积分点和沿着母线 ζ 方向的积分点个数往往不同,积分方式也是这两个方向上的组合。在 Diana 建模中,用户既可以直接采用 Create 创建实体单元再赋予实体单元的单元属性,也可以通过先建立平面单元,再通过 Extrude 和 Sweep 等一系列拉伸方式先创建实体单元再赋予单元属性。与其他单元不同的是,采用实体单元建模不需要对模型赋予几何尺寸特性,这是因为一方面实体单元建模中创立的点就是构成单元整体几何属性的几何点(尤其是同平面应力单元和壳单元相比,实体单元在一开始建模阶段就已确定截面形状和尺寸),另一方面,实体单元采用的建模单元形状均为立体形状,并不需要像其他单元那样赋予厚度方向的几何特性。

实体单元的节点位移主要为沿 X,Y,Z 方向的平动位移 U_x, U_y, U_z,应变变量不仅包括三个方向正应变 $\varepsilon_x, \varepsilon_y, \varepsilon_z$,也包括三个切应变 $\gamma_{xy}, \gamma_{xz}, \gamma_{yz}$。与之对应的应力变量包括三个方向的正应力和三个方向的切应力。

一种特殊的实体单元为组合实体单元。这种特殊单元由参考面、基本单元和组合单元构成。与常规实体单元不同的是,构成单元的每个节点只有一个沿厚度方向即 z 方向平动的自由度。这是因为每一个组合实体单元都有对应的基础单元,这些基础单元均为常规的非分层实体单元,这些单元沿着厚度方向只有单层数值积分。这些基础单元均由之前介绍的常规实体单元组成,组合单元本身没有力学特性,因此对整个有限元模型没有影响。

常见的实体单元类型见表 1-12。

表 1-12 常见的实体单元类型

单元名称	单元形状	单元类型	自由度总数	插值次数及积分方式	特性
TE12L	4节点四面体单元	普通实体单元	12	线性插值 体内1点数值积分	所有单元均为3D单元,单元上各节点位移变量只有沿着X,Y,Z三个方向的平动位移 Ux, Uy, Uz。各单元内的积分点个数均可通过减缩积分减少,有限元几何模型一旦建立成功,再添加实体单元,不需要再添加几何属性。三棱柱单元由顶面和侧面的各自积分点数和积分方式确定。且每种单元内的积分点个数除了默认设置之外还有可供选择的其他合适积分点数。但积分点个数均有上限,一旦超过上限,该单元便不可使用
TP18L	6节点三棱柱单元		18	$\xi\eta$ 方向线性面插值 ζ 方向线性插值 三角域内即 $\xi\eta$ 方向1个积分点 ζ 方向上2个积分点	
HX24L	8节点六面体单元		24	线性插值 体内 2×2×2 高斯积分	
CTE30	10节点四面体单元		30	二次插值 体内4点数值积分	
CPY39	13节点四面体单元		39	二次插值 体内13个点的数值积分	
CTP45	15节点三棱柱单元		45	二次插值 $\xi\eta$ 三角形域内4个积分点 ζ 方向上2个积分点	
CHX60	20节点六面体单元		60	二次插值 体内 3×3×3 高斯积分	
CTE48	16节点四面体单元		48	三次插值 体内64个点数值积分	
CTP72	24节点三棱柱单元		72	三次插值 三角形面 $\xi\eta$ 方向9个积分点, ζ 方向上4个积分点	
CHX96	32节点六面体单元		96	三次插值 体内 4×4×4 高斯积分	
HX25L	8节点六面体单元	橡胶特性单元	25	线性插值 体内 2×2×2 高斯积分	均由六面体单元组成,且积分方式均为高斯积分,适用于超弹性非线性分析
CHX64	20节点六面体单元		64	二次插值 体内 3×3×3 高斯积分	
组合实体单元名称	组合实体单元形状	组合实体单元特性	自由度数	对应的基础单元	
T3CMP	3节点曲面三角单元	由参考面、基础单元和组合单元组成。每个单元只有一个沿着 ζ 方向的平动自由度。组合单元本身没有	3	TP18L	
CT6CM	6节点三角曲面单元		6	CTP45	
CT9CM	9节点三角曲面单元		9	CTP72	
Q4CMP	4节点矩形曲面单元		4	HX24L	

(续表)

组合实体单元名称	组合实体单元形状	组合实体单元特性	自由度数	对应的基础单元
CQ8CM	8节点矩形曲面单元	力学特性，对整个有限元模型没有影响	8	CHX60
CQ12C	12节点矩形曲面单元		12	CHX96

在这里重点介绍一种常用的实体单元——CHX60，如图1.3-8所示，该单元由20个节点的六面体单元构成，位移插值曲线方程最高阶次数为2次，体内积分方式为3×3×3高斯积分。适用于模拟大坝、边坡或管廊等适用于实体单元建模的大体积工程结构，在大体积混凝土或工程方桩的耐久性及水化热反应分析中也有广泛的应用。由于大体积混凝土或方桩多数为六面体形状，而该单元恰为六面体单元，使用该单元往往具有良好的边界适应性和非线性计算收敛性。单元与单元接触面之间还可以添加面对面接触的界面单元用于模拟结构粘贴碳纤维布等公路桥梁加固工程，在采用实体单元建模的结构可靠度计算中具有广泛的应用。

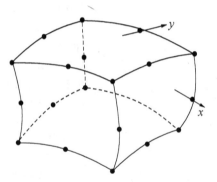

图1.3-8 CHX60实体单元示意图

1.3.8 钢筋单元（Reinforcements）

Diana中独有的嵌入式钢筋单元用于模拟纵筋、箍筋、架立钢筋、预应力钢绞线，故广泛适用于嵌入所有类型的单元中。所谓嵌入就是指不需用户手动建立钢筋与混凝土之间粘结，钢筋单元可以自动进入各种类型的混凝土单元中并且可与周边混凝土单元自动耦合，一起贡献刚度。这也是Diana软件相对于当前其他有限元软件独有的优势之一。Diana中的钢筋单元主要有两种，一种是用来模拟纵向钢筋和预应力钢绞线的Bar单元类型，另一种是用来模拟分布钢筋和钢筋网片的Grid单元类型，如图1.3-9及图1.3-10所示。对于Bar单元而言，需要选择钢筋材料本构类型和截面面积，同时通过建模点个数来确定Bar钢筋单元的形状。Grid单元通常用来模拟嵌入和各种类型单元的分布钢筋，Grid的定义方式通常有间距、直径和等效厚度两种。Bar钢筋单元的离散方式通常情况下有两种，一种是截面Section Wise类型，另一种为Element by Element类型，在结构的非线性分析中，采用前者往往会比后者具有更好的收敛性。

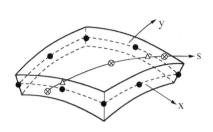 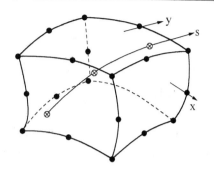

图 1.3-9　曲线 Bar 钢筋在曲壳单元中　　图 1.3-10　曲线 Bar 钢筋在实体单元中

预应力混凝土结构往往要求预应力筋的布筋形状与外荷载作用形式相匹配以储存预应力筋压应力从而抵消外荷载作用的影响，如纯弯结构要求在受压区布置直线筋，集中荷载需要布置折线形预应力筋，均布荷载则需要布置曲线形等效预应力筋，这时需要预应力筋相对于结构的中性轴有一个偏心距离，此时有两种处理方法，一种方式是直接通过建立 Bar 钢筋单元上各点坐标值来确定预应力筋的位置，这种方式适用于 Bar 单元单根预应力筋曲线形状阶数较低的情况；另一种是采用 Diana 软件自带的偏心距定义功能较为精确地模拟阶数较高曲线筋或含有转向块预应力筋各点作用位置。

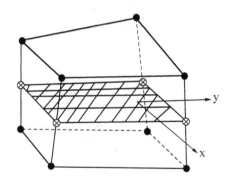 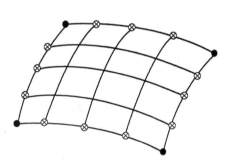

图 1.3-11　Grid 钢筋在实体单元中　　图 1.3-12　Grid 钢筋在平面应力单元中

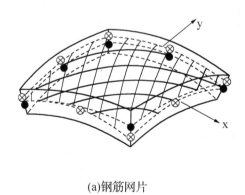 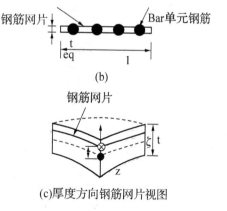

(a)钢筋网片　　　　　　　　　　(c)厚度方向钢筋网片视图

图 1.3-13　Grid 钢筋单元在曲壳单元中

钢筋单元的材料特性在第二章钢筋单元的材料属性中会详细介绍。

1.3.9 界面单元(Interface Elements)

在 Diana 软件中,界面单元的类型众多,功能也非常强大,足以满足各种单元类型下的界面连接。广义的界面类单元包括两个单元之间相互连接的界面单元(Interface)和兼具约束和双重特征的边界(Boundary interface)单元。其中根据分析类型的不同又可以将边界单元分为结构类边界单元(Structural interfaces)和热流边界(Heat Flow Boundary)单元。对上述各类界面单元均可在 Diana 中进行本构模型和几何特性的定义。本书主要介绍单元与单元相互连接的结构类界面单元。总体归纳起来主要有点对点连接的界面单元(1—1 Nodes)、点对实体连接(Point—Solid)界面单元、壳单元线对线连接(Line—Shell)的界面单元、线对实体连接的界面单元(line—Solid)以及实体单元之间采用面对面连接的界面单元,见表1-13。

表 1-13

界面单元接触方式	界面单元单元名称	界面单元接触特点	与界面相匹配母单元类型
点对点连接	N4IF	2D,1—1 nodes	2D梁单元,平面应力单元,平面应变单元
	N6IF	3D,1—1 nodes	3D梁单元,壳单元
线对线连接	CL12I	2D,3—3 nodes	平面应力单元,平面应变单元
	CL20I	2D,5—5 nodes	平面应力单元,平面应变单元
	CL24I	3D,3—3 nodes	曲壳单元
	CL32I	3D,4—4 nodes	曲壳单元
	L8IF	2D 2—2 nodes	平面应力单元,平面应变单元
	L16IF	3D,2—2 nodes	曲壳单元
	L20IF	3D,3—2 nodes	曲壳单元
点对实体连接	TE15IF	3D,1—4 nodes	线性四面体实体单元
	TP21IF	3D,1—6 nodes	线性三棱柱实体单元
	HX27IF	3D,1—8 nodes	线性六面体实体单元
	CTE33I	3D,1—10 nodes	二次四面体实体单元
	CTP48I	3D,1—15 nodes	二次三棱柱实体单元
	CHX63I	3D,1—20 nodes	二次六面体实体单元
线对实体连接	L12IF	3D,2—2 nodes	平面应力单元,平面应变单元
	CL18I	3D,3—3 nodes	实体单元
	TE18IF	3D,2—4 nodes	线性四面体实体单元
	TP24IF	3D,2—6 nodes	线性三棱柱实体单元
	HX30IF	3D,2—8 nodes	线性六面体实体单元(与线性直线接触)
	CTE39I	3D,3—10 nodes	二次四面体实体单元(与二次曲线接触)

(续表)

界面单元接触方式	界面单元单元名称	界面单元接触特点	与界面相匹配母单元类型
线对实体连接	CTP54I	3D,3—15 nodes	二次三棱柱实体单元(与二次曲线接触)
	CHX69I	3D,3—20 nodes	二次六面体实体单元(与二次曲线接触)
面对面连接	T18IF	3D,3—3 nodes	实体单元
	Q24IF	3D,4—4 nodes	实体单元
	CT36I	3D,6—6 nodes	实体单元(三角形面接触)
	CQ48I	3D,8—8 nodes	实体单元(矩形面接触)

界面单元中由于存在过渡单元,数量庞大,本书在这里挑选三种较为典型和常用的界面单元进行说明。

1. N4IF

N4IF 是一种同时存在于 9.4 版和 10.1 版本中的 2D 界面单元,如图 1.3-14 所示。这种界面单元通过点对点形成连接,该单元可通过控制材料的刚度从而实现模拟类似弹簧单元和单向受力桁架杆的作用与功能,在 Diana10.1 版本中还可以通过设置应力与应变关系生成材料的各项刚度曲线,是经常用于模拟 2D 结构模型中剪力墙构件中连系梁、新型材料的耗能杆件等抗震结构的一种 2D 界面单元,适用于桁架单元、2D 梁单元、平面应力单元和平面应变单元。

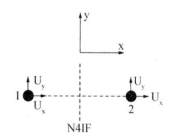

图 1.3-14 N4IF 界面单元示意图

N4IF 单元为点对点连接,其中沿着两点之间连线方向,即界面接触方向为界面单元的局部坐标 x 方向;垂直于接触方向,即界面单元方向为局部坐标 y 方向,接触点具有沿坐标系 X 及 Y 方向的平动位移 Ux,Uy。界面单元的材料本构特性有众多选择,常见的选择包括线弹性本构关系(Linear Elasticity)、非线弹性本构(Nonlinear Elisticity)关系以及库仑摩擦(Column friction)本构关系。每种材料本构关系都需要定义平面内法向抗压刚度(Normal Stiffness)、平面内抗剪刚度(Shear Stiffness in y)。根据不同界面单元所要模拟的对象不同力学性能的抗拉刚度可选择抗剪刚度为常数(No tension with constant shear stiffness)、抗拉刚度为零且抗剪刚度降低(No tension with reduced shear stiffness)以及用户自定义等三种模式,用户自定义模式针对抗剪刚度曲线和抗拉刚度曲线均有特殊需求的,通过选择编辑文本文件方式生成抗拉、抗压、抗剪刚度曲线进行用户自定义从而达到更为精细的模拟,然而从非线性计算收敛性角度而言,第三种方式的收敛性往往要低于前两种。其中线弹性本构关系和非线弹性本构关系截面材料的抗压刚度与抗剪刚度相互独立,而库仑摩擦本构模型中抗压刚度与抗剪刚度之间具有非独立性,通过库仑摩擦角来承接界面单元材料的抗压刚度和抗剪刚度关系。

以上关于 N4IF 的诸多材料特性不仅适用于该单元,也同样可以经过拓展后推广至所有的点对点接触的界面单元。

2. CL24I

CL24I 是一种专用于曲壳单元线对线连接的界面单元,通过组成壳单元之间边缘线上的 3 对节点相互接触形成线对线的界面单元,该单元采用二次插值,在 Diana10.1 中默认单元的数值积分方式为纵向(ξ 方向)采用 3 点牛顿科特斯积分,厚度方向采用 3 点辛普森积分。单元的局部坐标系 x 方向为单元内第一个节点指向第二个节点的方向,局部坐标系的 y 轴垂直于 x 轴。z 方向为平面外的厚度方向。该方向的单元位移变量主要为沿着 x,y,z 三个局部坐标轴方向的平动位移 U_x, U_y, U_z 和绕着 x 轴的转动位移 φ_x。局部坐标系的 x 轴方向为相互接触组成壳单元的边缘线(Edge)第一个节点指向第二个节点的方向。界面单元的局部坐标系朝向符合右手系定律。如图 1.3-15 所示。

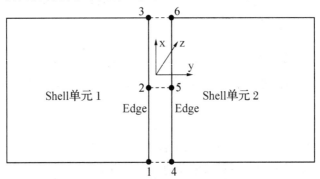

图 1.3-15 CL24I 界面单元接触方式

特别要注意的是,在新版的 Diana10.1 中,线对线接触的界面单元有两种定义方式,一种是方向向量平行于壳平面(Direction vector parallel to shell plane),另一种是方向向量垂直于壳平面(Direction vector normal to shell plane)。前者需要定义界面单元的接触方向(局部坐标系下 y 方向)对应的整体坐标系方向,后者需要定义局部坐标系 z 轴方向对应整体坐标系方向。之所以在 Diana 中确定,是因为局部坐标系中的 y 轴方向代表最基本的拉伸或压缩方向的单元刚度,一旦确定了 y 轴对应的整体坐标系的方向,其余两个局部坐标轴对应于整体坐标系下的方向的材料属性自然也就不言自明。

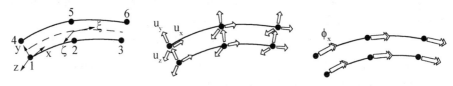

图 1.3-16 CL24I 单元位移变量

CL24I 的材料本构特性有众多选择,常见的有线弹性本构关系(Linear Elasticity)、非线弹性本构(Nonlinear Elisticity)关系、粘结滑移关系(Bond-slip)、非线弹性(Nonlinear elasticity)、非线弹性摩擦(Nonlinear elasticity)以及库仑摩擦(Column friction)等。CL24I 的单元接触方式为三维壳单元之间的线对线接触(3D line interface between shells)。与 N4IF 不同的是,CL24I 单元除了需要定义平面内法向抗压刚度、平面内抗剪刚度,还需要定义平面外的抗剪刚度。根据不同界面单元所要模拟的对象不同的力学性能,抗拉刚度可选择抗剪刚度为常数(No tension with constant shear stiffness)、抗拉

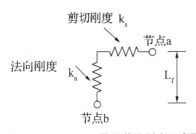

图 1.3-17 CL24I 界面单元刚度示意图

刚度为零且抗剪刚度降低(No tension with reduced shear stiffness)以及用户自定义等三种模式。用户自定义模式针对抗剪刚度曲线和抗拉刚度曲线均有特殊需求的,通过选择编辑文本文件方式生成抗拉、抗压、抗剪刚度曲线实现用户自定义从而达到更为精细的模拟。然而从非线性计算收敛性角度而言,第三种方式的收敛性往往要低于前两种。其中线弹性本构关系和非线弹性本构关系截面材料的抗压刚度与抗剪刚度相互独立,而库仑摩擦本构模型中抗压刚度与抗剪刚度之间具有非独立性,通过库仑摩擦角来承接界面单元材料的抗压刚度和抗剪刚度关系。

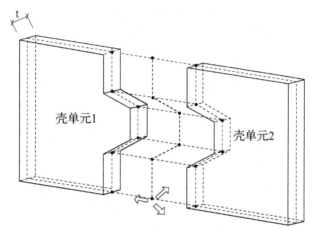

图 1.3-18 CL24I 界面单元力学行为示意图

3. CQ48I

CQ48I 作为一种实体面与面之间的连接单元,在两个相互接触的实体单元面之间定义面对面在接触的各个方向(法向、切向)的刚度形成界面单元。单元局部坐标系 x 轴为接触面上单元的第一个节点指向第二个节点的方向,y 方向为接触面法线方向,z 轴方向垂直于 x 与 y 轴确定的平面,即垂直于单元面的面外方向。

CQ48I 的材料特性常见的选项包括线弹性(Linear Elasticity)本构关系、非线弹性(Nonlinear Elasticity)本构关系以及库仑摩擦(Coulomb friction)本构关系。与 N4IF 不同的是,CQ48I 的单元接触方式为三维面对面接触(3D Surface interface)。CQ48I 单元除了需要定义平面内法向抗压刚度(Normal Stiffness)、平面内抗剪刚度,还需要定义平面外的抗剪刚度。根据不同界面单元所要模拟的对象不同的力学性能,抗拉刚度可选抗剪刚度为常数(No tension with constant shear stiffness)、抗拉刚度为零且抗剪刚度降低(No tension with reduced shear stiffness)以及用户自定义等三种模式。用户自定义模式针对抗剪刚度曲线和抗拉刚度曲线均有特殊需求的,通过采用界面操作编辑文本文件方式生成抗拉、抗压、抗剪刚度曲线从而实现用户自定义需求,以达到更为精细的模拟。然而从非线性计算收敛性

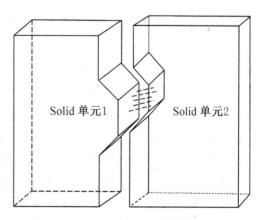

图 1.3-19 面对面接触图示

角度而言，第三种方式的收敛性往往要低于前两种。在 CQ48I 中线弹性本构关系和非线弹性本构关系截面材料的抗压刚度与抗剪刚度相互独立，而库仑摩擦本构模型中抗压刚度与抗剪刚度之间具有非独立性，通过库仑摩擦角和粘聚系数来承接界面单元材料的抗压刚度和抗剪刚度关系。

结合前文关于实体单元 CHX60 的介绍，该单元与 CH60X 单元是一对"黄金组合"。首先从单元形状而言，CQ48I 与 CHX60 这两种单元均为二次单元，位移插值函数均为二次，每个单元上均有 8 个节点。这就使得单元之间有较好的边界适应性，同时在对 CQ48I 单元添加合理的本构属性后也具有较好连接和接触效果，从而在计算中具有比较高的收敛性，特别适合模拟工程上含有 CFRP 碳纤维加固片等加固领域的工程结构。与此同时，该单元也同样适用于曲壳单元面之间的面对面接触问题。

1.3.10 接触单元(Contact Elements)

接触单元是一种特殊类型的界面单元，接触区域的形成必须由两个部分构成，一个是接触体(Contacters)，另一个是目标体(Targets)，接触方式可以是点对点、点对线、点对面接触，既有 2D 平面下的接触单元，也有 3D 条件下的接触单元。其中 2D 单元的位移变量只有沿着 X 轴和 Y 轴两个方向的平动变量 U_X，U_Y，3D 位移变量仅增加了一个沿着 Z 向的平动自由度 U_Z。与此对应的应力只有正应力，无切向剪应力。

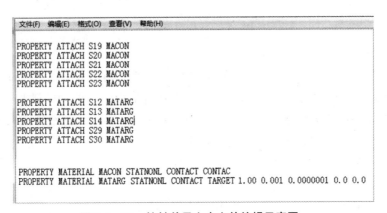

图 1.3-20 接触单元文本文件编辑示意图

在生成接触单元的具体操作流程中，Diana10.1 版本与之前的 9.4 版本有着极大的不同。以往的 9.4 版本命令流操作中只需先指定目标单元和接触单元的具体单元名称，再划分网格即可生成对应的接触单元。而在 Diana10.1 版的实际建模过程中，用户需要通过界面操作或 Python 语句指明哪些点、线或面作为接触单元，哪些点、线或面作为目标单元。网格划分前应设置好接触单元的形状。对于接触单元的材料本构特征，用户不需要输入具体的参数值，只需指定 Contacts 以表明这种材料是接触单元本构材料即可。而对于目标单元，用户则需要赋予目标单元名称、接触单元刺入的最大相对深度值(PENETR)、接触单元节点接触目标单元面后的相对距离，以及库仑摩擦系数和粘聚系数等一系列参数值。在实际接触过程中，接触单元能否顺利生成还与坐标轴的朝向有关。对于 2D 单元而言，y 轴的正向必须指向目标单元的外向；对于 3D 单元而言，z 轴的正向必须指向构成目标单元所在面的外法向。

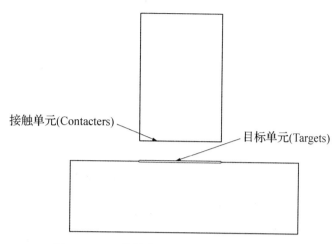

图 1.3‑21 接触单元与目标单元作用示意图

Diana 中接触单元类型众多,常见基本接触单元见表 1‑14。

表 1‑14 基本接触单元

单元名称	单元维数	单元形状	总自由度数	插值方式	特　点
L4CT	2D	2 节点直线单元	4	线性插值	单元上的每个节点只有沿着 X 和 Y 两个方向的平动自由度,且单元的节点必须位于 XOY 平面内
CL6CT		3 节点曲线单元	6	二次插值	
T9CT	3D	3 节点三角单元	9	线性插值	单元上的每个节点只有沿着 X,Y,Z 三个方向平动自由度,单元局部坐标系下 x 轴正向为单元第一个节点指向第二个节点的方向
Q12CT		4 节点四边形单元	12	线性插值	
CT18C		6 节点三角单元	18	二次插值	
CQ24C		8 节点四边形单元	24	二次插值	

1.3.11　弹簧单元(Spring Elements)

弹簧单元适用于两个点对点接触的有限元模型,分为位移类弹簧(Discrete Translation Spring/Dashpot)和转动类弹簧(Discrete Rotation Spring/Dashpot),其中位移类弹簧和转动类弹簧均可分为一点连接(one-node connection)和两点连接(two-node connection),名称分别为 SP1TR,SP2TR,SP1RO,SP2TR。与定义单元类型情况类似,Diana 10.1 版与 Diana 9.4 版定义弹簧单元的操作方法也有所不同。在 Diana10.1 界面中,弹簧单元隐藏在上方菜单栏几何连接(Edit connection property)的连接类型(Connection type)名目下。选择弹簧(Spring)模块后,用户需要确定添加弹簧单元的类型是位移类弹簧(Discrete Translation Spring/Dashpot)还是转动类弹簧(Discrete Rotation Spring/Dashpot),如图 1.3‑22所示。选择弹簧单元类型特征后,用户需要选择弹簧本构模型,一般有两种选择,即线弹性(Linear elasticity)本构模型和极限力(Ultimate forces)模型。前者需要定义弹簧刚度系数(Spring stiffness)和阻尼系数(Constant damping coefficient),如图 1.3‑23 所示;而后者除需定义上述内容外,还需要进一步定义力(位移类弹簧)或力矩(转动类弹簧)的最大值和最小值，如图 1.3-24 所示。

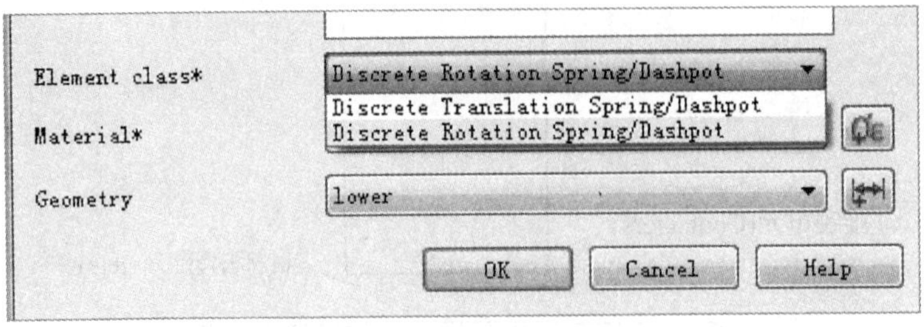

图 1.3-22　Diana10.1 中弹簧/阻尼单元类型

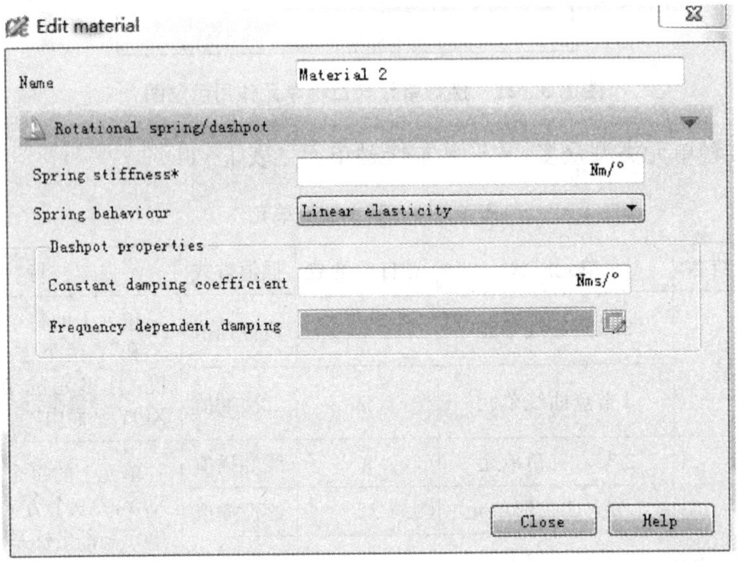

图 1.3-23　Diana10.1 中转动弹簧/阻尼单元线弹性模式下本构选择

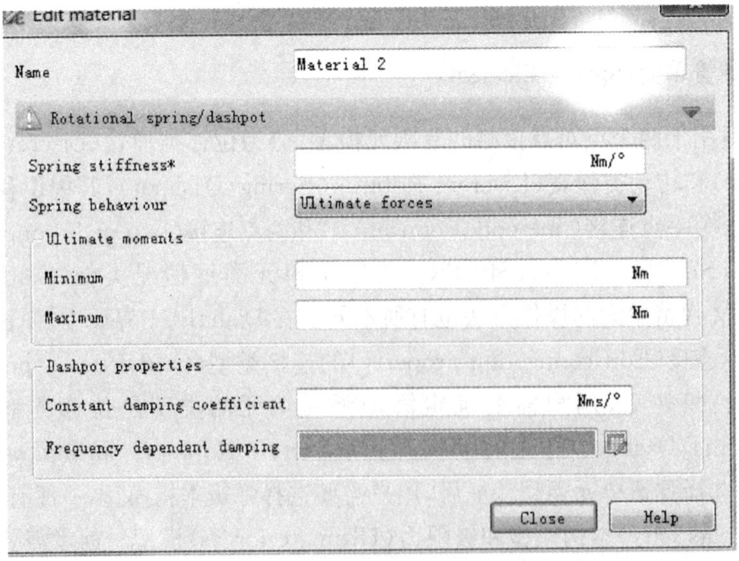

图 1.3-24　Diana10.1 中转动弹簧/阻尼单元极限力模式下本构选择

而位移类弹簧单元的本构采用力—位移关系进行描述,转动弹簧单元 SP2RO 的本构采用弯矩—转角关系进行描述,转动弹簧单元的转角通常用于对框剪结构节点核心区的剪切变形的模拟。因此不仅需要对模拟节点进行几何上简化,还需要进行力矩等效。

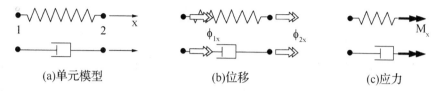

(a)单元模型　　　　(b)位移　　　　(c)应力

图 1.3－25　转动弹簧单元两点接触(SP2RO 单元)

具体建模时,将节点处的梁单元跟柱单元两个节点进行自由度耦合,包括水平方向和竖直方向,两个节点之间再通过 SP2RO 单元连接,从而达到传递弯矩的作用。建模时可以在需要添加弹簧单元的部位直接进行竖向自由度的耦合,以模拟剪力传递。

弹簧/阻尼单元适合在土木工程结构中模拟阻尼系数很大的隔震阻尼装置和隔震支座模型,另外一些抗震耗能结构的模拟也会用到这种单元。

1.4　Diana10.1 文件系统

在 Diana10.1 中,前处理建模包含 DianaIE 和 iDiana 两个不同模块,因此文件种类繁多,包括后缀格式为.dpf,.py,.dat,.bat 等二进制建模文件名、文本文件名和 Python 语言文件名。其中 DianaIE 在建模的可视化界面操作中采用 py 语言记录,并且兼具前处理、计算、后处理及历史记录文件功能于一身。Diana10.1 版各个后缀文件类型以及功能属性见表 1-15。

表 1-15　各个后缀文件类型以及功能属性

文件后缀名称	文件归属	文件类型	文件功能
.dpf	DianaIE	二进制文件	DianaIE 模型储存文件
.py	DianaIE	文本文件	记录操作的命令流
.dat	DianaIE,iDiana	文本文件	记录模型文件节点、单元、材料属性、荷载工况、单元数量及边界条件等参数特性的文本文件
.bat	iDiana	二进制文件	iDiana 模型参数储存文件及输入材料特性参数文本文件
.dcf	iDiana	文本文件	添加计算控制命令
.out	DianaIE,iDiana	文本文件	记录计算每一步的荷载以及输出计算结果
.G72	iDiana	二进制文件	前处理建模过程中形成的二进制文件

（续表）

文件后缀名称	文件归属	文件类型	文件功能
.V72	iDiana	二进制文件	计算结果生成后形成的后处理二进制文件
.his	iDiana	文本文件	记录建模过程中的历史记录文件
.ff	DianaIE, iDiana	二进制中转文件	计算过程中形成的二进制中转文件
.dnb	DianaIE	二进制文件	DianaIE 计算中生成的二进制文件

Diana9.4 和 Diana10.1 版中各类文件关系宏观逻辑如图 1.4-1 所示。

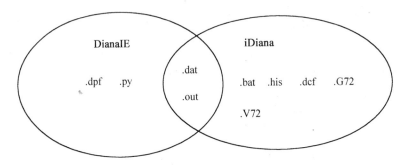

图 1.4-1　Diana9.4 和 Diana10.1 版中各类文件关系宏观逻辑图

DianaIE 文件中各个后缀文件类型转化示意如图 1.4-2 所示。

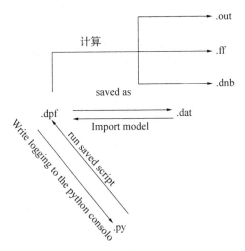

图 1.4-2　DianaIE 中各个后缀文件类型转化示意图

iDiana 各个后缀文件类型转化示意如图 1.4-3 所示。

Diana 软件中下的 DianaIE 模块和 iDiana 模块，有各自的文件打开、编辑、保存和关闭方式。无论是 DianaIE 文件还是 iDiana 文件，用户既可以采用应用菜单栏下的 File 文件相关操作打开文件，也可以直接通过键盘上 Ctrl 系列操作创立、打开、关闭、保存或者运行文件。对于老版的 Diana9.4 而言，当用户打开二进制模型数据库文件，也就是以 .bat 后缀结尾的二进制前处理建模文件时，不可直接采用 File→Open 的方式打开，而应该首先选择该

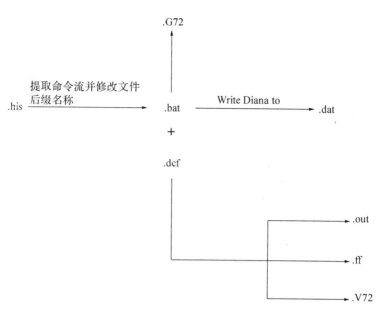

图 1.4-3　iDiana 模块各后缀文件类型转化示意图

.bat 文件所在的文件夹工作路径（Working directory），待将工作路径选取进 iDiana 之后，在模型树上寻找 Utility→Read→batch 后再在下方的命令流输入栏输入文件名称方可将文件打开。对于.dcf 计算控制文件，用户则需要在 Diana_w 的模块中选择该计算控制文件所在的路径后再添加进入，在计算的同时会生成二进制的中转.ff 文件和计算结果的.out 文件。为了不影响下一次的计算，用户在 iDiana 设置中可以选择让系统自动删除这些以.ff 结尾的二进制中转文件。对于 iDiana 模型数据库.dat 文件，用户可直接在应用菜单栏下用 File→Open 的方式打开。

而对于 DianaIE 而言，其操作界面对于文件的打开方式相对于老版本的 iDiana 操作界面不仅非常方便，而且打开路径清晰明了。首先，DianaIE 针对不同性质的文件统一在应用菜单栏下设置了不同打开方式，主要分为以下四大类：

（1）针对.dpf 二进制文件，采用应用菜单栏下 File→Open→双击该.dpf 文件名的方式即可打开.dpf 文件。当文件间需要转换的时候，直接采用 File→Open→Discard→点击下一个.dpf 文件即可实现不同.dpf 文件之间的转换。

（2）针对已经生成的模型数据库.dat 文件，在应用菜单栏下以 File→Import model 方式打开。

（3）采用 Python 编辑的以.py 结尾的模型命令流文件，在应用菜单栏下以 File→Run saved script 方式打开。

（4）针对 IGES，STEP IGES，STEP 等 CAD 模型文件，采用 File→Open→Import CAD/CSV file 操作路径即可打开。

1.5　Diana10.1 图形用户界面

Diana 的图形用户界面主要由以下板块组成：应用菜单栏、快捷工具栏图标、模型树、属性、python 语言命令流输入区、信息提示栏、GUI 可视化界面区。Diana10.1 图形用户界面各个板块位置如图 1.5-1 所示。

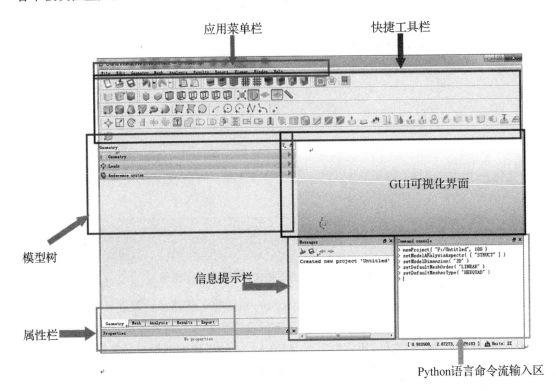

图 1.5-1　Diana10.1 中 DianaIE 用户图形界面

1. 应用菜单栏

应用菜单栏中包括用户所有需要的操作，菜单栏分为 File 栏、Edit 栏、Mesh 栏、Analysis 栏、Viewer 栏、Window 栏和 Help 帮助栏。下面重点介绍应用菜单栏中常用的几栏。

（1）File 栏

File 栏是文件操作方式栏，包括文件的打开、生成、创立、保存、建模。DianaIE 中文件的操作方式主要有创建二进制.dpf 文件(File→New)、打开二进制 dpf 文件(Open)、保存二进制 dpf 文件(Save/Save as)、导入文本文件(.dat 文件)并转换为二进制.dpf 文件(Import model)、读取 Python 语言的文本文件为.dpf 二进制文件(Run saved script)。Diana 菜单栏不仅可以打开或导入各种不同格式下的文件，还可以将 CAD 格式的文件导入并转化成GUI 用户可视化操作界面(run CAD/CAS file)。DianaIE 不同后缀名称的文件之间的转换方式前文已经介绍，这里不再赘述。

（2）Edit 栏

Edit 栏主要分为 Undo(撤回操作)、Redo(重复上述操作)、Preference(界面操作设置)、文件编辑项设置(Project settings)、亮度操作、工作框移动(Move working plane)、工作框尺寸调整(Working plane Grid)、两点间距离调整(Diatance between points)八类操作模式。其中 Preference 和 Project settings 为两种最常用的操作方式。通过 Preference 设置，用户

可以选择DianaIE界面外观颜色、背景颜色、节点颜色、线宽和节点尺寸等一系列设定方式，同时用户还可以在其界面下对模型几何颜色和尺寸相应设定。值得一提的是，在General选项设置中，用户可以决定在Diana重复计算中是否每次自动删除上一次生成的后缀名称为.ff的二进制中转文件Filos Files(Automatically remove Filos Files)，是否将界面操作自动记录成Python命令语句文件(Write logging to the python console)，以及Python命令流语句文件的储存位置(Logging Directory)。Preference界面下这三种操作对Diana10.1初学者至关重要，用户可以通过这三项设置在循环计算中自动减少不必要的计算中转文件，还可以根据生成的命令流Python文件学习Diana10.1中的Python语句编辑。

（3）Analysis栏

Analysis栏给出的操作功能众多，可以说是整个Diana10.1软件界面操作中最核心的部分。Analysis栏的子菜单栏中的Create，Modify，Analysis，Load，Supports，Functions，Materials以及Element geometries将众多功能基本覆盖。Create模块主要是点、线、面、体的创建，Diana中不仅可以创立直线，还可以创立曲线，还可以根据实际结构的需要创立多段直线、封闭直线和曲面形状。Modify模块主要是关于几何体形状的移动(Move Shape)、伸缩(Scale shape)、拉伸(Extrude)、扫略(Sweep)、镜像对称(Mirror Shape)、复制平移(Array Copy)等一系列的图形的变换操作。Laod模块主要是施加和定义荷载。Supports模块主要功能是对结构施加约束。在Material和Element geometries部分则是赋予模型中结构的本构及截面几何特性，与快捷栏中的黄、蓝、红三色图标定义效果有异曲同工之妙。

（4）Mesh栏

Mesh栏主要是对已经完成几何建模、材料属性赋予和约束荷载施加之后的结构模型网格属性的选择和定义。在Mesh状态栏下也同样存在着Load，Support，Material，Element geometries和Functions栏，只不过这些是在Mesh栏的模型树模块下对已生成网格单元后的结构模型进行修改和添加。

（5）Results栏

Results栏主要是对计算后的变形结果进行查看。用户既可以选择查看某一个视角下的结果(例如Normalized deformed results)，也可以查看全视角下的变形结果(Absolute deformed results)。此外还可以查看变形前后的单元网格结果(例如Undeformed/Deformed mesh feature edges)。

（6）Viewer栏

Viewer栏用于对计算后的各项结果进行选择性的查看。在Viewer栏中，用户既可以选择查看网格划分前的图形(View geometry)，又可以查看计算后的结果(View results)。还可以转换各种图形视角(如Fit all, perspective projection, show mesh seedings)。当计算完毕后，用户可以在Viewer中选择查看某一部分或形状的各项输出信息(Node/Shape/Face/vertex selection)、单元信息(Element selection)和钢筋单元的信息(Reinforcement selection)。

（7）Window栏

Window栏中包含窗口子菜单栏(Window panes)、工具栏(Tool bars)、模型部分(Mesh

sections)和网格部分(Mesh sections),这些部分包含了模型树和 Analysis 中各种功能。用户可以在该子菜单栏下选择显示哪些功能,忽略哪些功能。

(8) Help 栏

Help 栏中主要包含了 Diana 用户手册、发行注释(Release notes)、Diana 激活许可(Activate new Diana license)、Diana 更新许可(Update Diana license),以及关于 Diana10.1 版交互环境下的使用状态(About Diana Interactive Environment)。当使用 Diana 软件时间超过了交互环境下的许可时间时,用户需要申请重新激活软件方可继续使用。

2. 快捷工具栏图标

快捷键工具栏图标主要为各种图形按钮,用户通过点击图标按钮可快速进行点、线、面、体的几何建模,快速实现扫略(Sweep)、拉伸(Extrude)以及逻辑布尔运算加减的操作。同时还可以进行赋予混凝土和钢筋材料属性以及界面单元的生成等一系列操作。快捷工具栏图标中的功能与应用菜单栏的部分功能完全相同。快捷工具栏图标均为常用的一些界面操作图标。

3. 模型树

模型树作为用户 GUI 界面操作的主体展示功能之一,需要配合使用应用菜单栏和快捷工具栏图标。即用户一旦通过上述两个前处理建模方式添加或生成了某种操作,会在模型树下方展现相应的单元分组和操作结果,用户可以通过右击的方式对模型树进行名称修改以及属性的赋予。在 Analysis 中模型树可以直接添加分析类型(Anlysis type)和进行荷载模块(Load block)的设置。

4. 属性栏

属性栏作为界面操作的一种结果信息生成栏,是对前一步操作成功的一种反馈。DianaIE 属性窗口采用参数化控制。属性栏中各属性表格用于记录在建模中对应的每一个成功定义的属性所具有的数值或特征。Geometry(几何建模操作)、Mesh(划分操作)、分析(Analysis)、结果查看(Results)和输出(Export)模块内都有各自的属性列表。其中结果查看(Results)栏还可以根据用户个人的喜好在属性列表中进行后处理结果的设置。

5. Python 语言命令流输入区(Command console)

命令流栏采用 Python 语言记录和输入每一步操作生成的命令流信息,在软件中勾选"用 Python 语言记录每一步操作命令(Write loggings to python)"之后,用户在可视化 GUI 界面采用界面操作的每一个步骤都会在 Python 语言命令流输入区记录下来。用户也可以先在文本文件中编辑好 Python 语言命令流,然后将整段命令流复制粘贴到 Command console 区运行 Python 语言命令流。除此之外,由于 Python 功能的多样性和通用性,用户还可以采取界面操作和 Python 语言编辑混合使用的操作方式,例如可将事先编辑好的一部分 Python 语言粘贴到 Command console 区域进行运行,待出现 GUI 可视化操作界面后用户可以直接在可视化界面上进行 GUI 界面操作,这种方式可以轻而易举地实现 Python 语言和 GUI 用户可视化界面的转化。用户在使用 Python 命令流建立模型的同时,既可以通过前文所述的界面操作 File→Run saved script 方式打开后缀为 .py 的文件,也可以采用复制粘贴的方式将 Python 命令流复制粘贴到 Command console 区中,且后一种方式更容易逐条逐句查找命令流的错误。当 Python 语句中出现较为明显的错误时,计算会停止,在

Message 的信息提示栏会出现红色的针对 Python 语言命令流的报错提示,如图 1.5-2 所示,这时用户可以采取逐段或逐行提取 Python 语句并输入 Command console 区这种更有效的方式,以便查找错误之处与出错原因。

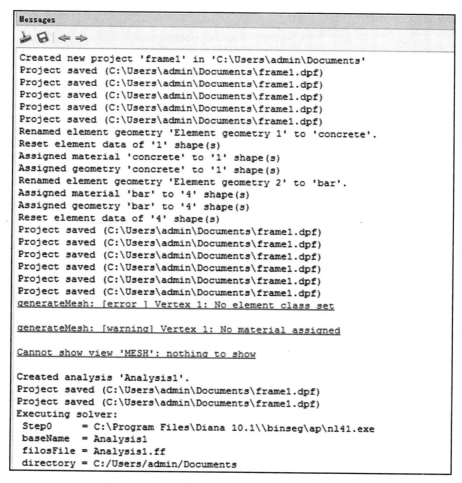

图 1.5-2　Message 的信息提示栏命令流报错提示

6. 信息提示栏(Message)

建模过程中所有的错误都可以在信息提示栏中找到。信息提示栏主要有两个功能,一个是提示各种文件在建模时阶段性状态(打开、导入、保存、关闭等),如果文件在上述的某一状态下发生错误,Diana 软件会自动在信息提示栏用红色的英文字母及时提示错误以便用户进行问题的查找,如图 1-36 所示;另一个功能是提示建模文件在建模阶段和计算阶段的错误,这时信息提示栏会根据错误的程度出现两种不同颜色的提示,一种是黄颜色的严重程度为警告(SEVERITY:WARNING)类错误,另一种是红颜色的严重程度为中断(SEVERITY:ABORT)类错误。如果文件在运行中的某行出现黄色字体的 WARNING 警告,说明该处的错误不影响计算顺利进行,尚可以完善和改进,但可能会对计算结果的精度造成影响。如果出现红颜色的提示,表明该处的问题非常严重(Severity),这时软件会接着以红色的 Abort 字体提示终止计算,有限元计算也随之终止。经常见到的中断计算的严重(Severity)错误是在非线性计算中出现迭代计算结果发散(Disconvergence)的错误。如图

1.5-3所示就是一种较为严重的错误。

```
Executing solver:
 Step0     = C:\Program Files\Diana 10.1\\binseg\ap\nl41.exe
 baseName  = Analysis3
 filosFile = Analysis3.ff
 directory = G:/例题

/DIANA/AP/NL41    23:14:14      0.03-CPU      0.06-IO     BEGIN
SEVERITY : ABORT
ERROR CODE: /DIANA/LB/AU32/0043
ERRORMSG.A: All licensed seats for module NONLIN are in use. See: http://localhost:1947
Extend your license or try again later.
DIANA-JOB ABORTED


Diana abort routine called

Trying to close things.
```

图 1.5-3　Diana10.1 信息提示栏错误提示

信息提示栏中建模部分和非线性计算部分存在着不同的报错形式。前处理建模部分主要是提示 Python 语句命令流中错误原因和出错结果,而非线性计算部分对错误的报错提示主要有以下几个方面:

（1）错误程度。提示是警告(WARNING)类型错误还是会导致计算中断的 ABORT 类型错误。

（2）错误代码。指明错误在 Diana 软件中的计算机代码信息,用户可以不予理会。

（3）错误信息。告诉用户错误的主要原因,这是用户需要重点关注和查找的信息。

（4）错误建议。指导用户可以进行何种操作修正这个错误。

7. GUI 图形可视化界面区

可视化界面区是 DianaIE 软件中最重要的一个模块,它提供几何建模施加约束、施加荷载、网格划分、查看后处理计算云图等图像的功能。相比 9.4 和 9.6 版本清一色黑色的 iDiana 的 GUI 界面风格,新版本的 DianaIE 的 GUI 界面部分有了非常重大的改进,用户既可以使用系统提供的可视化界面区默认的操作界面的颜色,也可以根据自己的喜好通过菜单栏→Edit→Preference settings 操作修改可视化界面区的背景颜色。同样的道理,混凝土单元颜色、钢筋单元颜色、界面单元颜色,甚至生成网格之后的网格单元颜色都可以在菜单栏的该路径下改变颜色。用户在应用菜单栏和快捷工具栏中的几何建模、修改、荷载添加、约束的每一步操作的结果都会在 GUI 可视化界面区呈现。网格划分成功以后,可视化界面会呈现出单元网格划分后的结构;非线性计算结束之后,同样会将变形图首先呈现在可视化界面区;当设置输出结果时,它还会把计算之后的结果以应力云图的方式一一呈现出来,对于具有多个荷载工况和多个荷载步计算的情况,用户在 DianaIE 界面查看结果应力云图将更加方便。

1.6　Diana 有限元分析流程

Diana 建模分析可分为以下几个步骤:
（1）建立结构的几何模型
建立结构尺寸模型,可以点击上方快捷菜单栏,也可以点击快捷键工具栏图标,还可以采用导入已编辑的 Python 语言的命令流文本文件的方式进行。组成几何模型的方式有点（Vertex）、线（Line）、面（Face）、体（Solid）四种。与 9.4 版本先建立点，后连接线、面、体的方式不同，Diana10.1 版本直接通过坐标输入建立线、面、体即可生成所需的几何模型。注意在建立单元面输入坐标时必须统一按照顺时针或逆时针方向依次输入坐标点，不可出现交叉建点（Intersect）的情况。

（2）赋予结构的材料属性和几何属性

点击快捷键或选中图形右击，选择如图 1.6-1 所示三种图标后会弹出对话框，可赋予模型材料属性和几何属性，其中黄色图标可以赋予混凝土的材料和几何属性，该图标主要用于赋予混凝土结构或砌体结构等单元的材料属性，蓝色图标用于配筋类型的本构属性设置（预应力筋单元、桩基础单元），红色图标用于边界单元和界面单元的单元类型本构属性设置。

图 1.6-1　各种材料几何属性快捷键图标

（3）施加荷载和边界约束条件

荷载的类型众多，主要有集中荷载（Force）、分布荷载（Distributed force）、预应力荷载（Postening）、点荷载（Point）。与其他软件不同的是，Diana 中可直接将预应力作为荷载添加，通过施加预应力荷载的方式（post-tensioning load），可以有效地模拟后张法施工工艺中预应力荷载作用。

（4）网格划分

Diana 单元网格划分主要分为两种类型：份数划分（Division）、尺寸划分（Element Size），在确定单元划分类型后，用户既可以对面（Face）进行网格划分，也可以对组成面的边（Edge）进行网格划分。网格划分既可以选择直接指定划分份数的 Division 方式，也可以选择指定单元尺寸的 Element Size 方式。Diana 网格划分方式为智能划分。基于非线性计算中收敛性考虑，用户在划分时优先考虑采用六面体或矩形（Hexa/Quad）的单元形状作为单元网格形状对边（Edge）进行网格划分时，设定的网格划分份数（Meshing Division）宜与各边长度比例保持一致。9.4 版本在 iDiana 中可以在前处理的任何阶段进行网格划分，而在 Diana10.1 版本 DianaIE 中，划分网格必须是有限元整个前处理建模流程的最后一步，且 Diana9.4 版本用户可以在生成单元网格前设定所需的单元类型，而 Diana10.1 版本只能通过确认网格形状和网格单元阶数，在网格成功生成之后才可以查看到单元类型。在 Diana9.4 版本中，用户无论是采用 GUI 界面操作方式还是命令流编辑方式均可进行网格划分操作，网格划分有以下几种步骤：

1）定义网格划分类型（边 Edge、面 Sheet、体 Solid）。

2）网格划分尺寸控制（Element size，Divisions）。

3）定义网格单元的形状（在 Diana 中，单元形状通常被限制为四边形、六面体单元或三角形、三棱锥单元）。

4）定义网格单元的中间节点位置（Mid-Side Location）。

5）生成网格（Meshing Generate）。

（5）添加荷载工况分析、计算并生成结果

添加荷载工况主要有三个模块。一是荷载的施加，主要包括点荷载（Point）、线荷载（Line）、面荷载（Face）、实体荷载（Solid）、温度荷载（Temperature）、支座移动荷载（Prescribed Deformation）、后张法预应力荷载（Post tensionning load）等主要荷载类型。无论施加哪一种类型的荷载，用户均要指定目标加载类型（Load Target Type）是点、线、边缘线、面、实体（Point、line、Edge、Face、Solid），加载类型（Load Type）是点荷载、线荷载、面荷载、体荷载，加载大小、方向和作用点这些几何要素。对于荷载作用点，在操作界面上通过鼠标点击进行拾取。在后张法预应力荷载中还需要指明一端/两端张拉（One/Both end）、锚固作用点、握裹系数（Wobble factor）、锚固回缩长度（Anchor retention length）以及库仑摩擦系数（Coulomb friction coefficient）等一系列参数，其中握裹系数和回缩长度等参数对于预应力筋随时间变化的衰减幅度有很大的影响。当用户施加了预应力筋荷载张拉力之后，Diana 计算中会自动根据上述参数扣除预应力损失。根据作者经验，一般地，握裹系数和回缩长度越大，预应力筋衰减的幅度越大。施加的各荷载工况往往采用荷载组合（Load combinations）的方式在结构分析中添加。而在结构非线性分析（Structural nonlinear analysis）中，用户需要设置迭代方法、最大迭代步数（Maximun Interations）、收敛准则（Convergence Norm）、收敛容差（Tolerance）以及最大计算次数。其中收敛容差默认值为0.01。当计算不收敛时，往往需要调整非线性计算的步长（Steps）、迭代方法和收敛准则（Convergence Norm）。收敛准则中，通常是力（Force）和能量（Energy）同时勾选，相比于只选择位移（Displacemnent）或力或能量这种单一选择的收敛标准，同时勾选的收敛方式在非线性迭代计算中往往最快，因为在实际多荷载步下的非线性迭代计算中，系统会同时对力和位移计算结果进行判定，只要力或位移中有一个达到了预定的收敛容差（Tolerance），该荷载步处的迭代计算即达到收敛。

（6）后处理结果查看

在后处理结果查看中用户往往可以查看某个方向应变云图、某个方向位移云图以及局部应力等主要的基本结果（All primaries），也可以在计算前添加荷载工况结束后的 OUTPUT 模块选项中选择用户自定义（User selection）输出某些结果，还可以在 OUTPUT 模块中 Device 部分选择输出类型为 Diana native 的 Result 查看方式或选择 Tabulated 选项以 .tb 后缀结尾的文本文件方式输出，来查看计算结果。

1.7 DianaIE Python 语言

在 Diana10.1 中，用户既可以采用 GUI 界面操作方式点击菜单栏或通过点击快捷键图标方式进行操作，还可以通过先编辑 Diana10.1 中特有的 Python 语言文本文件再导入 DianaIE 中进行建模。与单一鼠标点击操作的界面操作相比，这种使用 Python 语句编辑命令流建模的方式具有较强的通用性，尤其在土木工程有限元分析中，对于同种问题，用户不需要重复进行创建文件和建模操作，而只需要通过简单的复制、编辑或修改 Python 语句的命令流操作即可实现快速、批量化生成模型。与 iDiana9.4 中 .bat 二进制模型文件和 .dcf 计算控制文件中命令流采用繁琐晦涩的缩减语句相比，DianaIE 中的 Python 语句在命令流的语法和格式上较为随意,用户不需要查询前处理命令流语法手册(Synax Mannual)即可理解并自行编辑命令流。DianaIE 中常用的 py 语言命令流对应的操作翻译和说明见表1-16。

表1-16 py 语言命令流对应的操作翻译

Python 语言	操作
New Project("路径 A",模型范围)	在电脑的 A 路径上新创立一个文件,并设置模型长度范围
Set Model Analysis Aspects	设置分析类型
Set Model Dimension	确定有限元分析模型维数
Set Default Mesh Order	设置网格单元阶数
Set Default Mesher Type	设置所要划分的单元网格形状
set Default Mid Side Node Location	设置网格单元中间插值节点位置
Create veretex/line/Sheet/Solid	创立点、线、面、体
save Project	保存文件
set Parameter	对材料或几何属性设置材料参数或几何参数
add Material	添加材料属性
add Geometry	添加几何属性
create Line Connection	生成界面单元
attach To	赋予各种属性
set Element ClassType	设置单元类型
assign Material/Geometry	分配、赋予材料、几何属性
rename	重命名
add Geometry Load Combination	添加荷载组合
Set Geometry Load Combination Factor	对每一个荷载组合添加荷载系数
Set Mesher Type	设置网格划分形状类型
Set Mid Side Node Location	中点坐标位置设置
Set Element Size	网格单元尺寸设置
generate Mesh	生成网格
set Active Phase	阶段性分析激活
add Analysis Command	添加荷载工况分析模块
run Solver	运行计算
set Result Plot	查看结果的应力云图
set Result Case	查看荷载步对应的计算结果
array Copy	复制并且平移
rename	重新给予新的命名

所有的 GUI 图形界面交互环境(Diana Interative Environment)下生成的操作都可以以后缀.py 的文本文件的方式记录并保存下来,用户在生成图形界面时候有三种操作方式:① 在GUI 图像界面上直接操作;② 在 Command console 命令流区输入命令流;③ 预先将写好的含有 Python 命令流操作语句的.py 文本文件导入 DianaIE 中读取,或是将.py 文件中的 Python 语言的命令流语句直接复制粘贴到 Command console 命令流区。在 DianaIE 中,Python 语句就像在草稿纸上打草稿一样的随意,但用户可以通过修改或是删去不必要的冗余语句,形成规范式的 Python 批处理文件,这就保证了用户可以同时享用 Diana10.1 中的图形可视化特征和参数化建模这两大工具。在 DianaIE 的 Python 语言中,当括号内命令流语句为用户需要自行设置的语句时,通过"/"表示所需设置语句在对应 GUI 操作的模块间的从属关系。

1.8 Diana 单位

无论是在 Diana 软件中的 DianaIE 模块,还是保留的 iDiana 模块,Diana 软件的单位均是采用国际单位制。其中与土木工程有关的单位主要有长度单位(Length)、力学单位(Force)、质量单位(Mass)、时间单位(Time)、温度单位(Temperature)和角度单位(Angle)这几种主要的单位制。本节重点介绍 Diana10.1 的 DianaIE 模块下这几种单位以及单位之间相互转化方式。每改动一次单位制,长度、力和质量这三个单位中总有一个会以 derived 形式自动显示,在 DianaIE 中,可在 Geometry 下方的 Units 栏目属性下定义模型文件的单位制。在 DianaIE 中,默认的单位制为国际单位制,即 meter,kilogram,newton,second,kelvin,radian,unit 栏目属性下的国际单位制如图 1.8-1 所示。在 DianaIE 软件操作模块中,其他单位的单位制均可由上述几种单位制导出。

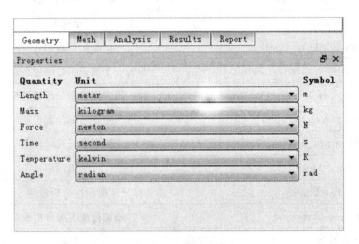

图 1.8-1 单位属性操作栏

(1) 长度单位

表征长度的单位主要有米(m)、分米(dm)、厘米(cm)、毫米(mm)四种形式,欧美单位制有英寸(Inch)、英尺(feet)、码(yard)、英里(mile)。其中国际单位制 m 和 mm 为 Diana 软件建模中最常用的两种单位。

(2) 力学单位

Diana 软件建模力学单位主要由 N 及 kN 组成,其中力的单位与长度单位之间的不同组合对弹性模量、密度、压强等单位量或数值均会产生影响,两种单位同时转化的情况下,Diana 中单位变化的同时数值也会相应调整。举例说明,当力的单位为 N,长度单位为 m 时,弹性模量单位为 N/m^2;当力的单位为 N,长度单位为 mm 时,这时弹性模量单位自动转化为 kg/mms^2;当力的单位为 kN,长度单位改为 mm 时,弹性模量单位为 kNs^2/mm^4。本书为了单位统一,也为了方便起见,后面例题长度单位一律采用 m。力学单位一律采用 N!

(3) 质量单位

质量单位的单位制种类众多,包括千克(Kilogram)、克(gram)、吨(ton)、盎司(ounce)、磅(pound)、千磅(kilo-pound),其中千克和吨为 DianaIE 建模中常用的单位。和长度单位及力的单位类似,当质量单位的单位制发生变化时,混凝土材料的本构模型弹性模量,各类本构模型密度的数值和单位均会同时发生变化,并且长度单位和力学单位中有一个会以 Derived 形式自动导出单位制;当质量单位为吨时候,力的单位在 DianaIE 中自动变化为为 kN,力的单位或长度单位制则不再以 derived 形式导出。

(4) 时间单位

时间单位是 Diana10.1 中常用的单位,尤其是在涉及混凝土收缩徐变、单元龄、加载龄期或混凝土水化热反应相关的时变计算时往往需要考虑时间单位的设置方式。时间单位制有年(year)、天(day)、小时(hour)、分(minute)、秒(second)。通常情况下时间单位的设置方式有两种,一种是以秒(s)为单位进行设置,另一种是以天(day)为基本设置单位,两种单位之间的转化除了会导致混凝土的单元龄、收缩固化曲线、加载龄期在单位制上的变化,还会导致与时间单位相关的密度单位和数值的同时变化,但这些变化对最终有限元计算结果不会造成影响。其中时间单位制转化结果如下所示

$$1 \text{ year} = 365 \text{ day} = 8\ 760 \text{ hours} = 525\ 600 \text{ min} = 31\ 536\ 000 \text{ s}$$

(5) 温度单位

作为与环境因素息息相关的指标,温度单位的表达方式主要有三种,一种是开氏温度(Kelvin),一种是摄氏温度(celsius),另一种是华氏温度(fahrenheit),其中前两种温度单位制是 Diana 软件中常用的单位制。在 Diana 各类规范本构定义中涉及到环境温度、热传导、热对流以及混凝土水化热温度反应中,通常默认温度的单位制为开氏温度 K,温度的默认值为 293.15 K(相当于 20℃)。

(6) 角度单位

在土木工程、岩土工程、地下工程等涉及采用库仑摩擦本构类型的模型中,角度也是一种常用的单位。角度单位的单位制有弧度(Radium)和角度(angle)两种。弧度和角度的转换关系为 1 弧度=57.3°。

第二章　Diana 材料模型和规范简介

本章针对土木工程结构中常用的混凝土、钢结构本构模型和世界各国相关规范,同时基于数值模拟经验,对软件中常用的收缩徐变长期性能、热对流反应和瑞雷阻尼模块进行介绍和比较。

2.1　Diana 各类材料模型简介

相比其他软件,Diana 软件的另一大特色是 Diana 具有广泛的、功能非常强大的诸多类型材料本构模型库,同时又吸收了欧洲 CEB-FIP 1990(MC1990)、fib 2010(MC2010)、荷兰规范、日本 JSCE 规范和美国的 ACI 等设计规范精髓,可有效模拟混凝土结构在线性、非线性、几何大变形、开裂模型、地震作用下的各种工程情况,以至于用户在绝大多数情况下不需再针对本构模型进行二次开发。

Diana 中的材料本构模型很多,主要有以下几大类型:

(1) 混凝土和砌体结构(Concrete and mansory)

该本构适用于模拟混凝土结构短期性能、早期阶段收缩和短期状态下裂缝开展的全过程。对裂缝开展全过程、裂缝位置的分布具有良好的模拟效果。但该本构对于徐变、单元龄等涉及长期性能参数并未加以考虑,因而无法像 Diana 中各规范那样做到对混凝土结构长期问题的有效模拟。总体而言,混凝土和砌体结构具有丰富的本构类型,包括线弹性材料各向同性(Linear elastic isotropic)、线弹性正交各向异性(Linear elastic orthotropic)、总应变裂缝模型(Total Strain based crack model)、多向固定裂缝模型(Multi-directional fixed crack)、塑性开裂模型(Crack and plasticity)和适用于多轴应力状态下的朗肯主应力(Rankine principal stress)模型等常用的本构类型。需要特别提到的是,在朗肯主应力模型中,拉伸截断曲线形式多为恒定或线性变化。而根据多轴应力状态曲线的不同,又可以分为朗肯塑性模型(Rankine plasticity)、朗肯冯米塞斯模型(Rankine von mises plasticity)和朗肯/德鲁克普拉格模型(Rankine/Drucker prager plasticity)。此外还有与剪切角度和粘滞系数相关的摩尔库仑-德鲁克普拉格模型(Mohr-Coulomb and Drucker-Prager);在滞回分析中可选择的前川—福浦混凝土模型(Maekawa-Fukuura concrete model)。在定义混凝土材料各向异性的问题中,还可以采用朗肯希尔各向异性模型(Rankine Hill anisotropy)。Diana10.1 中混凝土和砌体结构所有材料本构类型如图 2.1-1 所示。

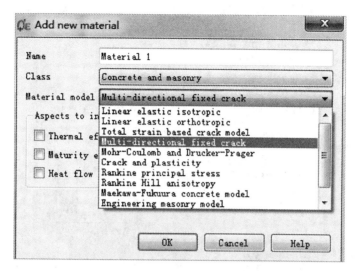

图 2.1-1　混凝土和砌体结构本构类型

各向同性线弹性材料本构参数主要确定弹性模量、泊松比和密度,由于材料为各向同性,输入上述参数值即表示材料性能指标在各个方向上均为相同数值。如图 2.1-2 所示。

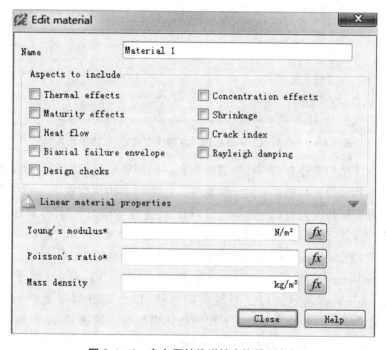

图 2.1-2　各向同性线弹性本构模型参数

正交各向异性准则,就是指材料的弹性模量、泊松比和剪切模量在材料互相垂直的两个方向上具有不同的性能指标。这个时候需要在各方向上都要输入弹性模量、泊松比以及剪切模量值。正交各向异性各项强度指标定义界面操作如图 2.1-3 所示。

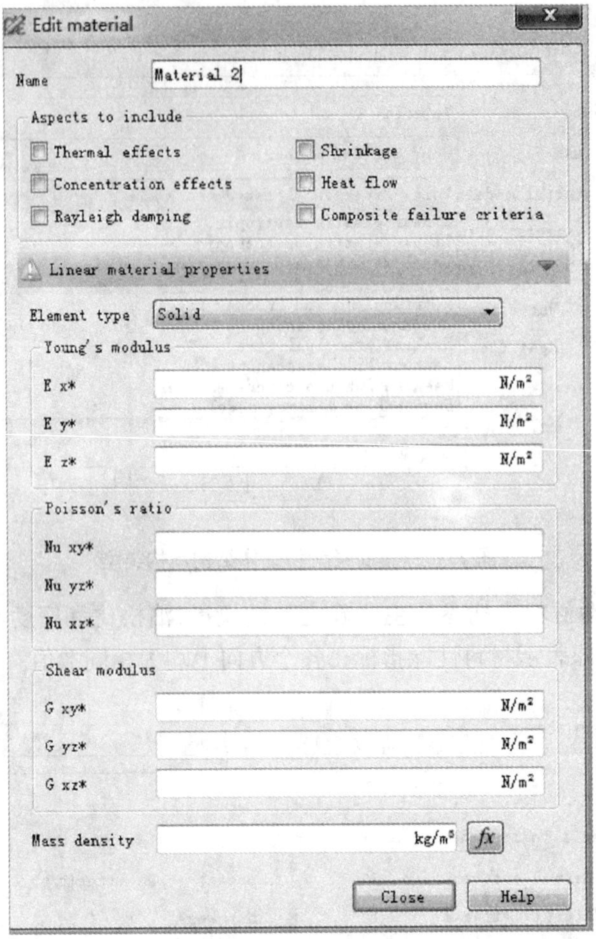

图 2.1-3　各向正交异性线弹性本构模型参数(实体单元为例)

混凝土和砌体结构中还有总应变裂缝模型和多向固定裂缝模型两种常用的本构模型,主要用于模拟结构弥散开裂模式下的开裂特性。这两种本构模型在下一章中会详细介绍,这里仅仅讨论界面操作设置方式。总体而言,相比线弹性本构关系而言,总应变裂缝模型不仅仅要输入弹性模量、泊松比以及剪切模量这些混凝土基本参量,同时还要输入整体应变裂缝下的裂缝特性(Total strain based crack model)、拉伸特性行为(Tensile Behavior)、剪切特性行为(Shear Behavior)以及压缩特性行为(Compressive Behavior),如图 2.1-4 所示。裂缝特性涉及对开裂方向是固定还是旋转的选择,以及对开裂带宽(Crack bandwidth specification)的设定。拉伸行为主要关于拉伸软化模型曲线的选择,断裂能、极限应变、残余拉伸应力参数值(Residual tensile strength)输入。残余拉伸应力的设置,用户还需要在泊松比折减(Poission's Ratio reduction)项目中选择折减类型(Reduction model)是无折减(No reduction)还是基于损伤(Damaged based)类型。如图 2.1-5 所示。

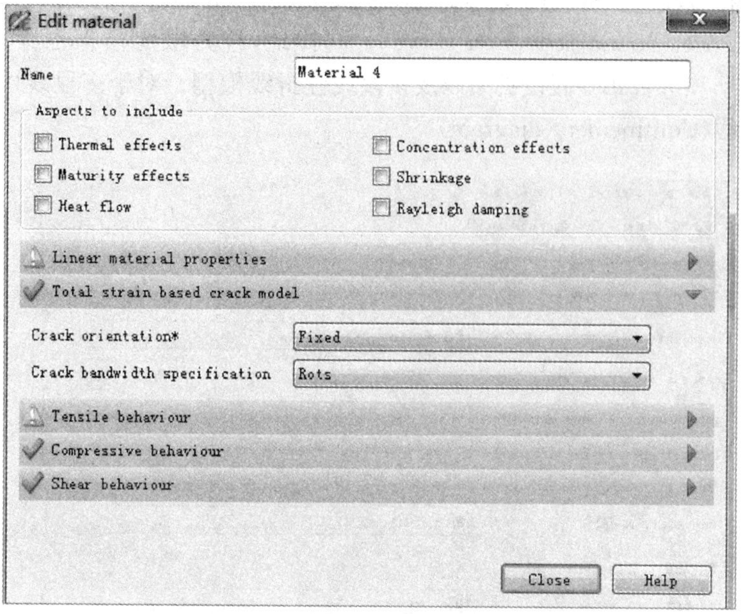

图 2.1-4　总应变裂缝本构模型操作界面

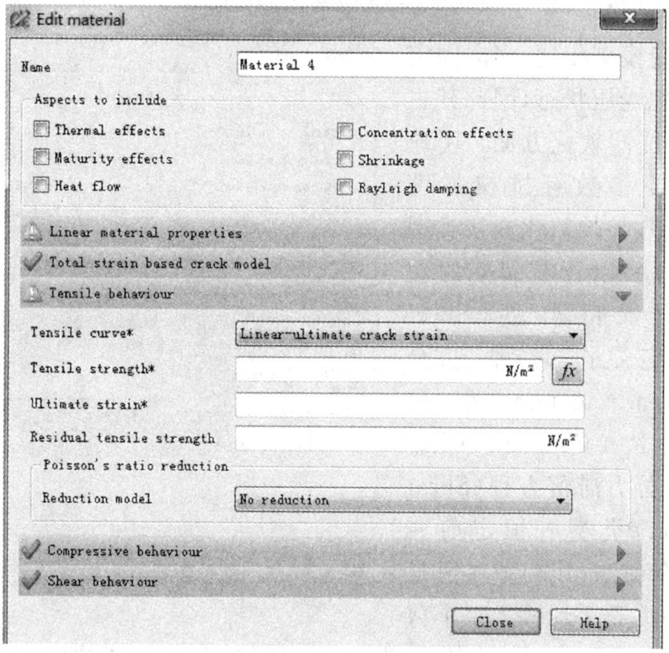

图 2.1-5　总应变裂缝本构模型下线性开裂拉伸软化模型操作界面

在总应变模型的抗压特性行为中，用户需要自行选择压缩曲线类型（Compression curve）、抗压强度（Compressive strength）、压缩断裂能（Compressive fracture energy）、残余压缩应力（Residual compressive strength），如图 2.1-6 所示。与拉伸软化模块不同的是，横向开裂状态下的残余抗压强度折减模型不仅存在传统的假设模型，而且根据规范和最新研究成果增加了其他类型，共有无折减（No reduction）模型、Vecchio 和 Collion1986 模型、Vecchio 和 Collion1993 模型、多段线性折线（Multi-linear diagram）和 JSCE2012 模型几种。用户需要在多段线性折线模型、Vecchio 和 Collion1993 模型中设置残余抗压强度折减曲线的下界(Lower bound reduction curve)。同样地,应力限定模型也存在不增加(No increase)、Selby 和 Vecchio 假设类型以及多段线性折线模型。对于多段线性类型,用户需要设置范围曲线(Confinement diagram)。

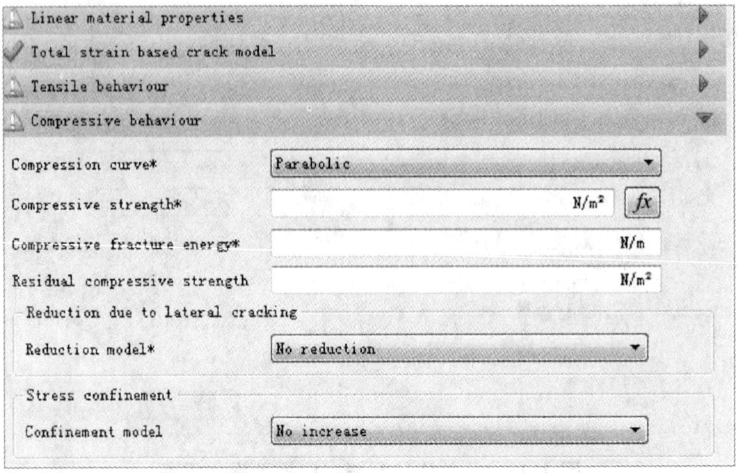

图 2.1-6　总应变裂缝本构模型下非线性压缩模型操作界面

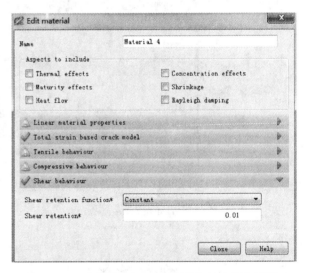

图 2.1-7　总应变裂缝本构模型剪切行为操作界面

同总应变裂缝模型相比,多向固定裂缝模型无法模拟受压状态,因此其主要的设置围绕着拉伸软化方面。用户需要输入的主要参数有抗拉强度(Tensile strength)、拉伸软化曲线类型(Tension softening)以及断裂能的确定。同时,剪力滞留类型(Shear retention)有两种定义方式,一种是完全剪力滞留(Full shear retention),另一种是恒定剪力滞留(Constant shear retention)。选择剪力滞留方程(Shear retention function)类型有恒定剪滞(constant)和完全剪滞(full)。在恒定剪滞类型下,用户需要手动输入一个剪滞系数 β(Factor Beta)。值得一提的是,当选择常用的 Hordijk 本构模型时,则需要输入两个用于表征拉应力与极限抗拉强度比值关系的参数 c1 和 c2,其中 c1 默认值为 3,c2 默认值为 6.93,如图 2.1-8 所示。软件中剪力滞留的参数默认值的详细内容会在后面章节中介绍,这里不再赘述。

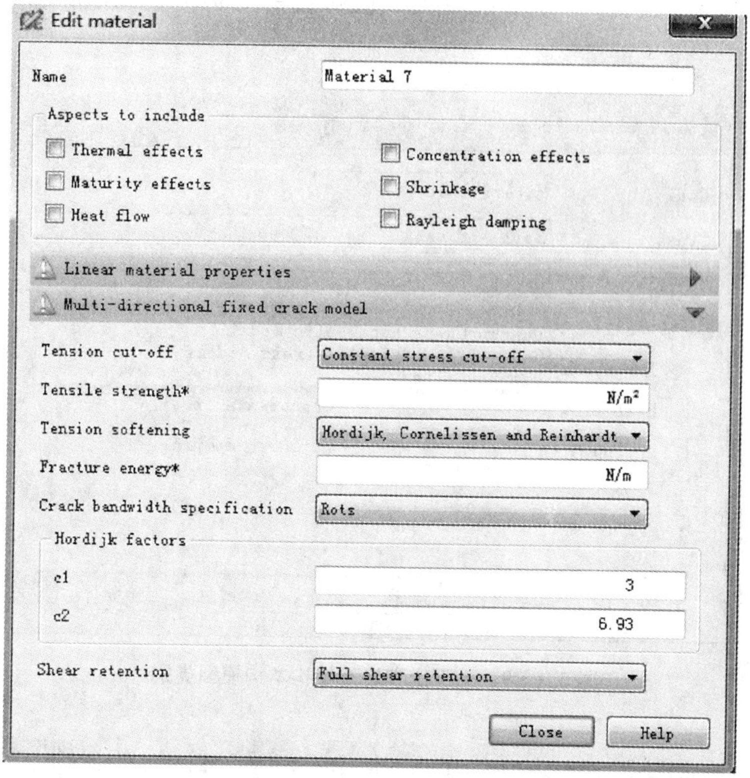

图 2.1-8　多向固定裂缝模型下 Hordijk 拉伸软化模型操作界面

（2）地下土和岩石本构结构模型

地下土和岩石结构被认为是一种半脆性结构材料，具有各向正交异性特征。与混凝土结构在线弹性状态、弹塑性状态甚至塑性状态下的计算相比，地下土和岩石结构往往需要进行塑性模型分析和计算。Diana 软件中地下土和岩石的本构模型很多来自于一些比较成熟的理论研究成果。此外，地下土和岩石结构还可以像混凝土那样进行瑞雷阻尼分析、热对流和热效应分析。与 Diana9.4 版稍有不同的是，该类材料的本构模型有时需要考虑初始应力。

地下土和岩石主要为正交异性材料，因此本构模型主要有以下四类：正交异性弹性材料（Orthotropic elasticity）、Duncan-Chang 双曲模型（Duncan-Chang hyperbolic）、Geotexile 简化应力模型（Geotexile-simple stress models）和用户自定义（User-supplied），如图 2.1-9 所示。其中较常用的是正交异性弹性本构。与混凝土和砌体结构的正交异性弹性本构类型设置方式类似，用户定义的弹性模量、泊松比和剪切刚度这些基本参量在正交多向异性的条件下在各正交方向的数值有所差别，因此需要确定各正交方向下的数值，如图 2.1-10 所示。除此之外，对初始应力的定义往往也是正交异性弹性本构材料中需要考虑的方面。这时用户需要考虑横向压力系数（Lateral pressure ratio），包括有效应力类型为各向同性（Effective stress-Istropic）还是各向正交异性（Effective stress-Orthotropic），或者是总应力下各向同性（Total stress-Istropic）还是各向正交异性（Total stress-Orthotropic）。

无论哪一种应力方式，对于各向同性材料，用户只需要设定一个应力的系数 K0，如图 2.1-11 所示；而对于各向正交异性材料，用户还需要输入横向分布系数的最大值 KT0max

和最小值 KTOmin 以及横向分布系数最大值对应的坐标方向(Direction of maximum pressure ratio)。

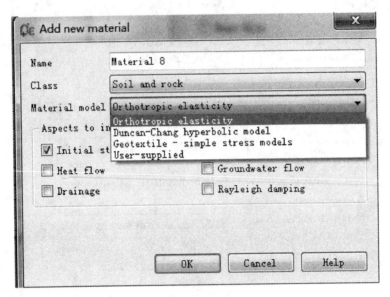

图 2.1-9　地下土和岩石结构本构模型类别

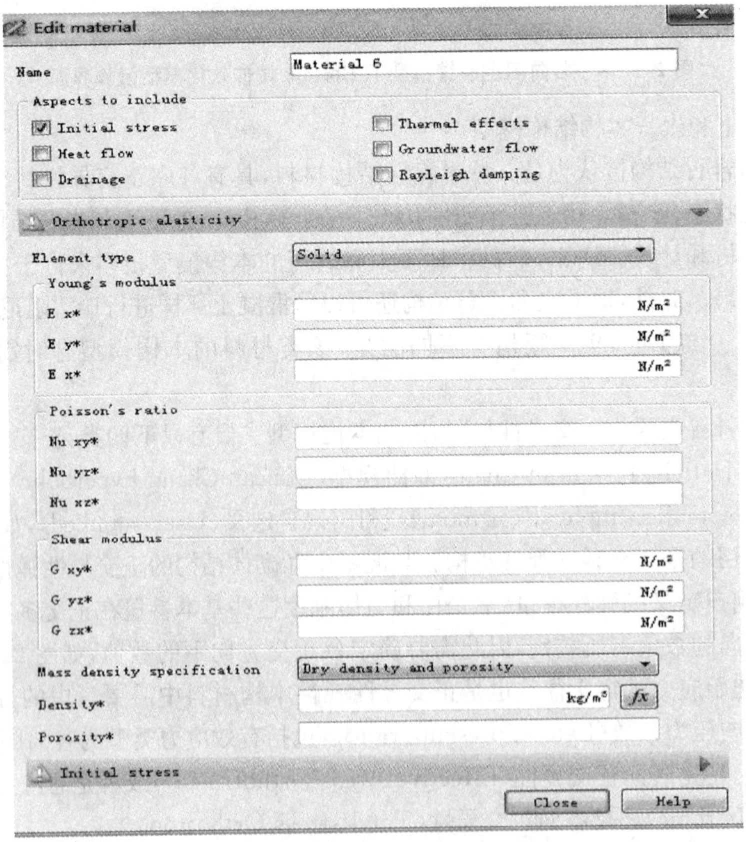

图 2.1-10　正交异性下的弹性本构参数定义界面

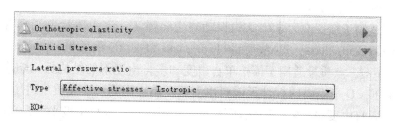

图 2.1‑11　各向同性有效应力初始应力的操作界面

(3) 组合结构和橡胶结构本构模型

组合结构和橡胶结构中不仅可以模拟线弹性各向同性材料和线弹性正交各向异性材料，还可以设置各向异性本构模型。对于各向异性材料而言，用户在 Diana10.1 中不仅要设置线弹性材料属性(Linear elastic property)，还需要设置每个坐标轴方向的正应力值和剪切应力值，如图 2.1‑12 所示。

图 2.1‑12　组合结构和橡胶结构各向异性材料属性设置界面

(4) 质量单元本构模型

质量单元适用于模拟动力学中结构振动响应问题,Diana 中质量单元具有多种不同本构类型:点质量(Point mass)、2D 线质量(Line mass 2D)、3D 线质量(Line mass 3D)、面质量(Surfass mass)。点质量适用于模拟单自由度和多自由度下的结构振动响应问题,2D 线质量适用于模拟 2D 平面内的杆件无限自由度振动响应问题,3D 线质量适用于模拟 3D 平面内的杆件无限自由度振动响应问题,面质量适用于模拟无限自由度平面振动响应问题,例如振动台的数值模拟常会用到面质量单元。

质量单元的定义,需要输入法向(Normal direction)和切向(Tangential direction)单位长度下的质量分布。图 2.1-13 为 3D 线质量定义操作界面。

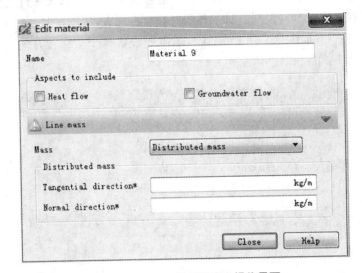

图 2.1-13　3D 线质量定义操作界面

(5) 弹簧和阻尼单元本构模型

弹簧单元的本构设置方式在 1.3 节中已有介绍,这里不再赘述。

(6) 界面单元本构模型

界面单元也同样具有内容丰富的本构单元库。界面单元中不仅有适用于点对点、点对线、点对面的线弹性本构材料(Linear elasticity)和非线弹性本构材料(Nonlinear elasticity),还具有可模拟已知固定位置处裂缝的离散开裂模型,这在下一节中会进行详细说明。一些受剪破坏结构与剪切角或摩擦因素有关,需要设置类似的界面单元情况,针对此类情况,Diana10.1 中相应增加了库仑摩擦(Coulomb friction)和非线弹性摩擦(Nonlinear elastic friction)两种与摩擦相关的本构关系。此外,考虑到采用桁架单元或梁单元模拟单根预应力钢绞线时无法像钢筋单元那样自动嵌入母体单元中,同时在实际情况中会有粘结滑移现象产生,Diana 软件提供了一款粘结滑移单元,如图 2.1-14 所示,可将离散成 truss 或者 beam 单元后的预应力筋与周围相连的母体单元产生耦合作用,并且将桁架或梁单元作为粘结滑移界面单元设置钢材的粘结滑移本构属性,再通过 DATA 中的 INTERF 选项中的积分转化选项将这些离散状态下的材料转化为粘结滑移状态。在采用粘贴钢板加固中往往也采用 Bondslip 粘结滑移本构模型。

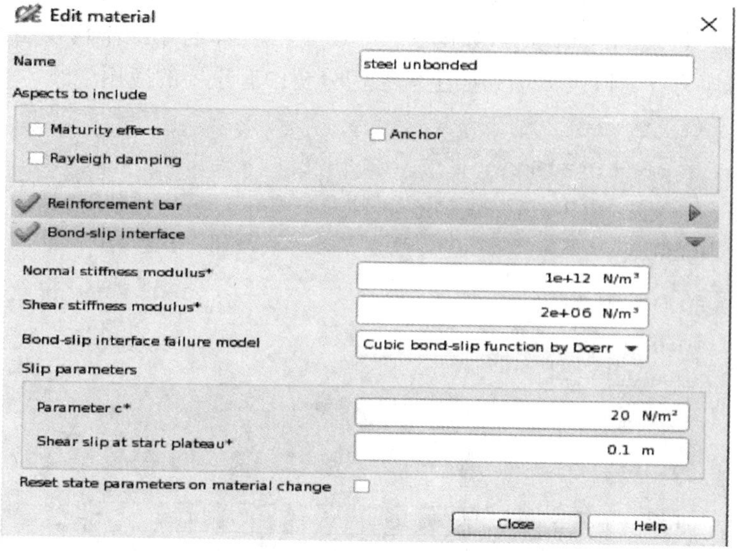

图 2.1-14　钢材粘结滑移本构属性定义界面

下面重点介绍两种常见界面单元材料图片:非线弹性本构材料和库仑摩擦本构材料—本构设置方式。非线弹性本构材料需要选择材料属性的类型(Type),同时根据维数的不同分别定义法向刚度模量(Normal stiffness modulus-z)、平面内切向刚度模量(Shear stiffness modulus-x),对于 3D 结构的界面单元本构类型,还需要定义平面外的切向刚度模量(Shear stiffness modulus-y)。非线弹性材料各项刚度指标定义的操作界面如图 2.1-15 所示。

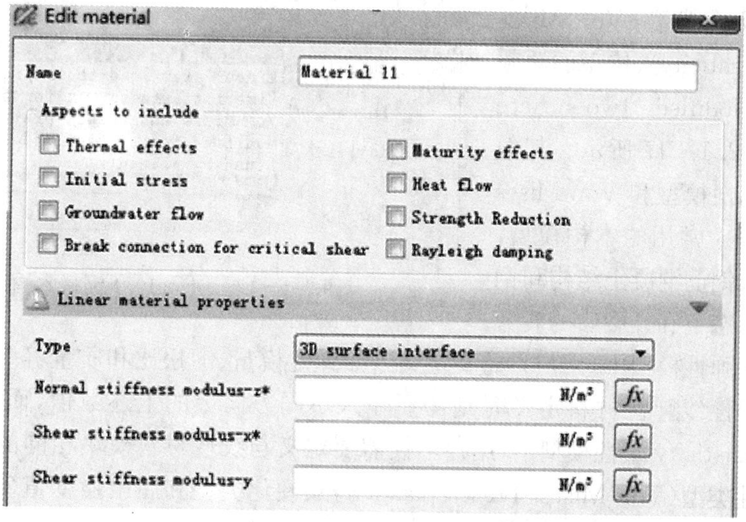

图 2.1-15　非线弹性材料各项强度指标定义界面

库仑摩擦本构模型主要应用于岩土领域的土体强度分析中。在非线弹性模型定义法向刚度模量和切线刚度模量的基础上,还需要定义库仑摩擦(Coulomb friction)有关的特性参数:粘滞系数(Cohension)、库仑摩擦角/内摩擦角(Friction angle)、膨胀角(Dilatancy angle),以及与粘滞力及摩擦力相关的硬化曲线。粘滞系数和内摩擦角是决定土体强度的重要因素。膨胀角则与体积应变有关,当膨胀角大于 0°时,体积应变增加,结构会产生膨胀。当库仑摩擦角角度较低时，土体强度较低，类似于Diana混凝土结构中的拉伸软化模型中的许用拉力值(Tension Cut-off)，如图2.1-16所示。库仑摩擦本构模型在土的抗剪强度指标、库仑摩擦力计算以及大坝和挡土墙结构分析中有重要的应用。

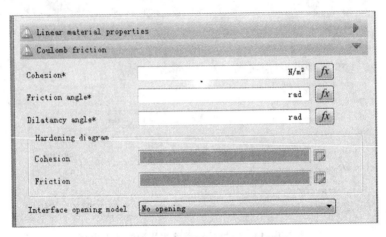

图2.1-16 库仑摩擦参数设置界面

(7) 钢材本构(Steel)

钢材本构模型主要包括线弹性各向同性材料（Linear elastic isotropic）、线弹性正交各向异性钢材（Linear elastic orthotropic）、冯·米塞斯塑性模型（Von Mises and Tresca plasticity）、修正的两截面模型（Modified two-surface model），如图2.1-17所示。其中线弹性各向同性模型和Von Mises模型为Diana中常用的本构模型，分别用于模拟钢筋混凝土结构和钢结构在弹性状态下和开裂后的疲劳

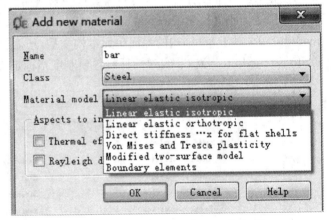

图2.1-17 Diana10.1钢材本构类型

等问题。在线弹性模型中，用户只需要定义诸如弹性模量、泊松比和质量等线性材料特征，在正交各向异性线弹性材料定义时则需要输入不同方向上的上述数值，而在Von Mises and Tresca plasticity塑性模型中，用户不仅需要定义与线性材料模块相同的基本参数，还需要输入塑性类型(Von Mises/Tresca plasticity)、硬化类型和屈服强度值(Yield stress)。关于Diana钢结构规范本构定义将在2.3节中进行介绍。

2.2 混凝土结构的裂缝模型

混凝土结构裂缝模型主要分为离散开裂(Discrete Cracking)及弥散开裂(Smeared-Cracking)。而弥散开裂种类众多，接下来就Diana10.1中混凝土开裂模型进行介绍。

1. 离散开裂(Discrete Cracking)

混凝土的开裂模型主要分为两种,一种是离散开裂模型(Discrete-Cracking),另一种是弥散开裂模型(Smeared-Cracking)。在 Diana10.1 中,用户通过定义材料本构特性实现裂缝模型的应用。这些特性主要包括弹性模量、泊松比、质量、拉伸截断(Tension cut-off)下的抗拉强度(Tensile strength)和抗压强度(Compressive strength)的定义,以及对拉伸软化关系(Tension softening)和剪力滞留(Shear Retention,以下简称剪滞)的设置。离散开裂模型本质为分离式单元与界面单元之间的一系列组合,通过单元之间的相对位移关系实现结构开裂行为的模拟,即在 Diana 中使用离散开裂模型对结构进行模拟时,通常把裂缝位置处的两个部分视为两个独立部分进行单独建模,中间通过添加线对线或者面对面的界面单元,并对这些界面单元的材料本构按照开裂的力学特性设置来完成对结构开裂的数值模拟。离散开裂模型中存在着一条主裂缝,在界面单元的法向应力达到抗拉强度时,主体单元之间出现相对位移,这时可认为混凝土产生了裂缝。离散开裂模型适用于查看结构在非线性计算后实际分布图形及局部应力状况,离散开裂之后,混凝土块之间会产生相对滑动位移以及相对滑移角度。离散开裂模式主要有五种拉伸软化模型:脆性拉伸软化模型、线性拉伸软化模型、JSCE 拉伸软化模型、Hordijk 拉伸曲线软化模型以及多段线性拉伸软化模型。其中 Hordijk 拉伸软化曲线模型为离散开裂模型中常用的拉伸软化模型,该模型考虑拉伸中的断裂能,对计算结果的收敛性有较大的帮助。五种离散开裂下的拉伸软化模型分别如图 2.2-2 所示。

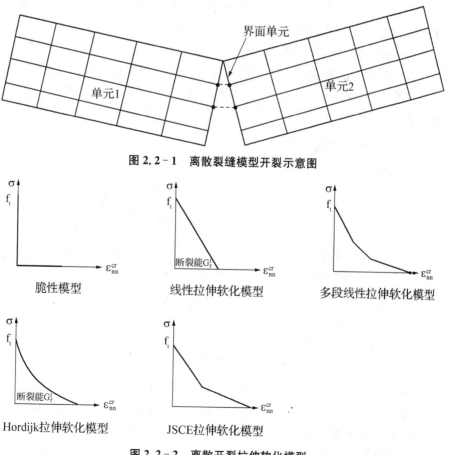

图 2.2-1 离散裂缝模型开裂示意图

图 2.2-2 离散开裂拉伸软化模型

除了采用界面单元模拟开裂特征,离散单元还可以模拟开裂后混凝土裂缝剪切相对位移与法向相对位移耦合的力学行为,这种行为称为裂缝剪切膨胀行为。由于在实际开裂结构中裂缝界面粗糙,裂缝切向滑移会引起法向的相对位移。离散开裂单元适用于模拟钢筋混凝土结构已知开裂位置处裂缝开裂前后的整个过程,但很多情况下开裂位置处会存在多条裂缝沿多向开展,该模型对开裂过程中方向和数量较多的裂缝开展过程的把握不如弥散开裂精准。

2. 弥散开裂(Smeared-Cracking)

弥散开裂(Smeared-Cracking)单元在建模中将诸如混凝土梁之类的开裂对象视为材料各向异性的连续体,通过弹性模量、抗压和抗拉强度等数值上的降低以及单元应变值来体现裂缝,适用于仅需要观测计算后结构的宏观特性,例如结构的整体位移、跨中位移或相应的结构荷载位移曲线,且开裂行为通过应力—应变关系确定。与离散开裂不同的是,弥散开裂的裂缝特征为裂缝主要分布在单元内的积分点位置,且与离散开裂相比,裂缝尺寸细小,无主裂缝,每一个单元里最多3条裂缝,裂缝沿主应力方向发展,呈现正交关系。由于弥散裂缝模型开裂时耗散的断裂能会受到单元尺寸的影响,即所谓的网格敏感性,因此在Diana软件中引入了一个裂缝长度尺度参数——裂缝带宽(Crack bandwidth)来解决这个问题。弥散开裂主要分为三类:多向固定裂缝模型(Multi-Directional Fixed Crack Model)、总应变裂缝模型(Total Strain Based Crack Models)和朗肯主应力模型(Rankine Principal Stress Model)。其中总应变裂缝模型是Diana钢筋混凝土开裂问题采用的主要本构模型。根据对裂缝开展的分析,总应变裂缝模型又可以分为正交固定裂缝模型(Fixed Orientation Based Crack Models)、正交旋转裂缝模型(Rotating Orientation Based Crack Models),以及裂缝发展不明显时采用正交旋转模型,裂缝状态明显时采用正交固定模型的混合裂缝模型(Rotating to Fixed Orientation Based Crack Models)。三种弥散开裂模型的范围以及适用特点如图2.2-3所示。

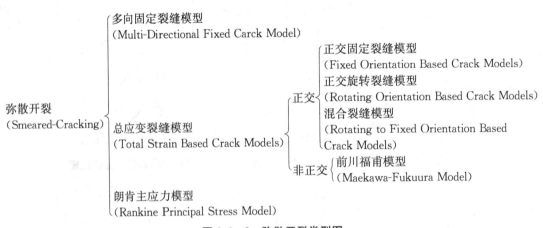

图2.2-3 弥散开裂类型图

(3) 多向固定裂缝模型(Multi-Directional Fixed Crack Model)

多向固定裂缝,是指同一个单元内同时存在几种不同方向的裂缝。多向固定裂缝模型核心思想是将应变分解成弹性阶段的应变 ε^e 和开裂应变 ε^{cr} 单独计算,即总应变为弹性应变和开裂应变两者之和。在Diana10.1中,多向固定裂缝模型的开裂后特征通过拉伸软化关系以及加载和卸载曲线的荷载割线刚度来确定,因此决定多向固定裂缝模型的特征主要有两个:拉伸行为(Tensile Behavior)、剪切行为(Shear Behavior)。其中拉伸关系用来模拟混凝土受拉及开裂下的力学行为特性,由于该模型不包含混凝土的压缩段,从而不考虑混凝土压缩特性。

弥散开裂下的各种模型拉伸软化曲线均较多,针对多向固定裂缝模型而言,主要有八类拉伸软化模型:脆性模型、线性软化模型、多段线性软化模型、Hordijk 非线性软化模型、基于断裂能的线性拉伸软化模型、指数软化模型、Moelands and Reinhardt 非线性软化模型、JSCE 拉伸软化模型以及 JSCE 应力刚化模型,分别命名为 TENSIO0—TENSIO7,如图 2.2-4 所示。

多向固定裂缝模型的八种拉伸软化模式中使用最多的是第五类拉伸软化曲线(TENSIO5)的 Hordijk 模型。这种本构模型考虑开裂后应力从极限拉应力衰减到 0 的下降阶段为非线性阶段,同时考虑在裂缝开展过程中存在的裂缝带宽和断裂能,可有效模拟混凝土在开裂状态下的应力变化。

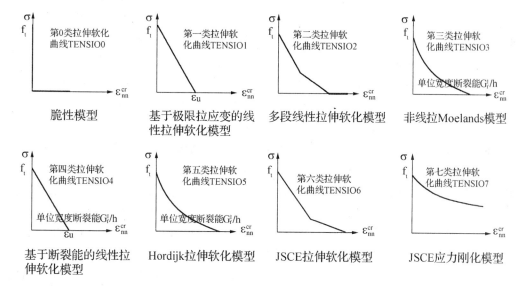

图 2.2-4 多向裂缝固定模型拉伸软化曲线

多向固定裂缝模型另一个需要定义的特性就是剪力滞留特性(Shear Retention,又称为剪力传递特性,以下简称为剪滞),这种剪滞特性只针对切向刚度。当混凝土结构发生开裂时,裂缝处材料的剪切刚度相对于初始刚度会有所下降,但是不会直接降低为0,这时除了需要在拉伸特性的 Tensile Behavior 模块下定义拉伸曲线类型(Tensile curve)、抗拉强度(Tensile strength)以及断裂能(fracture energy)外,还需要在 Diana 中定义泊松比衰减方式(Poisson's ratio reduction)和衰减模式(Reduction model)。通常剪滞系数用 β 表示,$0 \leqslant \beta \leqslant 1$。而在使用老版 iDiana 的命令流定义以后缀.bat 为首的二进制模型文件和以后缀.dat 文本为首的模型材料文件中的多向固定裂缝模型剪滞特征时,通常使用 TAUCRI 符号及其后面的数字来表征剪滞特性是完全剪滞(Full Shear Retention)或是恒定剪滞(Constant Shear Retention)。TAUCRI 0 代表完全剪滞模型,这种状态不需要再定义剪滞系数 β,即 β 值默认为1。TAUCRI 1 代表恒定的剪滞模型,这种状态需要定义一个剪滞系数。在 Diana10.1 中,当用

户选择恒定剪滞模型时，β 默认值为 0.01。这时在开裂过程中的裂缝剪切刚度可以表示为下式

$$D=\frac{\beta}{1-\beta}G$$

式中，D 代表开裂后的剪切刚度；β 是剪力滞留系数；G 代表初始剪切刚度。从公式可以明显看出多向固定裂缝模型开裂后在剪切方向的剪滞特征，即裂缝处材料剪切方向的单元刚度有所降低。

尽管多向固定裂缝模型可有效模拟混凝土在非线性计算中受拉开裂，但无法有效模拟非线性受压计算。因此本文重点介绍模拟钢筋混凝土结构短期作用下的裂缝开展最常用的一类模型——总应变裂缝模型。

（2）总应变裂缝模型(Total Strain based crack model)

根据裂缝方向与主应力方向是否始终保持一致，总应变裂缝模型又可分为正交固定裂缝模型(Fixed Orientation Based Crack Models)、正交旋转裂缝模型(Rotating Orientation Based Crack Models)、混合裂缝模型(Rotating to Fixed Orientation Based Crack Models)三种。在正交固定裂缝模型中，当拉应力达到断裂强度时裂缝开始出现，裂缝一旦出现便设定裂缝的方向不再随主拉应力的方向变化，因此称为"固定裂缝"。而在正交旋转裂缝模型中，随着荷载的增加，主拉应力的方向时刻变化，模型中设定裂缝的方向随着主拉应力的方向不断变化，看起来像是在"裂缝转动"。在总应变裂缝模型中，计算出来的应变值即为总体的应变值。

总应变裂缝模型的开裂后关系主要通过拉伸行为(Tensile Behavior)、剪切行为(Shear Behavior)和受压行为(Compressive Behavior)这三者共同确定。拉伸行为同多向固定裂缝模型相比，拉伸软化曲线增加了从初始承受拉应力到达到极限拉应力 f_t 阶段。此外，总应变裂缝模型具有更多的拉伸软化曲线，同时适用于更广泛的新型结构领域，例如可以采用总应变裂缝模型下纤维混凝土(Fiber-reinforced concrete)本构关系来模拟纤维混凝土结构。拉伸软化曲线类型一共有 16 种，包括脆性模型(brittle)、弹性拉伸模型(Elastic)、理想的双直线模型(ideal)、基于极限拉应变的线性拉伸软化模型(linear, ultimate strain based)、基于断裂能的线性拉伸软化模型(linear, fracture energy based)、基于总应变的多段线性拉伸软化模型(multi-linear total strain-based)、Hordijk 拉伸软化模型、指数模型(exponential)、CEB-FIP 欧洲 1990 规范模型(MC1990)、fib 2010 规范模型(MC2010)、JSCE 拉伸软化模型(JSCE softening)、JSCE 应力刚化模型(JSCE stiffening)、基于总应变模型的纤维增强模型(fiber reinforced total strain based)、基于裂纹开口的纤维增强模型(fiber reinforced crack opening based)和 Cervenka 拉伸软化模型，如图 2.2-5 所示。总体而言，基于整体位移的脆性和弹性模型不适用于非线性计算。

Hordijk 模型和指数模型为常用的拉伸软化模型，这种模型不仅在达到极限拉应力前的下降段为非线性曲线，能更好地适用于非线性分析计算，同时在材料的本构曲线中考虑了结构的断裂能和裂缝带宽（其中横纵坐标围成的面积为（裂缝处）单位宽度的断裂能）。此外，在诸如欧洲 1990 规范模型、欧洲 2010 规范模型以及 JSCE 模型中，软件会根据规范中的取值，自动地将这些数值嵌入到模型的本构曲线中。

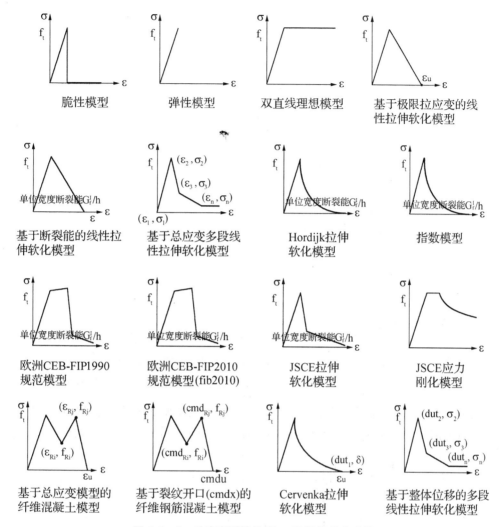

图 2.2-5 总应变裂缝模型 16 种拉伸软化曲线

基于开裂的纤维钢筋混凝土拉伸软化曲线也是近年来总应变裂缝模型中模拟纤维混凝土结构较为常用的一种拉伸软化模型。当用户选择设置纤维混凝土 CMOD 拉伸软化模型的本构参数时,除了要输入如图 2.2-6(a)所示的极限抗拉强度值(Tensile strength)fL 之外,还需要在如图 2.2-6(b)所示的界面框中输入表征纤维混凝土力学特性曲线的两个关键点处残余拉应力值(Residual Strength)和在该应力状态下对应的裂纹口张开值(Crack mouth opening at fR$_i$)坐标(ε_i,fR$_i$)和(ε_i,fR$_j$)和裂纹口张开极限值(Ultimate crack mouth opening)ε_u。

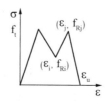

图 2.2-6(a) CEB-FIP 纤维钢筋混凝土拉伸软化曲线图

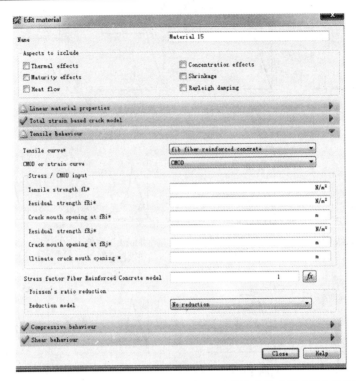

图 2.2-6(b)　纤维钢筋混凝土 CMOD 模式下拉伸软化曲线参数设置界面

当用户选择设置纤维混凝土总应变曲线下拉伸软化模型的本构参数时，除了要输入如图 2.2-6(a)所示的极限抗拉强度值(Tensile strength)fL 之外，还需要在如图 2.2-6(c)所

图 2.2-6(c)　纤维钢筋混凝土总应变模式下拉伸软化曲线参数设置界面

示的界面框中输入表征纤维混凝土力学特性曲线的两个关键点处残余拉应力值(Residual Strength)和曲线中对应该应力状态下的总应变值(Total strain at fRi/fRj)坐标(ε_i,fR$_i$)和(ε_j,fR$_j$)和总应变极限值(Ultimate crack mouth opening)ε_u。

总应变裂缝模型中的压缩行为,可模拟混凝土从受压到压碎的整个状态,尤其是具有丰富的非线性压缩模型,适用于大型结构在短期荷载作用下开裂的非线性计算,压缩模式除了经典的弹性(Elastic)、理想双直线(Ideal)模式和基于 Hognestad 理论的应力—应变压缩曲线之外,同样增加了 CEB-FIP 欧洲 1990 规范和 2010 规范的压缩曲线,除此之外常用的抛物线型也存在于总应变裂缝模型中。在 Diana10.1 的界面操作中,当用户已经确定了所采用的规范(例欧洲 1990 规范、EN1992 规范等)时,用户可以不需要设置受压特性,Diana 可自动设定所有的抗压参数。当采用非规范以外普通的本构模型时,用户必须手动输入每一个抗压特性参数以实现对应力—应变关系的模拟。总应变裂缝模型中的压缩本构不仅存在于各种模拟混凝土达到极限抗压强度(峰值应力)的上升过程,而且诸如 EN1992 规范受压模型、欧洲 CEB-FIP90 规范受压模型等还可以继续模拟从达到峰值应力下降至混凝土达到极限压应变的破坏阶段的应力—应变全曲线模型,在这一阶段裂缝继续扩展贯通,结构受到的破坏越来越严重,曲线出现拐点或收敛点,混凝土的抗压强度呈现非线性急剧下降。

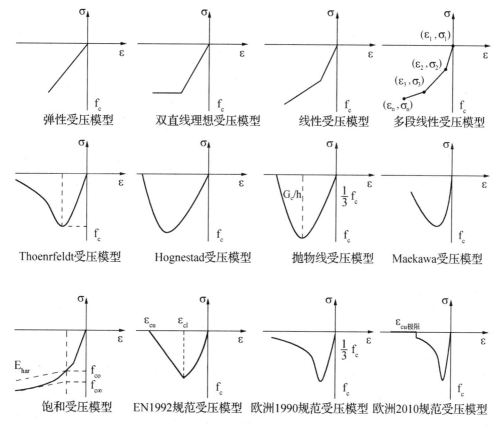

图 2.2-7 12 种总应变裂缝模型的受压模型

这里重点要提到的是以下三种模型:饱和受压模型、EN1992 规范模型以及欧洲的 2010 规范模型。饱和受压模型在考虑总应变模型下混凝土受压时,经历了一个线弹性变形阶段和一个非线性阶段,当抗压强度达到混凝土线弹性变形的比例极限点对应的最大抗压强度后即进入非线性上升阶段,这时非线性受压阶段的斜率即混凝土硬化期间的弹性模量。总应变裂缝模型的卸载采用割线形式卸载,因此在评估能量耗散时数值偏小,结果不够精确。

EN1992 规范相对于其他应力—应变本构关系而言,不仅存在裂缝稳定扩展的非线性上升段,而且还存在由于裂缝继续扩展、贯通造成结构破坏,抗压强度降低至 0 的线性下降段,即最终混凝土的极限压应变值为应力—应变曲线抗压强度值为 0 时对应的横坐标值。欧洲 2010 规范受压模型在欧洲 1990 受压模型基础上增加了极限压应变 ε_{cu},开裂后混凝土抗压强度先增加,达到峰值应力强度后呈现非线性下降,当混凝土抗压强度下降到越过拐点和收敛点后且数值足够小时陡然降低为零,此时应力—应变关系曲线上对应的横坐标值即 ε_{cu},之后一段直线表明在该模型中混凝土一直保持着内聚力耗尽、应力为零、应变急剧扩展的破坏状态。

在使用总应变裂缝模型中的固定(Fixed)裂缝模型时,同样需要考虑剪力滞留。而在正交旋转模型中,由于裂缝方向始终与主应力方向保持垂直,因此用户不需要输入剪力滞留系数 β。相较于多向固定裂缝模型下的完全剪力滞留和恒定剪力滞留模型而言,总应变裂缝模型具有较为丰富的剪力滞留函数(Shear retention function)模型。总体而言,总应变裂缝模型的剪力滞留有如下类型:

1. 恒定剪力滞留模型(Constant Shear Retention)。
2. 可变剪力滞留模型(Variable Shear Retention)。
3. 基于损伤模式下的剪力滞留模型(Damage Based Shear Retention)。
4. 基于骨料粒径的剪力滞留模型(Aggregate Size Based Shear Retention)。
5. 基于开裂正应变的剪力滞留模型(Normal Crack Strain Based Shear Retention)。
6. 前川剪力滞留曲线模型(Maekawa Shear Retention Curves)。
7. 马哈伊迪剪力滞留模型(Al-Mahaidi Shear Retention Function)。

上述这些剪力滞留模型可以适用于各种问题下的混凝土开裂。如考虑混凝土塑性损伤问题下的剪切刚度,可以采用基于损伤模式下的剪力滞留模型;在研究加载卸载等滞回分析下的剪切刚度退化问题可以采用前川剪力滞留函数模型,如考虑剪切刚度随着裂缝开口的增大而进一步降低时,用户可以采用基于骨料粒径的剪力滞留模型。

(3)朗肯主应力模型

短期混凝土开裂模型除了上述提到的几种模型之外,Diana 中还有一种朗肯主应力模型,这种模型只适用于模拟二维平面内混凝土双轴受压模型,即一个方向承受的压力会对另一个方向的应力产生影响。但朗肯主应力模型只适用于 2D 平面结构,对环境因素下的裂缝开展的模拟具有局限性。即对含有收缩徐变作用下的长期性能和有盐类环境侵蚀下的裂缝开展过程无法有效模拟。

此外,还有一种适用于非正交裂缝,模拟低周循环荷载在加载卸载滞回分析的弹塑性断裂模型——前川—福浦(Maekawa-Fukuura)模型。需要特别说明的是,这种模型在开裂前基于混凝土弹塑性损伤模型,开裂后可以归为总应变裂缝模型。本书在分类时,根据开裂后的特征将其归为总应变裂缝模型中的一种。前川—福浦裂缝类型为非正交裂缝。与总应变裂缝模型有所不同,前川模型允许单一积分点存在最多 6 条非正交裂缝。该模型在后续升级版本中正在不断研发和完善,并将在升级版软件中得到应用,限于本书篇幅的限制,这里就暂不作详细介绍。

弥散开裂模型下各类裂缝模型的适用特点见表2-1。

表2-1 弥散开裂模型下各类裂缝模型及适用特点

弥散开裂模型		适用特点
多向固定裂缝模型		主拉应力超过材料抗拉强度。 已有裂缝方向与当前主拉应力的方向需要超过临界角。
总应变裂缝模型	正交固定裂缝模型	一旦开裂，裂缝保持固定方向不变，不随主应力裂缝方向的改变而改变。但不能和徐变功能结合使用
	正交旋转裂缝模型	一旦开裂，裂缝位置随着裂缝主应力方向相对转动改变。不能和徐变功能结合使用
	混合裂缝模型	当裂缝走向不明朗时，采用正交旋转裂缝模型；当裂缝方向和裂缝位置明显时，转化为正交固定裂缝模型。但不能和徐变功能结合使用
	前川—福浦模型	开裂前基于混凝土弹塑性损伤模型，开裂后可归为总应变裂缝模型。单一积分点存在最多6条非正交裂缝，适用于模拟循环荷载作用下的滞回分析，可较为准确地反映能量耗散。
朗肯主应力模型		适用于双轴模型，并考虑一个方向受压对另一个方向产生的影响。只适用于2D建模，不能用于分析徐变收缩问题，也无法考虑诸如氯离子浓度等环境因素对模型的影响

注：在Diana10.1中，无论是上述哪一种裂缝模型，均只适用于混凝土短期性能下的非线性分析，对于含有收缩、徐变以及松弛状态下的长期分析以及与环境因素有关的耐久性分析均不可适用。

2.3 钢筋本构模型

Diana钢筋建模有两种方式，一种是离散式（Discrete）钢筋，一种是嵌入式（Embedded）钢筋。本文重点讨论嵌入式钢筋建模方式以及相应的配筋与桩基础（Reinforcements and pile foundations）本构模型设置。如前文所述，Diana中钢材模型较多，主要分为配筋和地基基础模型（Reinforcement and pile foundations）、钢结构本构模型（Steel）以及钢结构规范本构（Steel EN code）模型。其中每种模型又大致可分为各向同性的线弹性模型、冯·米塞斯（Von Mises）弹塑性模型以及完全塑性模型。钢结构规范又可以分为欧洲3号1993-1-1规范、荷兰6770规范、荷兰6720钢筋规范、荷兰6720的预应力索规范。本节重点讨论配筋和地基基础本构模型下钢材与钢结构规范下的本构模型将在2.5节中介绍。

1. 离散式钢筋

离散式钢筋采用Beam梁单元或者是Truss桁架杆单元建模，在Diana界面操作模块下的INTERF操作中将钢筋单元离散成Truss或者是Beam单元。离散式钢筋通过钢筋节点合并的方式与混凝土单元实现耦合作用，这种建模方式适用于分析单根预应力钢筋的粘结滑移问题，其中钢筋和混凝土之间通过创建界面单元进行连接。通过线对线、线对面、线对实体的界面单元实现钢筋条与其周边的母体单元连接。在定义离散式钢筋元本构模型时，用户首先应将这些梁单元或者是桁架单元赋予钢筋或者是钢绞线的本构特性，并且在

GUI 界面操作的 DATA 功能模块中实现这些已经定义钢筋属性的桁架单元和梁单元向嵌入式钢筋单元之间的转化。采用离散型钢筋建模具有较高的精度,但建模效率较低。

2. 嵌入式钢筋

嵌入式钢筋适用于大批量的钢筋单元建模问题,材料本构属性上属于配筋和桩基础类型(Reinforcements and pile Foundations)。嵌入式钢筋单元本身不具有自由度,只能通过单元上的坐标点及积分点与混凝土单元结合的方式贡献刚度。其中嵌入式钢筋根据建模功能、形状和几何属性赋予的不同,又可分为 Bar 钢筋单元和 Grid 钢筋网片单元,Bar 单元适合建立后张法预应力束、纵向钢筋等数量较多的纵筋;而钢筋网片适合箍筋或一个面上等间距分布的纵筋情况,有时候可以直接选中赋予钢筋网片材料和几何属性。嵌入式钢筋在模拟后张法施工(post-tensionng)中作为预应力,在添加前的材料本构定义中根据是否与母体单元粘结(bonded to mother element)可分为有粘结(bonded)和无粘结(Nobonded)两种类型。在有粘结的情况下,钢筋与嵌入的母体单元形成一体,共同变形,即钢筋对母体单元的刚度产生贡献;而无粘结时,钢筋对被嵌入的母体单元刚度没有贡献,钢筋应变和应力也不随母体变形而变化。

此外,Bar 单元适用于几乎所有 2D 和 3D 单元,具有良好的单元适应性。Bar 钢筋单元形状及钢筋单元的阶数随嵌入点的个数变化而变化。Diana10.1 版中 Bar 钢筋单元材料本构可选择配筋和桩基础(Reinforcements and pile foundations)、钢结构(Steel)、钢结构设计规范(Steel Design Codes)。对于非线性计算而言,钢性材料通常选用塑性屈服的冯·米塞斯(Von Mises)模型。

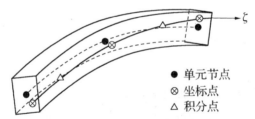

图 2.3-1 嵌入式钢筋线单元 Bar 单元嵌入梁单元中

Grid 钢筋主要模拟结构中的分布钢筋和钢筋网片。与 Bar 钢筋单元类似,Grid 钢筋也可有效地嵌入到各类 2D 和 3D 单元中,通过对弹性模量、泊松比以及塑性屈服值的一系列材料特征项的定义完成对钢筋网片的材料定义。对钢筋网片的建模既可以采用 Bar 单元逐个建模,也可以采用建模效率更高的 Grid 钢筋网片单元。

与 Bar 钢筋定义相比,采用 Grid 单元建立的钢筋网片的定义相对比较复杂。对于钢筋网片的厚度定义既可以通过在局部坐标系的两个正交方向输入等效厚度以表征配筋率的直接输入(Direct input)方式,又可以通过直径和间距(Diameter and spacing)的定义方式来定义钢筋网片的几何模型参数。在这两种方式下,用户不仅需要定义局部坐标系 x 和 y 两个方向上直径、间距及等效厚度各参数值还需要定义钢筋单元局部坐标系 x 轴方向与整体坐标系的关系。

无论是 Bar 钢筋还是 Grid 钢筋网片,在配筋和桩基础(Reinforcements and pile foundations)选择模块下都具有很丰富的材料模型(Material model),如图 2.3-2 所示,包括线弹性(Linear elasticity)本构模型、单轴非线性弹性模型(Uniaxial nonlinear elasticity)、冯米塞斯塑性屈服模型(Von Mises plasticity)、JSCE 塑性模型(JSCE plasticity)、粘结滑移模型(Bond-slip reinforcement)、欧洲 Eurocode 2 EN 1992-1-2 钢规模型、桩基础模型(Foundation pile)等。此外,Diana10.1 中还吸收了上世纪末一系列专门用于循环荷载下滞回分析的钢筋塑性材料模型的研究成果,如 Dodd Restrepo 塑性模型(Dodd Restrepo plasticity)、Monti-Nuti 塑性模型(Monti-Nuti plasticity)和 Menegotto-Pinto 塑性模型(Menegotto-Pinto plasticity)。其中线弹性本构模型和 Von Mises 塑性屈服模型为当前 Diana 的配筋和桩基础(Reinforcements and pile foundations)类型中最常用的本构模型。Von Mises 模型除了需要输入钢材的弹性模量和屈服强度值,还需要选择塑性硬化类型(Plastic hardening)。通过上述这些参数特征来确定钢材在塑性模型下的本构特性。无论是 Bar 钢筋单元还是 Grid 钢筋单元,嵌入式钢筋均可以嵌入任何类型的单元中,对梁单元、壳单元、平面单元及固体单元均适用。

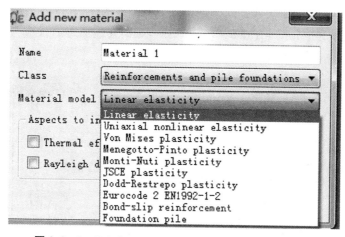

图 2.3-2　Reinforcement and pile foundations 本构类型

3. 嵌入式粘结滑移钢筋

与嵌入式钢筋先定义粘结滑移界面单元的老版 Diana 软件操作方式不同的是,新版的 Diana10.1 软件钢筋粘结滑移模型可以在嵌入式钢筋单元基础上直接定义粘结滑移特性,而不需要再建立界面单元。这就在实际建模过程中节省了时间,提高了建模效率,同时初学者也更方便入手。在定义钢筋的粘结滑移模型时,用户不仅要选择粘结滑移的失效模型(Bondslip failure model),还需要定义相关滑移参数。在定义嵌入式钢筋单元粘结滑移截面几何属性时,钢筋具有多种形式。桁架杆单元(Truss bondslip)力学行为的粘结滑移特性既可以作为一种选择,用户也可以将预应力筋选择为具有梁单元杆件力学行为的粘结滑移特性。此外 Diana10.1 软件具有较多形式的滑移截面形状,如圆形(Circle)、圆管状(Pipe)、矩形梁单元(Rectangular beam bondslip)粘结滑移特性和箱型梁单元(Box beam bondslip)粘结滑移特性。这都将在本书的第三章例题中将进行说明。

4. 钢拉索

与上述情况不同,钢拉索不与混凝土接触,通常采用普通桁架单元 L2TRU 或者 L7BEN 之类的梁单元进行模拟,采用 Prestress 方式对拉索施加力,与直接施加后张法预应力荷载(Post tensionning load)的单位不同,Prestress 的荷载单位为 N/m² 或 kN/m²,且施加方式是先按照桁架单元进行建模,然后在 DATA 界面下对桁架和梁单元进行离散化,定义钢材的本构和横截面几何尺寸属性,钢材的本构属性可以是钢本构(Steel)类型,也可以是钢结构规范本构(Steel codes)类型直接施加(Prestress)预应力荷载,还可以是配筋和桩基础本构(Reinforcements and pile foundations)类型中施加荷载。已经赋予本构属性的类型在添加荷载工况和施加荷载的时候再在端部施加 Prestress 类型荷载,用以模拟钢拉索的受力特性。钢拉索的模拟常用于斜拉桥中钢索的腐蚀、疲劳以及耐久性的研究。通常情况下,后张法横截面积较大的预应力钢绞线需要施加预应力荷载,而斜拉桥拉索以及预应力钢束等尺寸较小的结构则主要施加 Prestress 类型的预应力钢束荷载。还可以对钢拉索赋予 FRP 的本构属性近似模拟 FRP 筋力学特性。

预应力的荷载分为两种类型,一种是模拟后张法的预应力筋荷载(Post tensioning load),另一种是普通预应力荷载(Prestress),主要以应力的形式添加。荷载工况(Load case)选择 Prestress,目标加载类型(Load target type)为实体(Solid),荷载类型(Load type)为钢筋预应力荷载(Reinforcement bar prestress),对钢束施加普通预应力荷载的操作界面如图 2.3-3 所示。

图 2.3-3 Prestress 荷载施加方式

5. 以钢材为主体的本构定义

此外,当用户使用 Diana 软件进行钢结构的稳定问题、屈曲问题、后屈曲问题、钢材的疲劳问题甚至钢材防火问题分析的时候,往往需要将主体结构的材料属性定义为钢结构,而在后续的有限元分析流程中需要完成网格划分才能成功运行有限元计算,此时用户不可以右击选择或直接点击菜单栏下方的 而应该首先像赋予混凝土结构材料属性那样定义单元类型,接着再在下方 Material 模块中将材料属性定义为钢材本构属性。因为蓝色 Edit

Reinforcement property assignments 快捷键仅适用于定义钢筋混凝土结构中作为配筋使用的钢材材料属性,而不能定义以钢材为主体的结构。以实体单元为例,以钢结构为主体的钢材属性的具体定义流程如图 2.3-4 所示。Diana 中采用 Edit Reinforcement property assignments 的定义方式与采用 Edit Property assignments 的定义方式的区别就在于前者确认是配筋属性的情况下无法选择钢结构的单元类型,如图 2.3-5 所示。

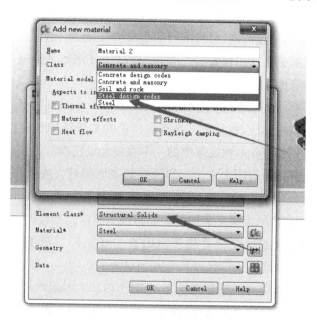

图 2.3-4　钢结构为主体的钢材属性的定义方式

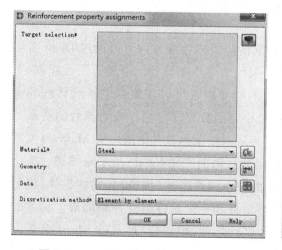 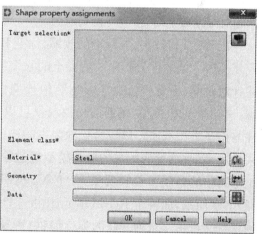

图 2.3-5　Reinforcement property assignments 与 Edit Property assignments 钢材定义方式区别

2.4　Diana 长期性能问题的本构模型介绍

所谓长期性能问题,是指钢筋混凝土结构在时变效应下的徐变、收缩和钢筋松弛问题。其中收缩和徐变是土木工程中混凝土材料常见的问题,属于混凝土材料固有特性。世界各国混凝土规范中考虑收缩徐变的本构模型主要基于以下几种：欧洲混凝土协会CEB-FIP系列规范CEB-FIP模型（包括CEB-FIP78和CEB-FIP90模型）、美国AASHTO模型、美国的ACI模型等。

在土木工程中，收缩、徐变等长期因素的作用，会使得结构中的构件的变形增加，在钢筋混凝土结构中还会发生应力重分布现象，在预应力结构中还会发生预应力的较大损失。因此近年来混凝土的收缩、徐变也成为一项越来越热门的研究课题。研究表明，混凝土的徐变不仅与时间有关，而且与加载龄期、材料本身的组成成分（水灰比、混凝土水泥用量）、结构构造尺寸和环境等因素有关。这些参数的影响在Diana10.1版本各种规范下的徐变收缩模型的本构参数中均有所考虑。

（1）结构构造尺寸。构件理论厚度（notional size of member h）是影响混凝土收缩、徐变的一个至关重要的内在参数，也是下文将重点讨论的关于结构长期性能非线性分析有限元Diana软件CEB-FIP1990本构规范中的一个重要的设定参数。构件理论厚度反映了构件暴露在空气部分的体积与表面积之比。当构件的表面积越大或体积越小，构件暴露在环境中的面积越大，水分失去得越快，徐变和收缩也相对严重。在Diana软件的CEB-FIP1990规范中，构件理论厚度定义为构件截面积的2倍除以构件的周长。

（2）环境因素。与收缩徐变相关的环境因素主要是环境温度（°C）和相对环境湿度（以％计数）。环境温度对混凝土的徐变具有相当显著的影响。一般说来，环境温度越高，混凝土的徐变越大。构件的相对环境湿度越大，混凝土内部的吸附水蒸发量越小，水泥在混凝土初期的水化反应中水化程度也越高，故徐变较小。与此同时，相对环境湿度针对混凝土结构早期的加载而言会产生更大的徐变。通常情况下，温度越高，相对环境湿度越低，混凝土的徐变现象越明显。而相对环境湿度对预应力的损失也具有很大影响。研究表明，当环境的相对湿度较低时，预应力损失越大。

在Diana软件中，混凝土的收缩、徐变和钢筋的应力松弛现象均可以通过Diana有限元建模嵌入到软件分析中。混凝土的收缩徐变采用的是率型模型。通过有限元建模，上述针对收缩、徐变和松弛等影响结构长期性能的时变因素可以在软件中得到体现和考虑。在Diana软件中，钢筋的松弛现象可以通过松弛函数自动考虑。松弛函数的选取往往与松弛类型有关。以应力松弛为主的松弛模型通过松弛函数并借助广义Maxwell模型来描述，以收缩徐变为主的松弛模型通过徐变函数并借助广义Kelvin模型来描述。类似于电路中的串联和并联现象，一个Maxwell体由一个弹性组件（弹簧）和粘性组件（阻尼器）串联而成，而一个Kelvin单元体由一个弹性组件和一个粘性组件并联而成。前者的物理意义可由一系列并联的弹性组件（弹簧）和粘性组件（阻尼器）形成的Maxwell单元来描述，如图2.4-1所示；后者的物理意义可由一系列串联的弹性组件（弹簧）和粘性组件（阻尼器）形成的Kelvin单元来描述，如图2.4-2所示。即Maxwell链由多个Maxwell体相互并联组成，Kelvin链由多个Kelvin体相互串联组成。由于Diana中Maxwell链基于混凝土非老化（Non-Aging）理论，而Kelvin链基于混凝土老化（Aging）理论，因此常常借助松弛函数及Maxwell链模型分析以钢筋松弛为主的钢筋混凝土结构时变问题，而借助收缩徐变函数和Kelvin链分析以收缩徐变为主的钢筋混凝土结构时变问题。对于

Diana 软件而言,要构成 Maxwell 链以及 Kelvin 链,模型单元的数量均不得少于 10 个。通常情况下,采用以钢筋松弛为主不考虑混凝土老化的 Maxwell 链模型比采用以收缩徐变为主基于混凝土老化理论的 Kelvin 链关于长期性能的非线性计算的结果要稍大,并且更容易达到模型单元所需的默认个数,而由于 Maxwell 链是以松弛为主的模型,相比 Kelvin 链,其钢筋的松弛结果会稍大于后者。因而用户在需要与长期性能有关的试验结果进行比对时,需慎重选择合适的模型。根据作者多年经验,Kelvin 链中的混凝土的弹性模量值是 Kelvin 链的重要参数之一,其数值又往往与单元龄、温度和相对环境湿度等因素息息相关。因此如何根据上述参数准确确定混凝土 28 天抗压强度是采用 Kelvin 模型数值模拟准确度的关键点之一。

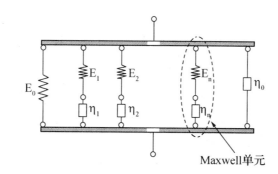

图 2.4-1 Maxwell 单元链

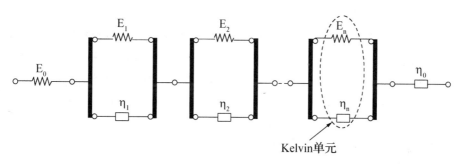

图 2.4-2 Kelvin 单元链

在率型模型中,松弛函数和徐变函数均以狄利克雷展开形式近似表示。其中松弛函数模型展开式为求和项内各项形式为指数函数与第 i 个单元体的松弛函数乘积。徐变函数模型展开式中求和项内的各项形式为指数函数除以第 i 个单元体的徐变函数。在广义 Maxwell 模型中,钢绞线松弛的应力—应变关系可表示为如下式(2-1)所示

$$\sigma(t) = \int_{-\infty}^{t} E(t,\tau)\varepsilon d\tau \tag{2-1}$$

其中,$E(t,\tau)$ 为加载龄期为 τ、计算龄期为 t 时松弛函数,松弛函数可用狄利克雷序列展开,展开式如(2-2)所示

$$E(t,\tau) = \sum_{i=0}^{n} E_i(\tau) e^{-\frac{t}{\lambda_i}} \tag{2-2}$$

其中,$E_i(\tau)$ 为 Maxwell 单元下的时变刚度,松弛时间则定义为

$$\lambda_i = \eta_i / E_i \qquad (2-3)$$

分别写出(2-2)式在 t 时刻和(t ＋Δt)时刻下时变松弛函数表达式,令 $t^* = t + \dfrac{\Delta t}{2}$。再将这两个表达式代入(2-1)式并经狄利克雷展开,则应力增量 Δσ 如式(2-4)所示

$$\Delta\sigma = \sum_{i=0}^{n} (1-e^{-\frac{\Delta t}{\lambda_i}}) \left(\frac{E(t^*)\lambda_i}{\Delta t} \overline{D}\Delta\varepsilon - \sigma_i(t) \right) \qquad (2-4)$$

在 Kelvin 链模型中,推导公式如式(2-5)所示

$$J(t,\tau) = \sum_{i=0}^{n} \frac{1}{E_i(\tau)} \left(1-e^{-\frac{t}{\lambda_i}}\right) = \frac{1}{E_0(\tau)} + \sum_{i=0}^{n} \frac{1}{E_i(\tau)} \left(1-e^{-\frac{t}{\lambda_i}}\right) \qquad (2-5)$$

其中,$\lambda_i = \dfrac{\eta_i}{E_i}$,$\eta_i$ 表示第 i 个 Kelvin 单元体的粘滞阻尼系数;E_i 表示第 i 个 Kelvin 单元体的弹性模量;$J(t,\tau)$ 为加载龄期为 τ、计算龄期为 t 时徐变函数。

从上式中不难看出,Kelvin 链计算公式本质上是通过徐变函数进行描述,该徐变函数可以用狄利克雷函数展开近似表示。每一个 Kelvin 单元链均可以通过收缩徐变试验数据或是规范中的最小二乘法拟合得到。

2.5　Diana 规范简介

在使用 Diana 软件时,既可以使用下文介绍的几种规范,也可以自定义规范,其中欧洲混凝土协会的欧洲 1990 规范(CEB‐FIP 1990)规范和 Diana10.1 新增的欧洲 2010 规范(fib Model Code for Concrete Structures 2010)为有限元建模中最常用的两种规范。本节将重点介绍欧洲 1990 规范、欧洲 2010 规范、美国 AASHTO 规范和日本 JSCE 这 4 种规范,并重点比较 4 种规范使用中的共性和特点。

Diana 混凝土规范种类如下:
(1) 欧洲 1990 规范(CEB‐FIP Model Code 1990)。
(2) 欧洲 2010 规范(fib Model Code for Concrete Structures 2010)。
(3) 美国 ACI 规范。
(4) 美国 AASHTO 规范。
(5) 日本 Japan Concrete Institute 规范(JCI)。
(6) 日本 Japan Society of Civil Engineers 规范(JSCE)。
(7) 韩国 Korea Concrete Institute (KCI) 2007 规范。
(8) 荷兰 NEN 6720/A4 规范。
(9) 丹麦结构安全联合委员会 JCSS Probabilistic Model Code 概率模型规范。

钢结构规范主要有:
(1) 欧洲 3 号 1993 钢结构规范(Eurocode 3 EN 1993‐1‐1)。
(2) 荷兰 6770 规范(Dutch NEN 6770)。
(3) 荷兰 6720 钢筋规范(Dutch NEN 6720 reinforcement steel)。

(4) 荷兰 6720 拉索预应力规范(Dutch NEN 6720 prestress cable)。

钢材本构定义类型,如图 2.5-1 所示。

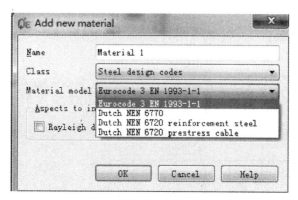

图 2.5-1 Diana 软件中钢结构规范类型选项

1. CEB-FIP1990 规范

混凝土 CEB-FIP1990 规范是在 CBE-FIP 1978 规范的基础上发展而来的。在 Diana 欧洲 1990 规范中,用户主要需要设置以下几大模块:涵盖内容(Aspect to include)、欧洲 CEB-FIP 1990 规范中混凝土内在特性(European CEB-FIP 1990)、直接输入的规范参数(Direct input)以及根据研究对象不同而选择的收缩徐变长期性能设置(徐变曲线类型、单元龄、固化收缩曲线)、热流或阻尼系数的相关模块参数。

涵盖内容模块提供弹性(Elasticity)、塑性(Plasticity)、收缩(Shrinkage)、徐变(Creep)、热流(Heat flow)等选择项,其中选择塑性计算时对模型的收敛性要求较高,而弹性计算不可以与收缩徐变等长期性能的因素同时勾选。规范参数中有诸多需要设置的因素,欧洲 1990 规范中的混凝土参数设置不仅涉及了诸如混凝土圆柱体抗压强度、弹性模量、泊松比、密度、28 天圆柱体抗压强度标准值、平均抗压强度等表征混凝土特性的基本项,同时还考虑了土木工程材料中的混凝土水泥硬化类型(Cement type)、构件理论厚度(Notional size of member)、骨料类型(Aggregate type)以及混凝土环境温度(Relative ambient humidity,RH)、相对湿度等环境因素的相关参量,而这些因素对于模拟结构长期性能问题具有重要意义。各参量符号名称、含义以及默认值见表 2-2(表格中注明用户自行设定的部分需要用户根据实际情况自行输入)。

表 2-2 CEB-FIP90 规范本构模型基本参数介绍

符号名称	含义	类型	默认值
Concrete class	混凝土抗压强度	European CEB-FIP 1990（欧洲混凝土协会 1990 规范模块）	用户自行选择
Cement type	水泥硬化类型		Normal and rapidly hardening
Ambient temperature	环境温度		20 ℃/293.15 K
Notional size of member h	构件的理论厚度		0.15 m
Relative ambient humidity RH in %	环境相对湿度		80%
Aggregate type	骨料类型		Quatrzite(硅酸岩)

(续表)

符号名称	含义	类型	默认值
Young's modulus	弹性模量	Direct Input（直接输入模块）	用户自行设定
Young's modulus at 28 days	28 天时的弹性模量		用户自行设定
Poisson's ratio	泊松比		0.3
Thermal expansion coefficient	热膨胀系数		1×10^{-5} 1/℃
Mass density	混凝土密度		2500 kg/m³
Characteristic strength at 28 days	28 天的混凝土抗压强度标准值		用户自行设定
Mean compressive strength at 28 days	混凝土抗压强度平均值		用户自行设定
Creep curve type	徐变曲线类型	Creep（徐变模块）	Non-aging（非老化龄期）
Concrete age at birth of element	混凝土的单元龄期		2419200s(28day)
Concrete age at end of curing period	混凝土收缩曲线固化周期结束时单元龄	Shrinkage（收缩模块）	86400s(1day)
Couductivity	热传导率	Heat flow（热流模块）	用户自行设定
Capacity	热容量		用户自行设定
Conductivity/capacity function	热传导/热容量方程		No dependency
Heat of hydration Method	水化热方式		Off
a-Factor for mass matrix	质量矩阵系数	Rayleigh damping（瑞雷阻尼模块）	用户自行设定
b-Factor for stiffness matrix	刚度矩阵系数		用户自行设定

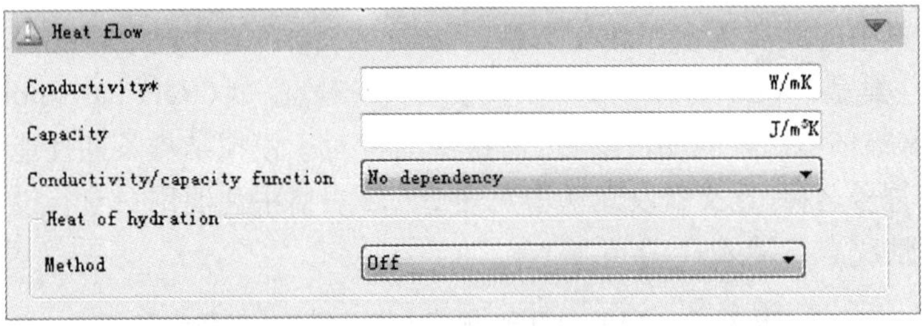

图 2.5-2　Diana10.1 CEB-FIP 1990 中热对流操作界面

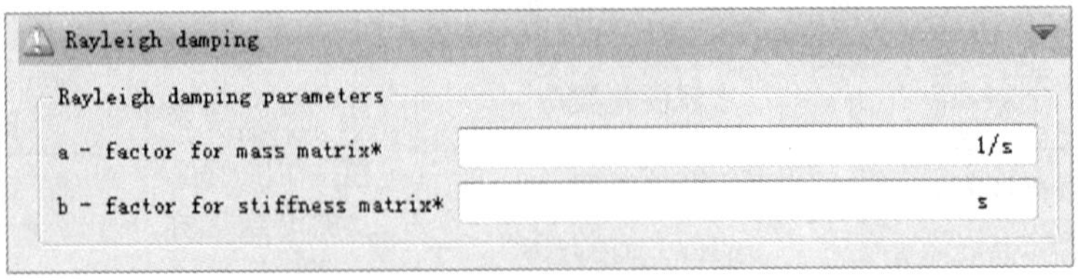

图 2.5-3　Diana10.1 CEB-FIP 1990 中瑞雷阻尼操作界面

以下几点需要特别说明：

(1) 在欧洲规范中,采用直径 152 mm、高 305 mm 的标准圆柱体试件确定的轴心抗压强度 f_{ck} 作为圆柱体抗压强度的标准值,所测的抗压强度数值即对应的混凝土强度标号后的数字,该数字对应的就是软件中混凝土抗压强度标准值(Characteristic strength at 28 days),例如 C30 对应的圆柱体抗压强度标准值为 30 N/m²。f_{ck} 与混凝土 28 天平均抗压强度值 f_{cm} (Mean compressive strength at 28days)存在以下关系

$$f_{cm} = f_{ck} + 8$$

需要使用试验测定的数值作为立方体抗压强度时,圆柱体抗压强度标准值与立方体抗压强度标准值之间的关系式根据混凝土强度等级可按下列分段公式确定

$$f_{ck} = \begin{cases} 0.79 f_{cu,k} & (混凝土强度 C60 以下) \\ 0.833 f_{cu,k} & (混凝土强度为 C60) \\ 0.857 f_{cu,k} & (混凝土强度为 C70) \\ 0.875 f_{cu,k} & (混凝土强度为 C80) \end{cases}$$

采用棱柱体轴心抗压强度测定的试件,首先应根据立方体抗压强度与棱柱体轴心抗压强度关系将轴心抗压强度转化为立方体抗压强度,再根据上述公式转换成圆柱体抗压强度输入软件中。

(2) 混凝土弹性模量,遵循以下公式

$$E = 2.15 \times 10^4 \times \alpha_E \times \left(\frac{f_{cm}}{10}\right)^{\frac{1}{3}} \text{(单位:MPa)}$$

当混凝土骨料类型为硅酸盐混凝土材料时,$\alpha_E = 1$;对于线性材料,新型计算状态下计算出来的弹性模量还需在此基础上进行折减,折减系数为 0.85。

2. fib Model Code for Concrete Structuraes 2010

fib Model Code for Concrete Structures 2010 规范(MC2010)是 Diana10.1 版本中新增加的规范,如图 2.5-4 所示。与 CEB-FIP 1990 规范相比,fib Model Code for Concrete Structures 2010 规范(以下简称 2010 规范)的一大好处是减少了用户需要直接输入参数的

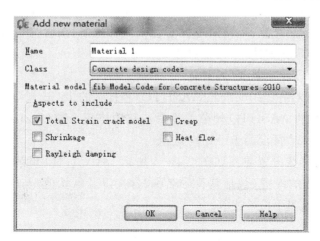

图 2.5-4 fib Model Code for Concrete Structuraes 2010 版本构界面

Direct Input 模块,即一旦确定了混凝土的强度等级,这些在 CEB-FIP 1990 规范本构中需要手动输入的诸如弹性模量、泊松比、密度、热膨胀系数、抗压强度平均值等参数会由软件自动确定,在批量化分析中免去了大量手动输入工作,如图 2.5-5 所示。但由于土木工程材料具有一定程度的离散型,构件尺寸、材料的弹性模量和强度的数值往往在一定范围内波动,采用该规范模型下的本构定义无法根据实际情况手动输入现场测定的数值,此时往往会产生一定程度的误差,这种情况在需要同试验状态下的结果对比时是一个不利的影响。

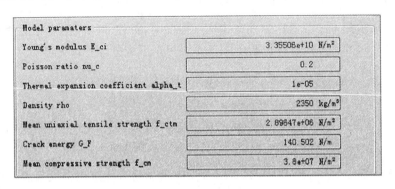

图 2.5-5 fib Model Code for Concrete Structuraes 2010
规范中软件系统自动确定的材料参数

2010 规范的另一个特点是认定构件理论尺寸和环境因素是与收缩、徐变息息相关的内因和外因,并将收缩、徐变与环境温度、环境相对湿度、构件的理论尺寸放在一个模块下考虑相互之间的作用。

2010 规范的第三个特点是设置了极限状态下的材料安全系数(Material safety factors, ULS),包括弹性模量安全系数、轴向拉伸强度安全系数以及平均抗压强度安全系数。通过考虑承载力极限状态下的材料弹性模量、抗拉强度和抗压强度的变化,设置了材料的安全储备系数(Material Safety Factors),即材料取得的设计值。通过对这三个参数的取值可以全面且较精准地模拟钢筋混凝土结构在承载力极限状态下的受力特性。由于混凝土材料的设计值低于标准值,因此在输入材料安全储备系数时均应选择大于 1 的数值。

此外,2010 规范的热对流、瑞雷阻尼设置包括其符号名称、含义以及默认值与 CEB-FIP90 规范相同,这里不再赘述。

3. 美国 AASHTO 和日本 JSCE 规范

美国的 AASHTO 规范(American AASHTO)和日本的 JSCE 规范也是 Diana 规范库里常用的规范。美国的 AASHTO 规范采用骨料来源修正系数(Correction factor K1 for source of aggregate)来表征混凝土中骨料类型。

AASHTO 规范参数设置界面如图 2.5-6 所示。AASHTO 规范在收缩、徐变、热传导、瑞雷阻尼系数方面的设置(包括其符号名称、含义以及默认值)与 CEB-FIP90 规范、2010 欧洲规范相同,这里不再赘述。

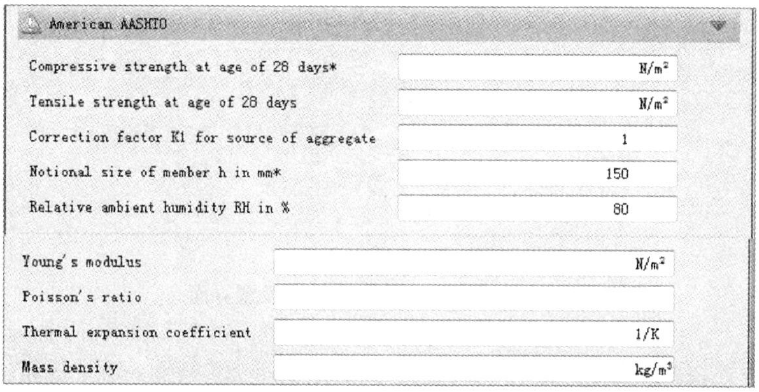

图 2.5-6 Diana10.1 中美国 AASHTO 规范参数设置界面

与欧洲规范和美国 AASHTO 规范不同的是，JSCE 规范不适用于收缩徐变等结构长期性能分析。在 JSCE 规范中，混凝土强度取值为 91 天后的混凝土立方体抗压强度标准值(Characteristic strength after 91 days)，弹性模量取值则为 91 天时的弹性模量(Modulus of elasticity at age of 91 days)。在 Diana10.1 模块的 JSCE 规范中，并未涉及针对收缩、徐变等长期性能的选项，诸如单元龄、构件理论厚度、收缩固化曲线时间等一系列影响收缩、徐变的因素也并未在软件中有体现。图 2.5-7 是 JSCE 规范本构类型设置界面。

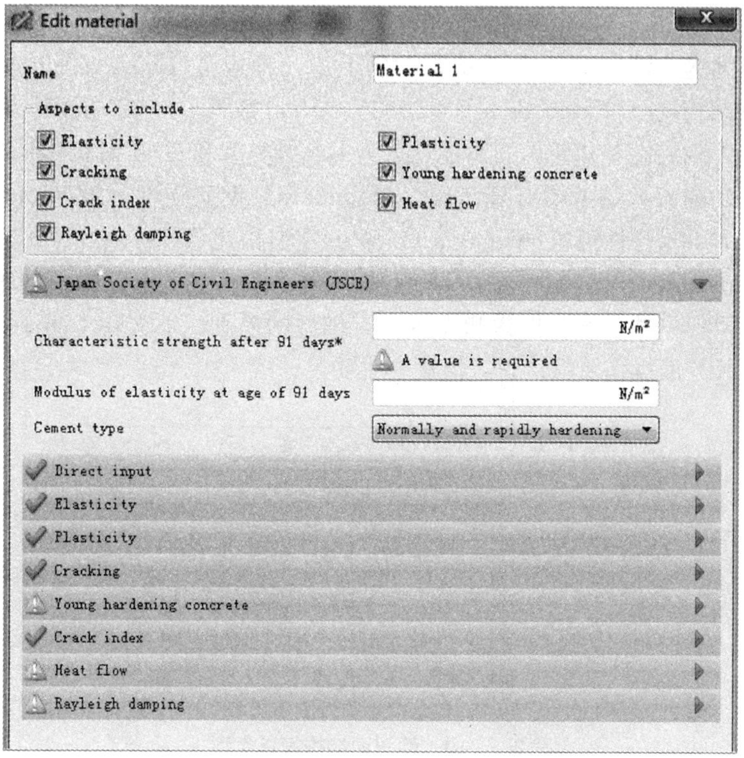

图 2.5-7 JSCE 规范本构类型设置界面

JSCE 规范中热对流、瑞雷阻尼系数的设置与其他规范操作完全一致，如图 2.5-8、图 2.5-9 所示。

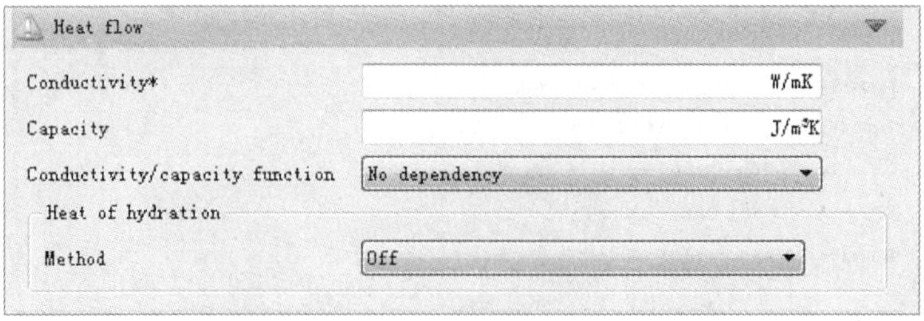

图 2.5-8 JSCE 热对流本构设置界面

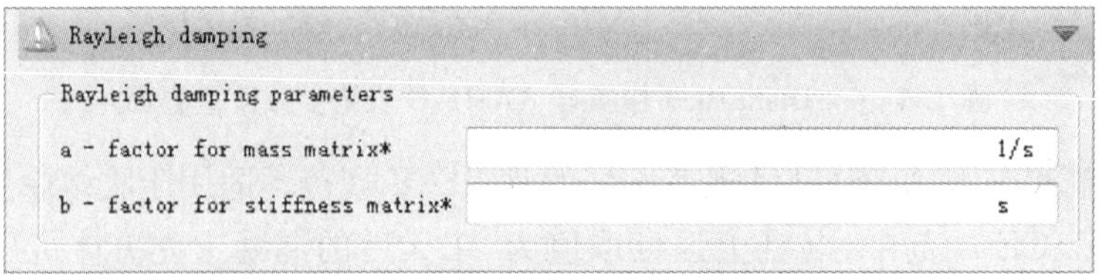

图 2.5-9 JSCE 瑞雷阻尼本构设置界面

4. 钢结构规范本构模型

钢材料本构模型也是 Diana 中常用的本构模型,其中以荷兰钢结构规范本构模型种类最多。荷兰的 Dutch 钢结构规范不仅有常用的钢结构规范(Dutch NEN 6770)和钢筋规范(Dutch NEN 6720 reinforcement steel),也有适用于斜拉桥结构钢拉索的 Dutch NEN 6720 prestress cable 规范。除了 Diana9.4 版本中的上述几种规范,Diana10.1 中还新增加了欧洲 3 号 1993 钢结构规范(Eurocode 3 EN 1993-1-1),如图 2.5-10 所示。无论上述哪一种规范,材料的本构属性和几何界面属性均分为 Bar 线类型和 Grid 钢筋网片类型两种定义方式,材料模块的输入参数主要有规范类型(Standard)、钢筋强度等级(Grade)、名义厚度(Nominal thickness)、弹性模量(Young's modulus)、泊松比(Poisson's ratio)、热膨胀系数(Thermal expansion coefficient)、密度(Density)等基本表征参数。根据不同钢结构规范使用的主要对象不同,一些参数的选择和设置会有细微的差别。下文就 Diana10.1 中各种钢结构规范的 GUI 界面操作及默认设置值逐一说明。

图 2.5-10 欧洲钢结构规范类型

一旦生成钢结构的几何模型对象,选择好钢结构规范,对应的默认值就会自动生成,用户可自行进行选择。以 Eurocode 3 EN 1993-1-1 为例,各项材料默认参数如图 2.5-11 所示。

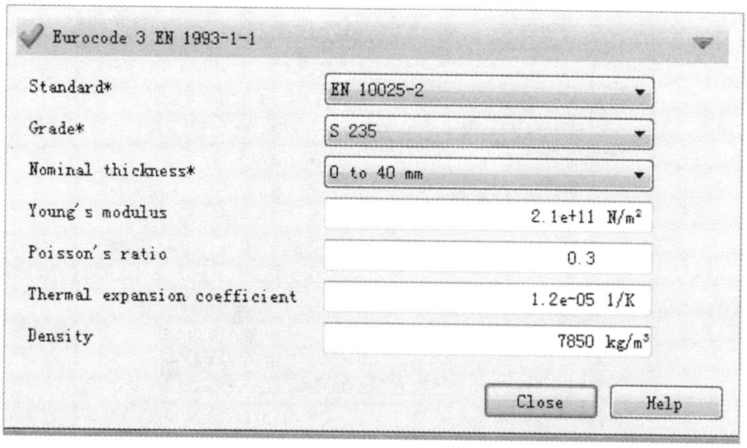

图 2.5-11　Eurocode 3 EN 1993 1-1 规范默认值

在荷兰钢结构 6770 规范中，用户需要定义钢结构的强度等级（Grade）、名义厚度范围（Nominal thickness）、弹性模量、泊松比、热膨胀系数（Thermal expansion coefficient）以及钢材的密度，各种参数的默认值如图 2.5-12 所示。

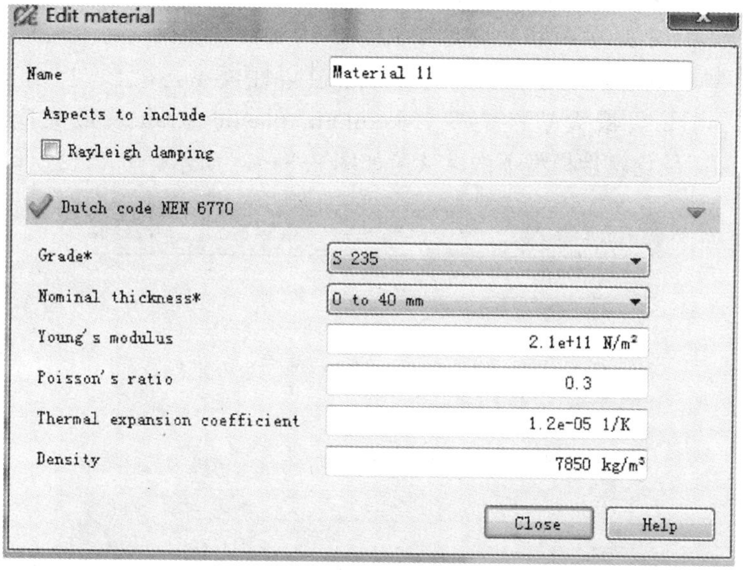

图 2.5-12　荷兰 NEN 6770 钢结构规范操作界面及默认值

在荷兰钢结构 6720 钢筋规范中，用户的操作方式与配筋和桩基础类型中的本构定义方式类似，首先需要考虑有无粘结（bonding），即选择钢筋是否与母体单元相连。除了要定义钢材等级（FeB）和弹性模量，还要选择配筋的本构模型为线弹性（Elasticity）还是理想塑性模型（Ideal plasticity）。如果用户选择理想塑性模型，还需要定义屈服强度值。Dutch NEN 6720 规范在理想塑性模型下的各参数默认值如图 2.5-13 所示。

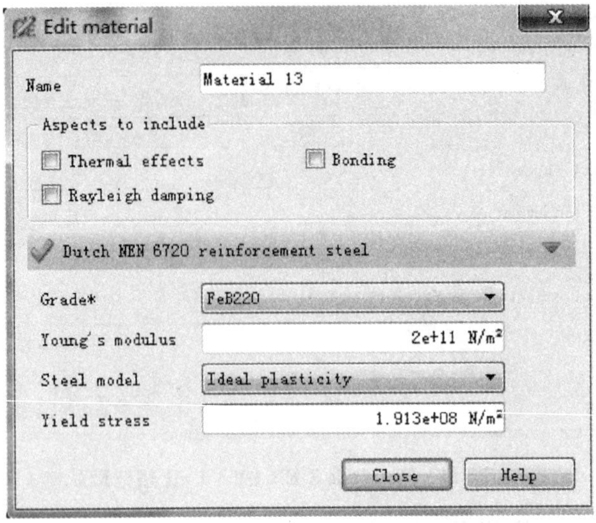

图 2.5‐13　Dutch NEN 6720 钢配筋规范参数默认值

　　Dutch NEN 6720 prestress cable 规范模型参数的设置方式与 Dutch NEN 6720 钢筋规范非常相似，只在强度等级和钢材模型上有细微差别。该模型主要为模拟斜拉桥类结构中钢拉索钢材的振动、腐蚀、疲劳和断裂的力学特性，因此强度等级较高。此外，钢材本构模型用硬化塑性模型（Hardening plasticity）取代理想塑性模型，并且增加了初始塑性应变（Initial plastic strain）、最大塑性应变（Maximum plastic strain）、最大应力（Maximum stress）三个方面。硬化塑性模型下的默认值如图 2.5‐14 所示。

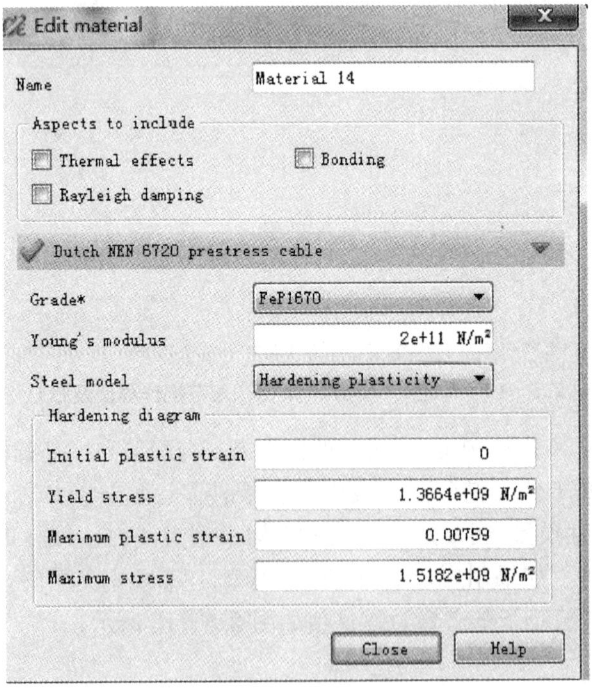

图 2.5‐14　Dutch NEN 6720 prestress cable 规范参数默认值

第三章 Diana 土木工程非线性建模案例分析

3.1 案例一：平面预应力框架非线性分析

本模型为一个简易的二维平面两层混凝土框架，框架与基础之间采用刚接框架。框架梁和框架柱均配有普通钢筋，截面面积分别为 100 mm² 和 120 mm²，每层框架梁均配有一根预应力筋，预应力筋直径均为 15.24 mm，屈服强度为 1860 MPa，弹性模量为 1.95×10^{11} Pa。梁和柱混凝土弹性模量为 3.45×10^{10} Pa，抗拉强度均为 2.64×10^{6} N/m²，抗压强度为 32.5 MPa，其中框架梁截面尺寸 500 mm×500 mm，框架柱截面尺寸 350 mm×400 mm，底层框架距离底面 4.6 m，顶层框架与底层框架间距离为 4 m，框架柱间距为 6.1 m。预应力筋初始张拉力均为 500 kN，每层框架顶端作用集中荷载各 10 kN，顶部梁作用均布荷载 20 kN/m，框架梁和柱均采用梁单元 L7BEN 建模，混凝土本构模型采用总应变裂缝模型。对两层框架采用非线性分析。框架平面图（不含梁柱纵筋）如图 3.1-1 所示。

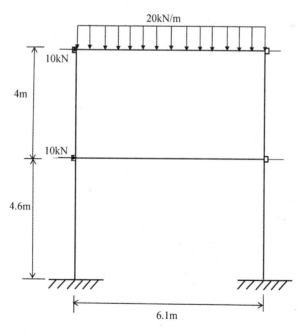

图 3.1-1　框架平面图

注：作为 Diana10.1 算例例题，本书例题分析条件全部是假定值，分析结果妥当性另当别论，望谅解。

学习要义：

(1) 学习创立直线(Line)和多段直线(Polyline)的几何建模方式。

(2) 学习平面梁单元类型选择、总应变裂缝模型下各本构参数的设置和梁单元截面几何特性赋予方式。

(3) 学习平移复制(Array Copy)直线、移动(Move)直线等简单的直线修改界面操作。

(4) 学习普通 Bar 纵向钢筋单元及预应力筋 Bar 单元本构模型参数设置。

(5) 学习施加集中荷载、均布荷载、预应力筋荷载和固端约束等界面操作。

(6) 学习荷载组合添加方法。

(7) 掌握非线性计算荷载工况的添加和计算。

首先，点击 DianaIE，弹出 New project 对话框，在电脑 G 盘的工作路径下创立一个新文件夹，将文件夹名命名为例题，本节模型文件名(Project name)命名为 Frame，该模型文件即为.dpf 格式的二进制文件。该文件夹保存在路径中。选择分析类型为结构分析(Structural)，模型尺寸最大范围为 100 m，默认智能划分生成的网格形状类型为六面体/四边形单元(Hexa/Quad)，默认单元网格阶数为线性单元网格(Linear)，如图 3.1-2 所示。

图 3.1-2 New project 操作界面

点击 OK，生成 GUI 操作界面，建立有限元几何模型。点击 Add a line，添加两点坐标 (0,0,0) 和 (6.1,0,0)，命名为 beam1，点击 OK 生成直线。

点击模型树 Geometry 栏下方的 beam1，右击，选择 Select 选项，这时可看到 GUI 界面中 Beam/几何模型被选中，再在 GUI 图形可视的操作界面中直接右击，选择 Move Shape，弹出如图 3.1-3 所示的移动图形的对话框。在 Displacement 位移框中选择沿着 Y 方向正向平移 4.6 m。

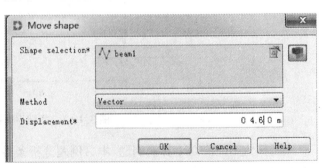

图 3.1-3 Move Shape 操作界面

点击选中 Geometry 栏下方的 beam1,右击选择 Select,再用鼠标右键点击 GUI 可视化界面区,选择 Array copy(平移复制),弹出如图 3.1-4 所示的 Array copy 操作界面,相对平移位移(Relative Displacement)选择(0 4 0),即沿着 Y 方向正向平移 4 m,复制份数(Number of copies)选择 1 份,生成 beam2。

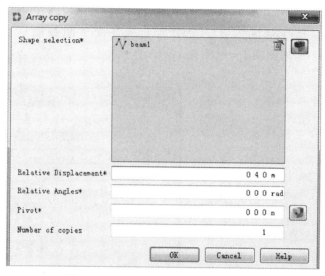

图 3.1-4　beam1 Array copy 操作界面

选择快捷工具栏图标添加多段线(Add a polyline),如图 3.1-5 所示,取消多段闭合线(Closed)的勾选。拾取两根梁上左端对应位置的点生成坐标(0,8.6,0),(0,4.6,0),再手动输入添加(0,0,0),点击 OK,命名 column1,生成第一根柱,column1 各点坐标如图 3.1-6 所示。

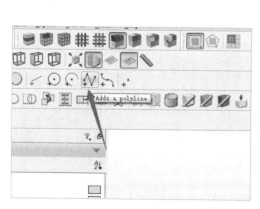

图 3.1-5　Add a polyline 快捷图标按钮

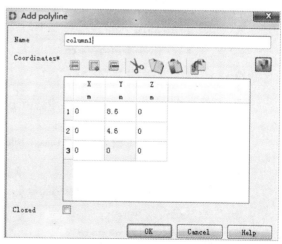

图 3.1-6　column1 各点坐标

再用上述 Array copy 的方式沿 X 正向平移柱,平移距离为 6.1 m,生成 column2,生成的整体几何模型如图 3.1-7 所示。

在 Geometry 栏的 Shape 下方右击选择 beam1 和 beam2,点击 Select 选项拾取这两根线,右击进行属性赋予(Property assignments)。先赋予梁单元属性和材料属性,单元类型(Element class)选择 2D 第二类梁单元(Class-II Beams 2D),梁单元的材料属性选择混凝土和砌体类型(Concrete and mansory),材料的本构模型(Material model)选择总应变裂缝模型(Total strain based crack model),如图 3.1-8 所示。混凝土的弹性模量为 $3.45 \times 10^{10} \text{ N/m}^2$,泊松比为 0.15,密度为 2500 kg/m^3,裂缝方向选择旋转裂缝模型(Rotating),拉伸软化曲线(Tension Softening)选择经典的 Hordijk 软化曲线,抗拉强度为 $2.64 \times 10^6 \text{ N/m}^2$,第一类拉伸断裂能(Mode-I tensile fracture energy)为 200 N/m,抗压强度曲线选择抛物线形(Parabolic),抗压强度为 32.5 MPa,非线性受压断裂能(Compressive fracture energy)为40 000 N/m,如图 3.1-9、图 3.1-10 所示。

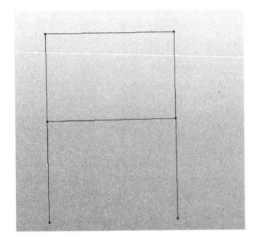

图 3.1-7 整体几何模型

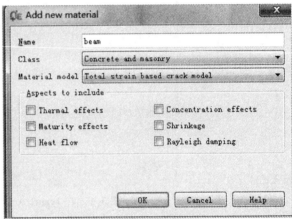

图 3.1-8 单元属性赋予窗口

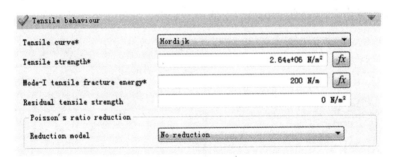

图 3.1-9 Tensile behaviour 属性赋予界面

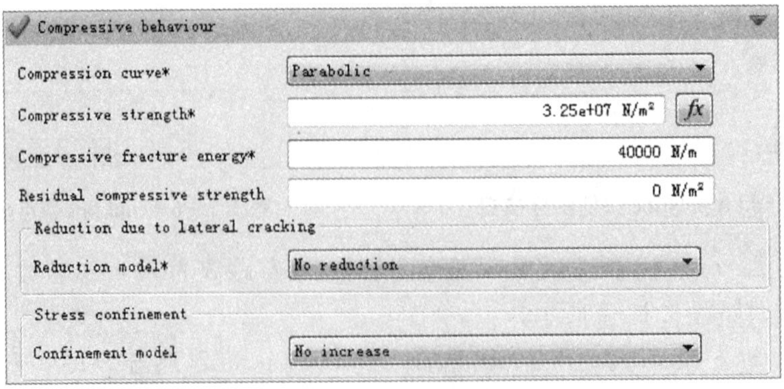

图 3.1-10 Compressive behaviour 属性赋予界面

编辑截面几何特性,命名为 beam,选择梁端截面为矩形截面(Rectangle),截面尺寸(Dimensions of a filled rectangle)高、宽均为 0.5 m,如图 3.1-11 所示。

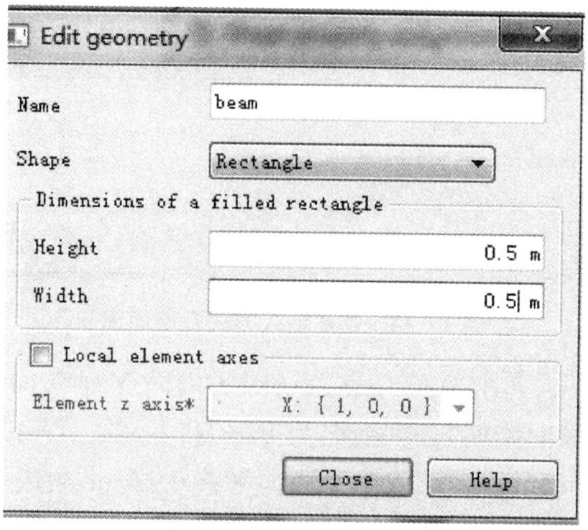

图 3.1-11 编辑截面几何特性窗口

建立好梁组的材料属性和截面几何特性之后,用同样的方式赋予柱单元的材料属性。材料名称和几何名称均命名为 column,点击 OK,生成柱组的各项属性。其中柱的材料属性定义与梁相同,截面的几何特性定义为高度 0.35 m,宽度 0.4 m。

点击快捷菜单栏图标 Add a line 创立直线,命名为 bar1,输入第一层钢绞线所在两点坐标(0,4.5,0),(6.1,4.5,0)。在 GUI 图形界面区选中 bar1,右击选中 Array copy,采用复制平移方式沿 Y 正向平移 4 m,生成第二层预应力筋,名称为 bar2。平移操作如图 3.1-12 所示。

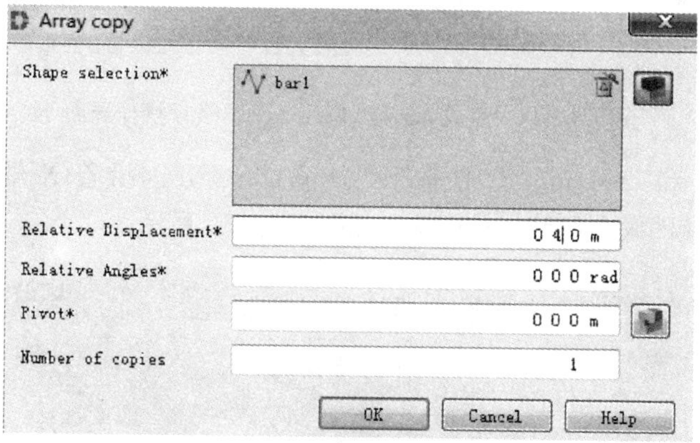

图 3.1-12 复制平移 bar1 生成 bar2

选中 bar1 和 bar2,赋予 bar1 和 bar2 预应力筋的材料属性。钢筋本构类型为配筋和桩基础(Reinforcements and pile foundations)类型下的 Von Mises 模型,输入预应力钢绞线弹性模量为 1.95×10^{11} Pa,屈服强度值为 1860 MPa。在输入预应力筋的截面几何特性时,钢筋类型选择嵌入(Embedded)方式,预应力钢绞线的截面面积(Cross-section area of bar)为 1.4×10^{-4} m²,如图 3.1-13 所示。为使得非线性计算更容易收敛,本构属性定义界面最下方的离散化方法(Discretization Method)选择 Section wise 类型。

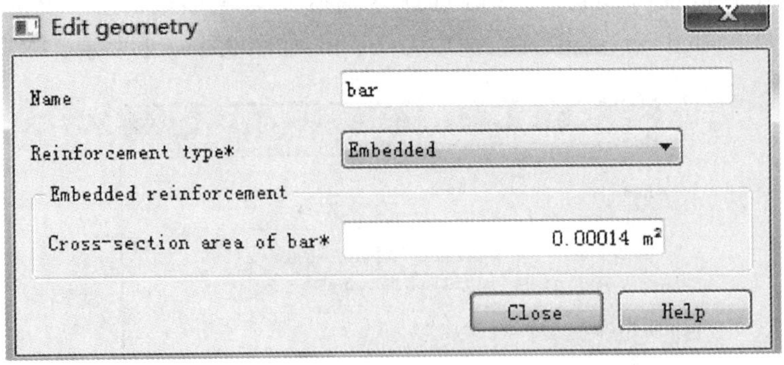

图 3.1-13　钢筋嵌入方式及截面面积

选择 beam1 和 beam2，右击选择 Array copy，方向向 Y 轴负向平移 0.15 m，复制份数为 1 份，生成框架梁的纵向配筋，分别命名为 bar3 和 bar4。Array copy 的界面操作如图 3.1-14 所示。再采用相同的方式沿着 Y 轴正向平移 0.15 m，生成 bar5 和 bar6。

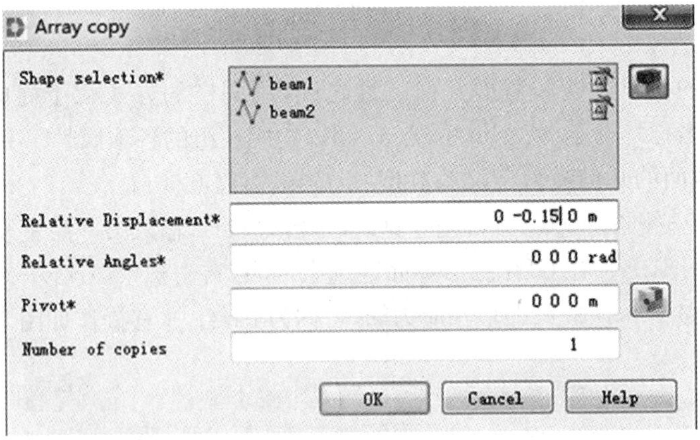

图 3.1-14　生成 bar3 和 bar4 Array copy 操作界面

选中 column1 和 column2，右击选择 Array copy，分别沿着 X 轴正向和负向平移 0.12 m，生成框架柱的纵向配筋。具体操作如图 3.1-15 所示。

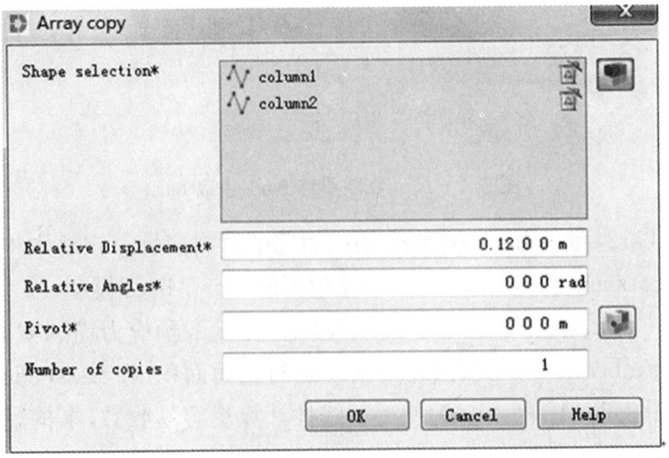

图 3.1-15　生成纵向配筋的复制平移操作界面

待生成框架梁和框架柱的纵向钢筋几何模型之后,分别针对梁和柱的纵向钢筋赋予纵筋的本构特性,框架梁纵向钢筋和框架柱纵向钢筋选择配筋和桩基础模块(Reinforcements and pile foundations),本构类型均定义为线弹性,本构名称分别命名为 barlong1 和 barlong2,各纵向钢筋钢筋组的弹性模量均为 2.1×10^{11} N/m²,其中框架梁纵筋的截面几何属性为嵌入式(Embeded),截面面积为 1×10^{-4} m²,框架柱纵筋的截面面积为 1.2×10^{-4} m²。截面几何属性定义分别如图 3.1-16 和 3.1-17 所示。

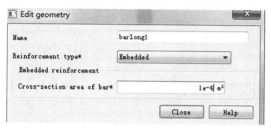
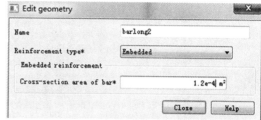

图 3.1-16　barlong1 截面几何特性定义界面　　图 3.1-17　barlong2 截面几何特性定义界面

添加约束。选择菜单栏 Geometry→Analysis→Attach supoport 施加约束。创立约束 co1,约束目标类型(Support target type)选择对点施加约束(Point),将基础的两个点施加 X 和 Y 两个方向平动和转动约束。约束定义方式和生成约束如图 3.1-18 和 3.1-19 所示。

图 3.1-18　约束定义操作界面

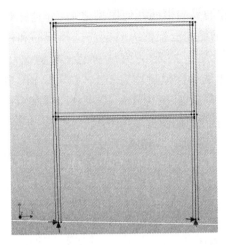

图 3.1-19　生成约束

施加荷载。点击 Load 栏下方的 Define a global load 图标,如图 3.1-20 所示,添加重力荷载的荷载工况,命名为 gravity,荷载类型选择为恒载(Dead load)。

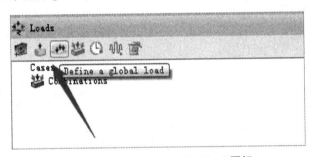

图 3.1-20　Define a global load 图标

施加集中点荷载的荷载工况,命名为 Force。加载目标类型(Load target type)选择为施加点荷载(Point),加载类型(Load type)为施加集中点荷载(Force),两个点的加载大小均为 10 kN,方向均沿 X 轴正向,如图 3.1-21 所示。

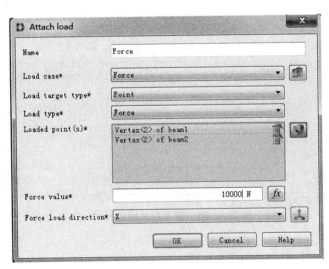

图 3.1-21　点荷载定义界面

添加预应力荷载的荷载工况,命名为 bar。预应力荷载加载目标类型(Load target type)选择实体(Solid)类型,加载类型(Load type)选择预应力筋荷载(Post tensioning load),张拉类型(Tension type)为两端张拉(Both ends),两层预应力筋大小均为 500 kN,两端锚固端回缩长度(First/Second anchor retention length)均设置为 0.01 m,库仑摩擦系数(Coulomb friction coefficient)为 0.22,握裹系数(Wobble factor)为 0.01/m。注意这里的实体含义与前文所述的实体单元(Solid)含义不同,之前预应力筋已经添加了作为预应力钢绞线的几何属性和材料属性,因此这里的实体应为已经赋予材料属性和截面几何属性的几何单元体。

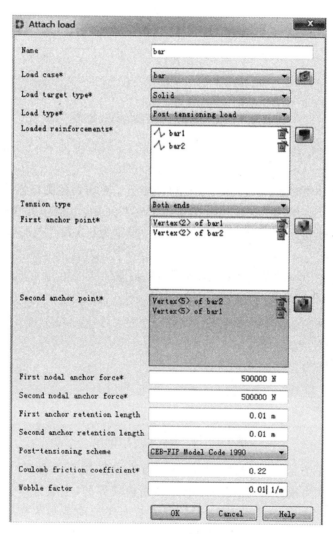

图 3.1-22　添加预应力荷载的荷载工况

继续添加荷载工况,命名为 pressure,施加均布荷载。加载目标类型(Load target type)是直线类型(Line),加载类型为均布荷载(Distributed force),方向为 Y 方向负向,大小为 20 kN/m,如图 3.1-23 所示。

添加荷载工况组合。点击 Load 栏下方 Combinations 图标,右击设置荷载工况组合表,将重力荷载与预应力筋荷载作为荷载组合 1(Load Combination1)。框架模型应先施加均布

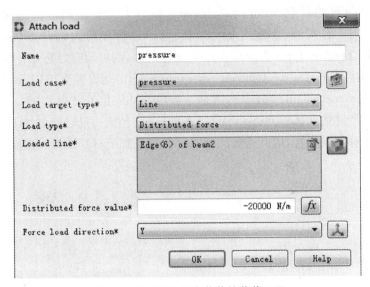

图 3.1-23　添加均布荷载的荷载工况

荷载,均布荷载作为荷载组合 2(Load Combination2)。最后施加 X 方向正向的集中荷载,作为荷载组合 3(Load Combination3)。

接下来进入前处理的最后一个部分——网格划分。点击黄色圆柱体的快捷图标按钮设置图形的划分属性(Set mesh properties of a shape),Operation 操作类型选择线(Edge),划分类型选择单元尺寸划分(Element size),单元尺寸设定为 0.1 m。点击 Generate mesh of a shape 按钮生成网格。

查看 meshing 栏下方的单元类型,确认是预先希望的网格单元 L7BEN,此时确定最后网格划分成功。如图 3.1-24 所示。

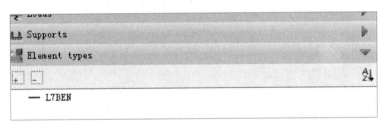

图 3.1-24　网格单元 L7BEN

在 Analysis 栏下方添加非线性计算模块。首先点击左下角含有红色加号的小图标 Add an analysis 添加生成分析,弹出 Analysis 分析模块界面,生成计算分析模块。右击选择结构非线性分析(Structural nonlinear)。首先右击添加荷载步为初始荷载步(Ececute Steps→Start Steps),命名为 Bar,在初始模块属性中选择荷载组合 1 作为初始荷载步,荷载步长因子(User specified sizes)为 1,用以模拟自重和预应力筋共同作用下结构的初始状态。最大迭代步数(Maximum Interations)设定为 20 步,收敛准则(Convergence Norm)选择力(Force)或位移(Displacement)收敛准则。添加物理非线性属性(Physical nonlinear)用于模拟有粘结预应力筋。

右击添加荷载步(Load Step),用同样方式添加荷载组合 2。荷载步、迭代方法和收敛准则的设置与上述相同。

点击 Run analysis 开始计算。计算结束后，点击 Output→Total displacement→TDty 和 Output→Total displacement→TDtx 查看 Y 方向和施加水平荷载后 X 方向的位移云图，其中施加竖向荷载前施加预应力后初始状态下 Y 方向的位移云图如图 3.1-25 所示，这时可看出有明显的反拱现象。

图 3.1-25 施加预应力后的初始状态

施加荷载后的 Y 方向位移云图如图 3.1-26 所示

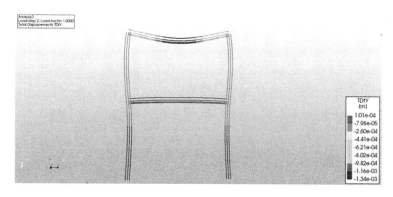

图 3.1-26 施加竖向荷载后 Y 方向的位移图

施加水平方向的点荷载后 X 方向的位移云图如图 3.1-27 所示。

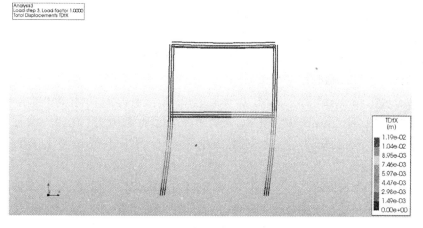

图 3.1-27 施加水平方向的点荷载后 X 方向的位移图

查看钢筋拉力值,施加初始预应力荷载和施加竖向均布荷载后的预应力筋拉力云图分别如图 3.1-28 和 3.1-29 所示。

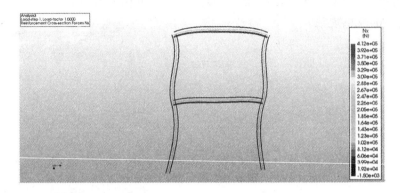

图 3.1-28　施加初始预应力荷载的预应力筋拉力云图

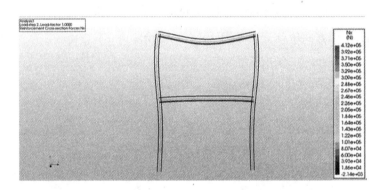

图 3.1-29　施加竖向均布荷载后的预应力筋拉力云图

附上 Python 命令流语句

```
newProject( "Frame", 100 )
setModelAnalysisAspects( [ "STRUCT" ] )
setModelDimension( "2D" )
setDefaultMeshOrder( "LINEAR" )
setDefaultMesherType( "HEXQUAD" )
createLine( "beam1", [ 0, 0, 0 ], [ 6.1, 0, 0 ] )
translate( [ "beam1" ], [ 0, 4.6, 0 ] )
arrayCopy( [ "beam1" ], [ 0, 4, 0 ], [ 0, 0, 0 ], [ 0, 0, 0 ], 1 )
createPolyline( "column1", [[ 0, 8.6, 0 ],[ 0, 4.6, 0 ],[ 0, 0, 0 ]], False )
arrayCopy( [ "column1" ], [ 0, 6.1, 0 ], [ 0, 0, 0 ], [ 0, 0, 0 ], 1 )
undo( 1 )
arrayCopy( [ "column1" ], [ 6.1, 0, 0 ], [ 0, 0, 0 ], [ 0, 0, 0 ], 1 )
saveProject(  )
addMaterial( "beam", "CONCR", "TSCR", [] )
setParameter( "MATERIAL", "beam", "LINEAR /ELASTI /YOUNG", 3.45e+10 )
```

```
setParameter( "MATERIAL", "beam", "LINEAR/ELASTI/POISON", 0.15 )
setParameter( "MATERIAL", "beam", "LINEAR/MASS/DENSIT", 2500 )
setParameter( "MATERIAL", "beam", "MODTYP/TOTCRK", "ROTATE" )
setParameter( "MATERIAL", "beam", "TENSIL/TENCRV", "HORDYK" )
setParameter( "MATERIAL", "beam", "COMPRS/COMCRV", "PARABO" )
setParameter( "MATERIAL", "beam", "COMPRS/GC", 200 )
setParameter( "MATERIAL", "beam", "COMPRS/RESCST", 0 )
setParameter( "MATERIAL", "beam", "TENSIL/TENSTR", 2640000 )
setParameter( "MATERIAL", "beam", "TENSIL/GF1", 200 )
setParameter( "MATERIAL", "beam", "TENSIL/RESTST", 0 )
setParameter( "MATERIAL", "beam", "LINEAR/ELASTI/YOUNG", 3.45e+10 )
setParameter( "MATERIAL", "beam", "COMPRS/COMSTR", 32500000 )
setParameter( "MATERIAL", "beam", "COMPRS/GC", 200 )
setParameter( "MATERIAL", "beam", "COMPRS/GC", 40000 )
setParameter( "MATERIAL", "beam", "COMPRS/COMSTR", 32500000 )
addGeometry( "Element geometry 1", "LINE", "CLS2B2", [] )
rename( "GEOMET", "Element geometry 1", "beam" )
setParameter( "GEOMET", "beam", "SHAPE/RECTAN", [ 0.5, 0.5 ] )
setParameter( "GEOMET", "beam", "SHAPE/RECTAN", [ 0.5, 0.5 ] )
setParameter( "GEOMET", "beam", "SHAPE/RECTAN", [ 0.5, 0.5 ] )
clearReinforcementAspects( [ "beam1", "beam2" ] )
setElementClassType( "SHAPE", [ "beam1", "beam2" ], "CLS2B2" )
assignMaterial( "beam", "SHAPE", [ "beam1", "beam2" ] )
assignGeometry( "beam", "SHAPE", [ "beam1", "beam2" ] )
resetElementData( "SHAPE", [ "beam1", "beam2" ] )
saveProject( )
addMaterial( "column", "CONCDC", "MC1990", [ "CRACKI", "ELASTI", "PLASTI" ] )
remove( "MATERIAL", "column" )
addMaterial( "column", "CONCR", "TSCR", [] )
setParameter( "MATERIAL", "column", "LINEAR/ELASTI/YOUNG", 3.45e+10 )
setParameter( "MATERIAL", "column", "LINEAR/ELASTI/POISON", 0.15 )
setParameter( "MATERIAL", "column", "LINEAR/MASS/DENSIT", 2500 )
setParameter( "MATERIAL", "column", "MODTYP/TOTCRK", "ROTFIX" )
setParameter( "MATERIAL", "column", "MODTYP/TOTCRK", "ROTATE" )
setParameter( "MATERIAL", "column", "TENSIL/TENCRV", "HORDYK" )
setParameter( "MATERIAL", "column", "TENSIL/TENSTR", 2640000 )
setParameter( "MATERIAL", "column", "TENSIL/GF1", 200 )
setParameter( "MATERIAL", "column", "TENSIL/RESTST", 0 )
setParameter( "MATERIAL", "column", "COMPRS/COMCRV", "PARABO" )
```

```
setParameter( "MATERIAL", "column", "COMPRS /COMSTR", 32500000 )
setParameter( "MATERIAL", "column", "COMPRS /GC", 40000 )
setParameter( "MATERIAL", "column", "COMPRS /RESCST", 0 )
setParameter( "MATERIAL", "column", "COMPRS /RESCST", 0 )
addGeometry( "Element geometry 2", "LINE", "CLS2B2", [] )
rename( "GEOMET", "Element geometry 2", "column" )
setParameter( "GEOMET", "column", "SHAPE /RECTAN", [ 0.35, 0.4 ] )
clearReinforcementAspects( [ "column1", "column2" ] )
setElementClassType( "SHAPE", [ "column1", "column2" ], "CLS2B2" )
assignMaterial( "column", "SHAPE", [ "column1", "column2" ] )
assignGeometry( "column", "SHAPE", [ "column1", "column2" ] )
resetElementData( "SHAPE", [ "column1", "column2" ] )
createLine( "bar1", [ 0, 4.5, 0 ], [ 6.1, 4.5, 0 ] )
saveProject( )
arrayCopy( [ "bar1" ], [ 0, 4, 0 ], [ 0, 0, 0 ], [ 0, 0, 0 ], 1 )
saveProject( )
addMaterial( "bar", "REINFO", "VMISES", [] )
setParameter( "MATERIAL", "bar", "LINEAR /ELASTI /YOUNG", 1.95e+11 )
setParameter( "MATERIAL", "bar", "LINEAR /ELASTI /YOUNG", 1.95e+11 )
setParameter( "MATERIAL", "bar", "PLASTI /HARDI1 /YLDSTR", 1.86e+09 )
setMaterialAspects( "bar", [ "NOBOND" ] )
addGeometry( "Element geometry 3", "RELINE", "REBAR", [] )
rename( "GEOMET", "Element geometry 3", "bar" )
setParameter( "GEOMET", "bar", "REIEMB /CROSSE", 0.00014 )
setReinforcementAspects( [ "bar1", "bar2" ] )
assignMaterial( "bar", "SHAPE", [ "bar1", "bar2" ] )
assignGeometry( "bar", "SHAPE", [ "bar1", "bar2" ] )
resetElementData( "SHAPE", [ "bar1", "bar2" ] )
setReinforcementDiscretization( [ "bar1", "bar2" ], "SECTION" )
arrayCopy( [ "beam1", "beam2" ], [ 0, -0.15, 0 ], [ 0, 0, 0 ], [ 0, 0, 0 ], 1 )
arrayCopy( [ "beam1", "beam2" ], [ 0, 0.15, 0 ], [ 0, 0, 0 ], [ 0, 0, 0 ], 1 )
renameShape( "beam3", "bar3" )
renameShape( "beam4", "bar4" )
renameShape( "beam5", "bar5" )
renameShape( "beam6", "bar6" )
arrayCopy( [ "column1", "column2" ], [ 0.12, 0, 0 ], [ 0, 0, 0 ], [ 0, 0, 0 ], 1 )
renameShape( "column3", "bar7" )
renameShape ( "column4", "bar8" )
arrayCopy( [ "column1", "column2" ], [ -0.12, 0, 0 ], [ 0, 0, 0 ], [ 0, 0, 0 ], 1 )
```

```
renameShape( "column3", "bar9" )
renameShape( "column4", "bar10" )
addMaterial( "barlong1", "REINFO", "LINEAR", [ ] )
setParameter( "MATERIAL", "barlong1", "LINEAR/ELASTI/YOUNG", 2.1e+11 )
addGeometry( "Element geometry 4", "RELINE", "REBAR", [ ] )
rename( "GEOMET", "Element geometry 4", "barlong1" )
setParameter( "GEOMET", "barlong1", "REIEMB/CROSSE", 0.0001 )
setReinforcementAspects( [ "bar3", "bar4", "bar5", "bar6" ] )
assignMaterial( "barlong1", "SHAPE", [ "bar3", "bar4", "bar5", "bar6" ] )
assignGeometry( "barlong1", "SHAPE", [ "bar3", "bar4", "bar5", "bar6" ] )
resetElementData( "SHAPE", [ "bar3", "bar4", "bar5", "bar6" ] )
setReinforcementDiscretization ( [ "bar3", "bar4", "bar5", "bar6" ], "SECTION" )
saveProject( )
addMaterial( "barlong2", "REINFO", "LINEAR", [ ] )
setParameter( "MATERIAL", "barlong2", "LINEAR/ELASTI/YOUNG", 2.1e+11 )
addGeometry( "Element geometry 5", "RELINE", "REBAR", [ ] )
rename( "GEOMET", "Element geometry 5", "barlong2" )
setParameter( "GEOMET", "barlong2", "REIEMB/CROSSE", 0.00012 )
setParameter( "GEOMET", "barlong2", "REIEMB/CROSSE", 0.00012 )
setReinforcementAspects( [ "bar7", "bar8", "bar9", "bar10" ] )
assignMaterial( "barlong2", "SHAPE", [ "bar7", "bar8", "bar9", "bar10" ] )
assignGeometry( "barlong2", "SHAPE", [ "bar7", "bar8", "bar9", "bar10" ] )
resetElementData( "SHAPE", [ "bar7", "bar8", "bar9", "bar10" ] )
setReinforcementDiscretization( [ "bar7", "bar8", "bar9", "bar10" ], "SECTION" )
addSet( "GEOMETRYSUPPORTSET", "co1" )
createPointSupport( "co1", "co1" )
setParameter( "GEOMETRYSUPPORT", "co1", "AXES", [ 1, 2 ] )
setParameter( "GEOMETRYSUPPORT", "co1", "TRANSL", [ 1, 1, 0 ] )
setParameter( "GEOMETRYSUPPORT", "co1", "ROTATI", [ 1, 1, 0 ] )
attach( "GEOMETRYSUPPORT", "co1", "column1", [[ 0, 0, 0 ]] )
attach( "GEOMETRYSUPPORT", "co1", "column2", [[ 6.1, 0, 0 ]] )
saveProject( )
addSet( "GEOMETRYLOADSET", "gravity" )
createModelLoad( "gravity", "gravity" )
saveProject( )
addSet( "GEOMETRYLOADSET", "Geometry load case 2" )
rename( "GEOMETRYLOADSET", "Geometry load case 2", "Force" )
createPointLoad( "Force", "Force" )
```

```
setParameter( "GEOMETRYLOAD", "Force", "FORCE /VALUE", 10000 )
setParameter( "GEOMETRYLOAD", "Force", "FORCE /DIRECT", 1 )
attach( "GEOMETRYLOAD", "Force", "beam1", [[ 0, 4.6, 0 ]] )
attach( "GEOMETRYLOAD", "Force", "beam2", [[ 0, 8.6, 0 ]] )
saveProject( )
addSet( "GEOMETRYLOADSET", "Geometry load case 3" )
rename( "GEOMETRYLOADSET", "Geometry load case 3", "bar" )
createBodyLoad( "bar", "bar" )
setParameter( "GEOMETRYLOAD", "bar", "LODTYP", "POSTEN" )
setParameter( "GEOMETRYLOAD", "bar", "POSTEN /BOTHEN /FORCE1", 500000 )
setParameter( "GEOMETRYLOAD", "bar", "POSTEN /BOTHEN /FORCE2", 500000 )
setParameter( "GEOMETRYLOAD", "bar", "POSTEN /BOTHEN /RETLE1", 0.01 )
setParameter( "GEOMETRYLOAD", "bar", "POSTEN /BOTHEN /RETLE2", 0.01 )
setParameter( "GEOMETRYLOAD", "bar", "POSTEN /SHEAR", 0.22 )
setParameter( "GEOMETRYLOAD", "bar", "POSTEN /WOBBLE", 0.01 )
attach( "GEOMETRYLOAD", "bar", [ "bar1", "bar2" ] )
attachTo( "GEOMETRYLOAD", "bar", "POSTEN /BOTHEN /PNTS1", "bar1", [[ 0, 4.6, 0]] )
attachTo( "GEOMETRYLOAD", "bar", "POSTEN /BOTHEN /PNTS1", "bar2", [[ 0, 8.6, 0]] )
attachTo( "GEOMETRYLOAD", "bar", "POSTEN /BOTHEN /PNTS2", "bar1", [[ 6.1, 4.6, 0
]] )
attachTo( "GEOMETRYLOAD", "bar", "POSTEN /BOTHEN /PNTS2", "bar2", [[ 6.1, 8.6,
0 ]] )
saveProject( )
addSet( "GEOMETRYLOADSET", "Geometry load case 4" )
addSet( "GEOMETRYLOADSET", "pressure" )
remove( "GEOMETRYLOADSET", [ "Geometry load case 4" ] )
createLineLoad( "pressure", "pressure" )
setParameter( "GEOMETRYLOAD", "pressure", "FORCE /VALUE", -20000 )
setParameter( "GEOMETRYLOAD", "pressure", "FORCE /DIRECT", 2 )
attach( "GEOMETRYLOAD", "pressure", "beam2", [[ 3.05, 8.6, 0 ]] )
saveProject( )
setDefaultGeometryLoadCombinations( )
setGeometryLoadCombinationFactor( "Geometry load combination 1", "gravity", 1 )
remove( "GEOMETRYLOADCOMBINATION", "Geometry load combination 1" )
remove( "GEOMETRYLOADCOMBINATION", "Geometry load combination 2" )
remove( "GEOMETRYLOADCOMBINATION", "Geometry load combination 3" )
setGeometryLoadCombinationFactor( "Geometry load combination 4", "bar", 1 )
setGeometryLoadCombinationFactor( "Geometry load combination 4", "gravity", 1 )
setGeometryLoadCombinationFactor( "Geometry load combination 4", "pressure", 1 )
```

```
remove( "GEOMETRYLOADCOMBINATION", "Geometry load combination 4" )
addGeometryLoadCombination( "" )
setGeometryLoadCombinationFactor( "Geometry load combination 1", "gravity", 1 )
setGeometryLoadCombinationFactor( "Geometry load combination 1", "bar", 1 )
addGeometryLoadCombination( "" )
setGeometryLoadCombinationFactor( "Geometry load combination 2", "pressure", 1 )
addGeometryLoadCombination( "" )
setGeometryLoadCombinationFactor( "Geometry load combination 3", "Force", 1 )
saveProject( )
setElementSize( "beam1", 1, [[ 3.05, 4.6, 0 ]], 0.1, 0, True )
setElementSize( "beam2", 1, [[ 3.05, 8.6, 0 ]], 0.1, 0, True )
setElementSize( "column1", 1, [[ 0, 6.6, 0 ],[ 0, 2.3, 0 ]], 0.1, 0, True )
setElementSize( "column2", 1, [[ 6.1, 6.6, 0 ],[ 6.1, 2.3, 0 ]], 0.1, 0, True )
saveProject( )
generateMesh( [] )
hideView( "GEOM" )
showView( "MESH" )
addAnalysis( "Analysis3" )
addAnalysisCommand( "Analysis3", "NONLIN", "Structural nonlinear" )
renameAnalysis( "Analysis3", "Analysis3" )
renameAnalysisCommand ( " Analysis3", " Structural nonlinear", " Structural nonlinear" )
addAnalysisCommandDetail ( " Analysis3", " Structural nonlinear", " MODEL /EVALUA /REINFO /INTERF" )
setAnalysisCommandDetail ( " Analysis3", " Structural nonlinear", " MODEL /EVALUA /REINFO /INTERF", True )
removeAnalysisCommandDetail( "Analysis3", "Structural nonlinear", "EXECUT(1)" )
setAnalysisCommandDetail ( " Analysis3", " Structural nonlinear", " EXECUT /EXETYP", "LOAD" )
renameAnalysisCommandDetail( "Analysis3", "Structural nonlinear", "EXECUT(1)", "bar" )
setAnalysisCommandDetail( "Analysis3", "Structural nonlinear", "EXECUT(1) /ITERAT /MAXITE", 20 )
setAnalysisCommandDetail( "Analysis3", "Structural nonlinear", "EXECUT(1) /ITERAT /CONVER /DISPLA /NOCONV", "CONTIN" )
setAnalysisCommandDetail( "Analysis3", "Structural nonlinear", "EXECUT(1) /ITERAT /CONVER /FORCE /NOCONV", "CONTIN" )
setAnalysisCommandDetail ( " Analysis3", " Structural nonlinear", " EXECUT /
```

```
        EXETYP", "LOAD" )
    renameAnalysisCommandDetail( "Analysis3", "Structural nonlinear", "EXECUT
(2)", "PRESSURE" )
    setAnalysisCommandDetail( "Analysis3", "Structural nonlinear", "EXECUT(2)/
LOAD /LOADNR", 2 )
    setAnalysisCommandDetail( "Analysis3", "Structural nonlinear", "EXECUT(2)/
ITERAT /MAXITE", 20 )
    setAnalysisCommandDetail( "Analysis3", "Structural nonlinear", "EXECUT(2)/
ITERAT /CONVER /DISPLA /NOCONV", "CONTIN" )
    setAnalysisCommandDetail( "Analysis3", "Structural nonlinear", "EXECUT(2)/
ITERAT /CONVER /FORCE /NOCONV", "CONTIN" )
    setAnalysisCommandDetail( "Analysis3", "Structural nonlinear", "EXECUT/
EXETYP", "LOAD" )
    renameAnalysisCommandDetail( "Analysis3", "Structural nonlinear", "EXECUT
(3)", "LOAD" )
    setAnalysisCommandDetail( "Analysis3", "Structural nonlinear", "EXECUT(3)/
LOAD /LOADNR", 3 )
    setAnalysisCommandDetail( "Analysis3", "Structural nonlinear", "EXECUT(3)/
ITERAT /MAXITE", 20 )
    saveProject( )
    runSolver( "Analysis3" )
    showView( "RESULT" )
    show( "SUPPORTSET", [ "co1" ] )
    rename( "SUPPORTSET", "co1", "co1" )
    setResultPlot( "contours", "Total Displacements /node", "TDtZ" )
    setResultPlot( "contours", "Total Displacements /node", "TDtY" )
    setResultPlot( "contours", "Total Displacements /node", "TDtX" )
    setResultPlot( "contours", "Reinforcement Cross - section Forces /node", "Nx" )
    setResultCase( [ "Analysis3", "Output", "Load - step 2, Load - factor 1.0000" ] )
    setResultCase( [ "Analysis3", "Output", "Load - step 1, Load - factor 1.0000" ] )
    setResultCase( [ "Analysis3", "Output", "Load - step 2, Load - factor 1.0000" ] )
    setResultCase( [ "Analysis3", "Output", "Load - step 3, Load - factor 1.0000" ] )
```

3.2 案例二:箱梁体外粘贴钢板加固案例

本节案例为一段混凝土简支箱梁简化模型,箱梁长度为 10 m,顶板宽度 6 m,顶板及腹板厚度均为 0.4 m,底板厚度 0.3 m,底板宽度 4 m,整个箱梁最大高度为 2.4 m,混凝土箱梁截面尺寸如图 3.2-1(a)所示。其中箱梁混凝土强度等级为 C50,采用欧洲 CEB - FIP 1990 规范中的收缩徐变模型,顶板和底板的纵筋直径均为 8 mm,底板纵筋间距

1.2 m,顶板纵筋间距 1.2 m。对混凝土箱梁底部采用尺寸为 10 m×4 m×0.01 m 的钢板进行粘贴加固,箱梁和钢板均采用二次实体单元 CHX60 建模。钢板与混凝土箱梁之间接触采用面对面接触的界面单元 CQ48I,粘贴接触面的本构类型为粘结滑移(Bond-slip)类型。接触方式如图 3.2-24(b)所示。整个箱梁顶板承受竖向向下均布荷载,大小 50 kN/m² 采用阶段性分析方法来模拟钢板粘贴加固过程。第一阶段模拟简支箱梁的长期性能劣化过程,时间荷载步设定为 100 年,假设到 100 年时箱梁的跨中挠度超过了许可控制值。第二阶段激活钢板单元,启动加固阶段,待计算完毕后,对比粘贴钢板前后的箱梁位移,以验证粘贴钢板加固效果。此外,本案例还将展示提取 GUI 操作界面生成的 py 文件并修改其中 Python 语言命令流语句中参数生成新的 py 文件,并将这些文件导入 DianaIE 界面中自动计算并生成计算结果。

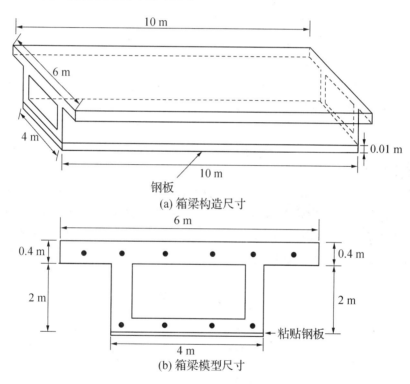

图 3.2-1 箱梁模型尺寸

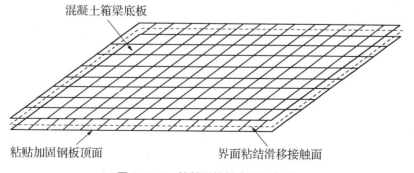

图 3.2-2 接触面接触类型示意图

注：正常情况下，当桥梁挠度超过一定容许值（例如：正常使用极限状态下约为桥梁跨径的 1/200～1/250）则应当采取加固补强措施。本算例为方便向读者展示阶段性分析操作流程，粗略选择 100 年作为粘贴钢板加固补强起始点，并忽略了对钢板加固后继续承受时变效应下的挠度控制效果的模拟。采用二次受力加固的阶段性分析方法进行分析。如有不符合实际之处，请读者谅解。

学习要义：

（1）学习在几何建模中对图形轮廓的布尔加减运算。

（2）掌握采用拉伸操作建立箱梁三维实体模型和采用参考点法建立立体模型。

（3）学习 CEB-FIP 1990 规范中收缩徐变参数设置。

（4）学习使用 Property assignments 定义钢材本构特性。

（5）熟悉实体单元接触面间定义面对面接触的界面单元（Surface to Surface）的方法。

（6）掌握面对面界面单元粘结滑移本构类型中多段线（Multi-Linear）类型下的应力—滑移双线型模型材料本构参数定义。

（7）学习箱梁钢板粘贴加固过程中添加阶段性分析及应力单元激活操作方式。

（8）学习复制、粘贴、修改和编辑 Python 语言并将其导入软件生成 GUI 界面图像和 .dfp 文件的方法。

首先打开 DianaIE 界面，点击菜单栏 File→New，弹出 New project 对话框，在 3.1 节的 G 盘例题文件夹中创立新的 dpf 文件，命名为 Girder，分析类型选择结构（Structural）类型，模型维数选择三维（Three dimensional），模型最大尺寸范围（Model size）为 100 m，对于该模型的默认网格单元形状类型（Default mesher tyor）为六面体/四边形（Hexa/Quad）单元，网格单元的默认阶数（Default mesh order）为二阶（Quadratic），单元中间节点位置的确定方式（Mid-side node location）为线性插值（Linear interpolation）。如图 3.2-3 所示。

图 3.2-3　New project 对话框

点击菜单栏下方的快捷工具栏 Add a New Sheet，创立 Sheet 1，如图 3.2-4 所示，依次输入各点坐标创立箱梁截面外部轮廓面，其中各点坐标见表 3-1。

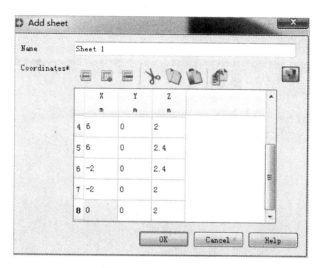

图 3.2-4　创立 Sheet 1

表 3-1　外轮廓面坐标

1	(0,0,0)
2	(4,0,0)
3	(4,0,2)
4	(6,0,2)
5	(6,0,2.4)
6	(−2,0,2.4)
7	(−2,0,2)
8	(0,0,2)

点击 OK，生成如图 3.2-5 所示的外部轮廓面形状。

创立内轮廓面。依次输入坐标 (0.4,0,0.3)，(3.6,0,0.3)，(3.6,0,2)，(0.4,0,2) 创立 Sheet 2，建立内轮廓面，如图 3.2-6 所示。

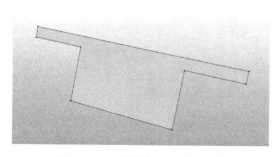

图 3.2-5　箱梁截面外部轮廓面

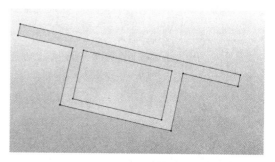

图 3.2-6　创立 Sheet 2

鼠标左键点击选中操作界面的图形,点击界面上方快捷工具栏 Subtract two or more shapes,位置如图 3.2-7 所示,进行布尔减法运算。弹出 Subtract 操作界面,目标选项(Target selection)为 Sheet 1,工具选项(Tool selection)为 Sheet 2,进行几何图形的布尔减法运算,操作界面如图 3.2-8 所示。

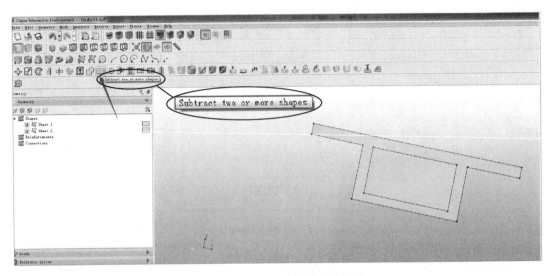

图 3.2-7　Subtract 快捷菜单位置

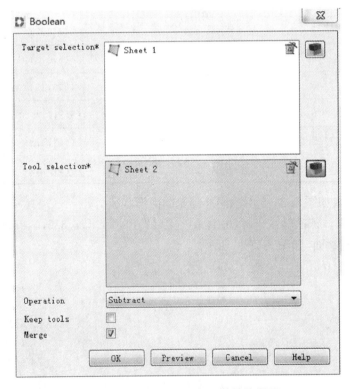

图 3.2-8　布尔减法逻辑运算操作界面

点击 OK,生成如图 3.2-9 所示的结果,完成箱梁截面模型的建立。

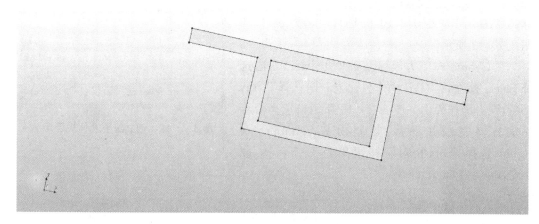

图 3.2‑9　布尔减法运算后的箱梁截面模型

点击快捷工具栏图标 ▣（Extrude a shape），将几何截面沿着 Y 方向拉伸为箱梁立体图形，拉伸距离为 10 m，如图 3.2‑10 所示。

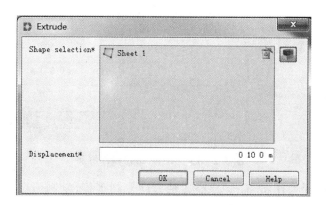

图 3.2‑10　Extrude a shape 拉伸操作

点击 OK，生成箱梁立体图形如图 3.2‑11 所示。

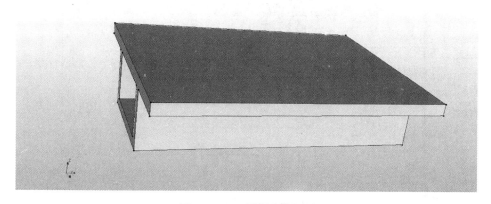

图 3.2‑11　箱梁立体图形

输入底板纵筋的坐标。点击快捷工具栏图标 Add a line 创立直线,输入第一根纵筋几何模型坐标(0.2,0,0.15)和(0,2,10,0.15),命名为 bar1,如图 3.2-12 所示。

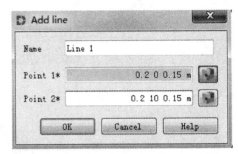

在 Geometry 下方鼠标左键单击 bar1,右击选择 Select 操作,在 GUI 操作界面上右击选择 Array copy,弹出操作界面如图 3.2-13 所示。选择沿着 X 轴方向复制并平移,复制份数(Number of copies)选择 3 份,相对平移距离为 1.2 m,生成 bar2,bar3 和 bar4。

图 3.2-12 创立纵筋几何模型示意图

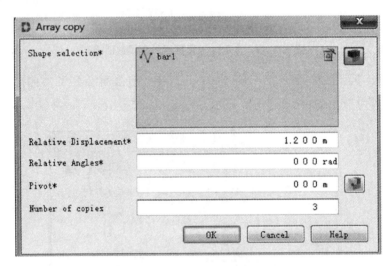

图 3.2-13 Array copy 操作界面

再使用上述相同操作将 bar1 沿着 Z 轴方向复制平移,复制份数为 1 份,平移距离为 2 m,点击 OK,生成 bar5。再针对 bar5 重复上述复制平移操作,方向为沿着 X 方向,平移距离为 1.2 m,复制份数为 4 份,如图 3.2-14 所示。

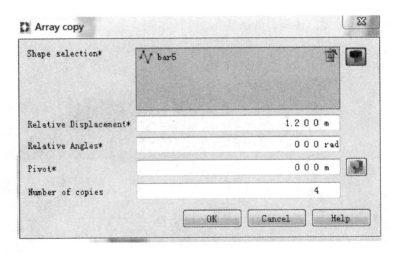

图 3.2-14 复制平移 bar5

再点击鼠标左键选中 bar5,右击选择 Select 选项,沿着 X 轴负向复制平移 1 份,距离为 1.2 m,点击 OK 生成 bar10。此时已完成整个箱架模型的建模。生成 bar10 后的立体图体,如图 3.2-15 所示。

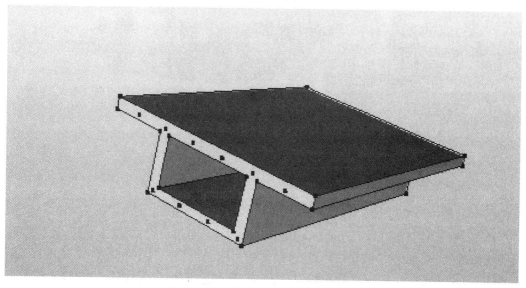

图 3.2-15　生成 bar10 后的立体图形

选中整个箱梁结构,右击选择 Move shape,将模型沿着 Z 轴向上平移 0.01 m。如图 3.2-16 所示,点击快捷栏图标 Add a block solid 创立立体 Block 1,输入起始参考点位置(0,0,0),并且依次定义 X,Y 和 Z 三个方向的尺寸为 4 m,10 m,0.01 m,生成加固钢板几何模型,如图 3.2-17 所示,生成如图 3.2-18 所示立体模型。放大界面,可以看到粘贴钢板加固层。

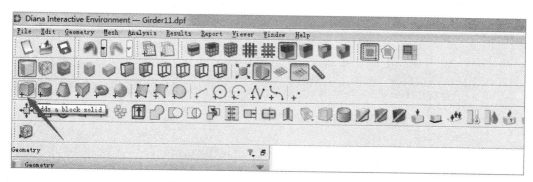

图 3.2-16　Adds a block solid 快捷图标

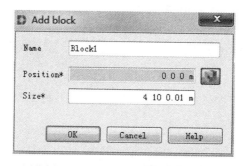

图 3.2-17　创立钢板几何模型

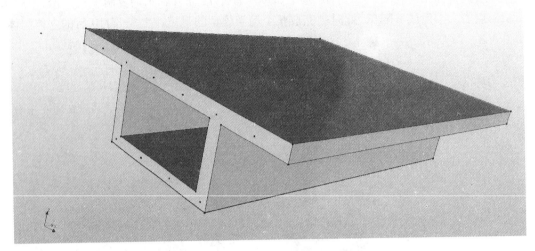

图 3.2-18　箱梁桥和加固钢板几何模型

单击 Sheet 1，右击选择 Property assignments 对箱梁进行材料属性的定义，选择单元类型为实体单元（Structural Solid）类型，材料本构为欧洲规范 CEB-FIP 1990 模型，命名为 concrete，勾选徐变、收缩和开裂指数三个选项，环境温度设定为 20℃，相对环境湿度为 60%，构件理论厚度采用默认值 0.15 m，如图 3.2-19 所示。混凝土弹性模量为 3.8926×10^{10} N/m²，28 天弹性模量设定为 3.15×10^{10} N/m²，混凝土泊松比为 0.15，CEB-FIP 1990 规范中抗压强度平均值设定为 58 MPa，如图 3.2-20 所示。

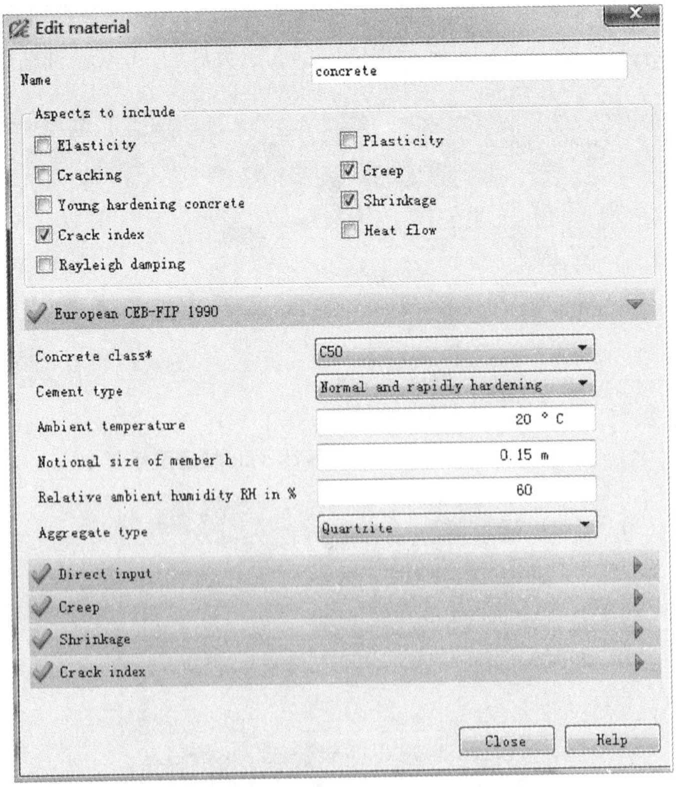

图 3.2-19　箱梁 CEB-FIP 1990 材料属性定义界面

图 3.2‑20　箱梁 CEB‑FIP 1990 材料属性定义界面

松弛模型选择基于混凝土老化理论(Aging)以收缩徐变为主的 Kelvin 链模型,徐变混凝土单元龄(Concrete age at birth of element)设定为 28 天(2 419 200 s),收缩固化曲线设定为 1 天(86 400 s)。由于实体单元不需要截面几何特性定义,点击 OK,关闭混凝土本构界面定义对话框,完成材料属性定义。

定义纵筋属性,选中所有纵筋,右击选择 steel,弹出钢筋材料属性定义的操作界面,命名为 bar。对 bar 直线筋组进行材料属性定义。在本案例中,纵筋本构模型选择线弹性且各向同性(Linear elasticity isotropic),设定纵筋的弹性模量为 $2.1 \times 10^{11}\,\text{N/m}^2$,钢材泊松比为 0.3,密度为 7800 kg/m³。定义纵筋的截面几何特性(Edit geometry),纵筋采用嵌入方式(Embedded)与混凝土进行耦合,面积为 $\frac{\pi}{4} \times 8^2 \approx 50.265\,\text{mm}^2$,如图 3.2‑21 所示。纵筋离散化方法(Discretization method)选择截面智能化方法(Section wise)。

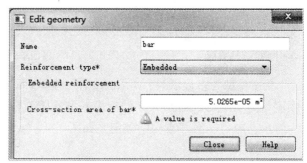

图 3.2‑21　纵筋截面几何特性定义

点击箱梁模型,右击 Hide 隐去,在 Geometry 栏下方取消 bar 钢筋组勾选,只保留底层 Block1,图形显示如图 3.2-22 所示。先进行界面单元材料和几何特性的定义。

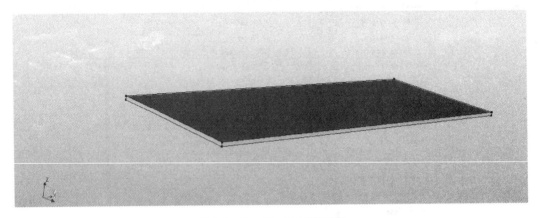

图 3.2-22　Block1 图形显示

点击快捷菜单栏中的 Edit connection property assignments,即 图标,进行界面单元本构属性的定义。弹出界面单元操作界面,命名为 int,连接类型选择界面单元(Interface elements),连接模式选择自动连接(Auto connect),界面单元接触类型(Selection type)为面(Face)接触,单元类型(Element class)为结构类界面单元(Structural element)。点击下方的 Material,弹出界面单元材料属性本构定义,由于采用钢板粘贴加固,因此界面单元接触材料类型选择粘结滑移(Bondslip)类本构,如图 3.2-23 所示。线性材料属性选择为 3D 面接触(3D surface interface),单元的法向刚度(Normal stiffness modulus - z)取 $3.65\times 10^{16}\,\text{N/m}^3$,x 向和 y 向的切向刚度(Shear stiffness modulus - x/y)均为 $3.65\times 10^{12}\,\text{N/m}^3$,如图 3.2-24 所示。

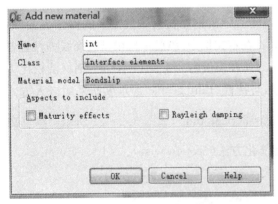

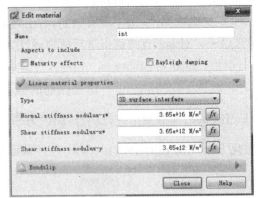

图 3.2-23　本构类型选择界面　　　　图 3.2-24　界面单元材料属性刚度定义

在粘结滑移模块,选滑移类型为多段线性(Multi-Linear)模型,即应力—滑移三直线模型。该模型传递法向刚度与剪切刚度。对于界面的应力—滑移模型,其滑移应力—滑移长度本构图如图 3.2-25 所示。应力—滑移本构参数定义如图 3.2-26 所示。

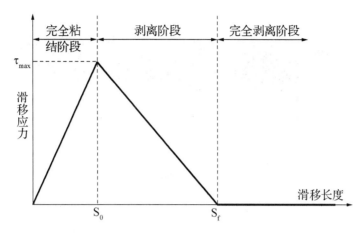

图 3.2‑25 应力—滑移模型

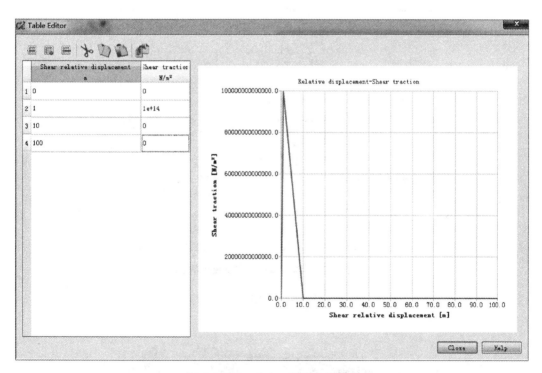

图 3.2‑26 应力—滑移本构参数

点击 OK,完成界面单元本构定义,如图 3.2‑27 所示。红色区域代表生成的定义的界面单元。

选中 Block1,右击 Property assignments 对钢板进行钢材料属性定义,钢板单元类型选择实体单元(Structural Solids)。点击 Material,弹出材料属性定义界面,选择钢材(Steel)类型,材料类型为各向同性线弹性(Linear elasticity isotropic),各向材料特性参数定义与纵筋相同。点击 OK,完成钢材属性定义。

添加约束,点击 Geometry 栏下方 图标定义重力荷载,荷载类型选为恒载(Dead weight)。创立荷载工况 1,命名为 load,加载目标类型(Load target type)为面荷载,荷载形式

图 3.2-27 面对面接触单元定义

(Load type)为均布面荷载(Distributed force),大小为 50 kN/m²,方向 Z 轴向下。如图 3.2-28 所示。

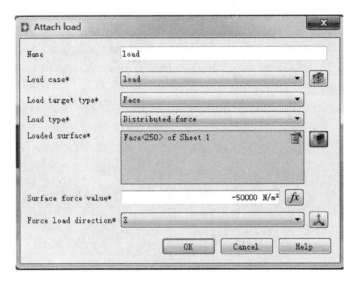

图 3.2-28 均布面荷载定义操作界面

点击 OK,生成如图 3.2-29 所示的操作界面。

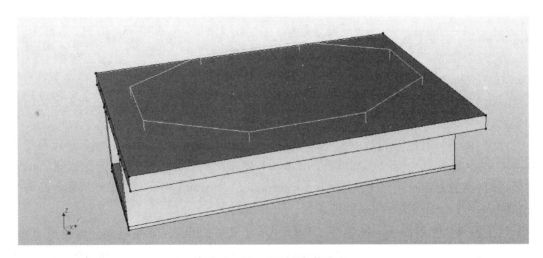

图 3.2-29 均布面荷载形成

点击 Combination,同时添加重力荷载和均布荷载作为荷载组合 1。点击该荷载组合,再点击快捷菜单栏图标 (Edit time dependency factors),定义含有时变特性的时间—荷载系数函数关系式。由于施加的是恒定的长期荷载,因此选择力不随时间变化,load factor =1。最大时间荷载步设定为 500 年(1.5768×10^{10} s)。如图 3.2-30 所示。

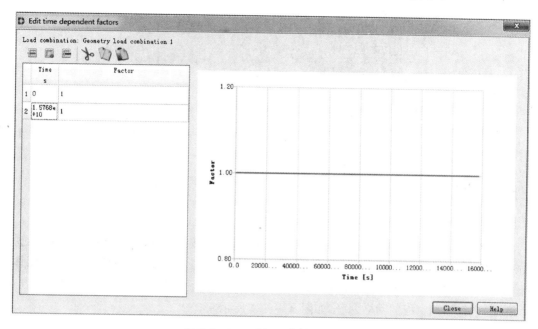

图 3.2-30　时间—荷载系数定义

添加约束。点击菜单栏 Geometry→Analysis→Attach support,对 Y=0 m 的箱梁桥截面添加 X,Y,Z 三个方向的平动约束,命名为 co1。支座目标类型(Support target type)为面约束,如图 3.2-31 所示。

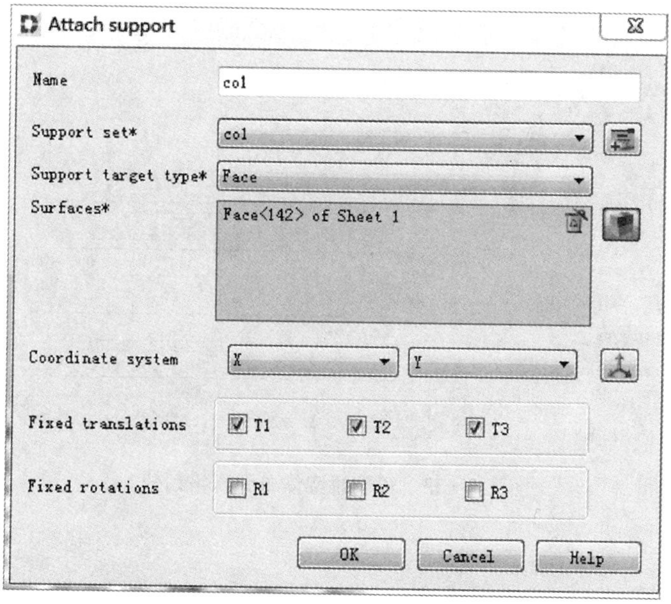

图 3.2-31　箱梁 Y=0 所在截面约束信息

用同样的方法对 Y=10 m 所在的箱梁截面施加简支约束,命名为 co2,施加约束的形式与 co1 完全相同。点击 OK,查看 GUI 操作界面生成的约束,如图 3.2-32 所示。

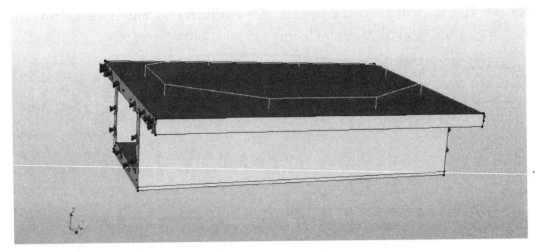

图 3.2-32 箱梁简支约束

点击快捷工具栏图标 Set mesh properties of a shape 按钮,进行单元网格的定义。选择 Sheet 1,划分方式选择单元尺寸划分类型,单元网格尺寸大小为 0.5 m,网格形状为六面体/四边形,单元中间节点位置处的确定方式为线性插值。如图 3.2-33 所示。再用同样方式对 block 进行粘贴钢板的网格划分,如图 3.2-34 所示。

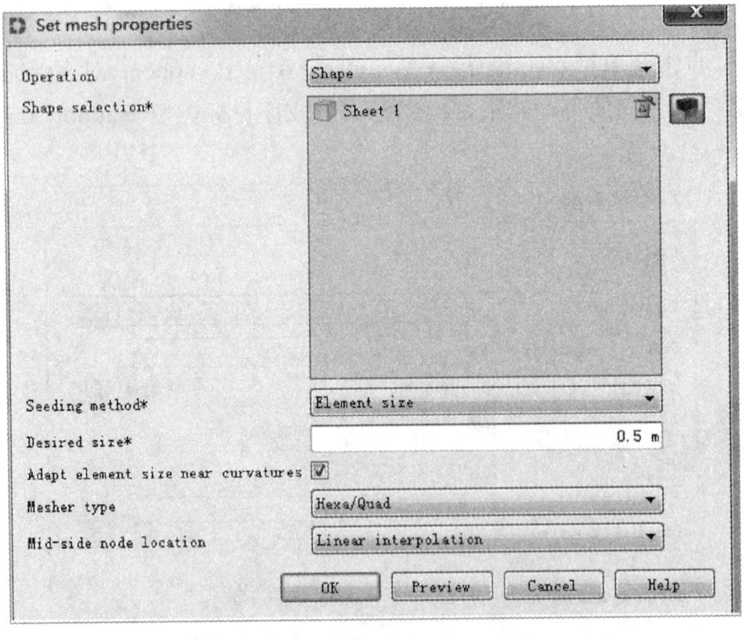

图 3.2-33 箱梁桥实体单元网格划分

图 3.2‑34　粘贴钢板网格划分

最后对界面单元进行网格划分，如图 3.2‑35 所示。

图 3.2‑35　界面单元网格划分

点击快捷图标中的 Generate mesh of a shape 按钮，生成如图 3.2‑36 所示的单元网格。查看 Mesh 栏中 Element types 下方的单元名称，确认是需要划分的实体网格单元和界面网格单元（CHX60 和 CQ48I）。

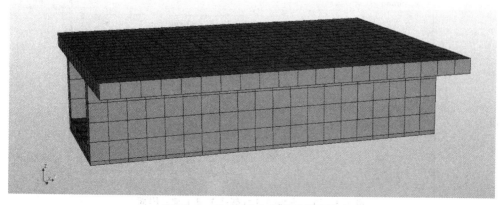

图 3.2‑36　划分之后的单元网格

点击属性栏 Analysis 模块中的 ■ (Add analysis command)生成 Analysis 分析模块,点击 Analysis,右击 Add command→Phased 生成阶段性分析第一阶段,取消勾选 Block 1 单元,表示激活第一阶段除钢板单元之外的所有单元。如图 3.2-37 所示。

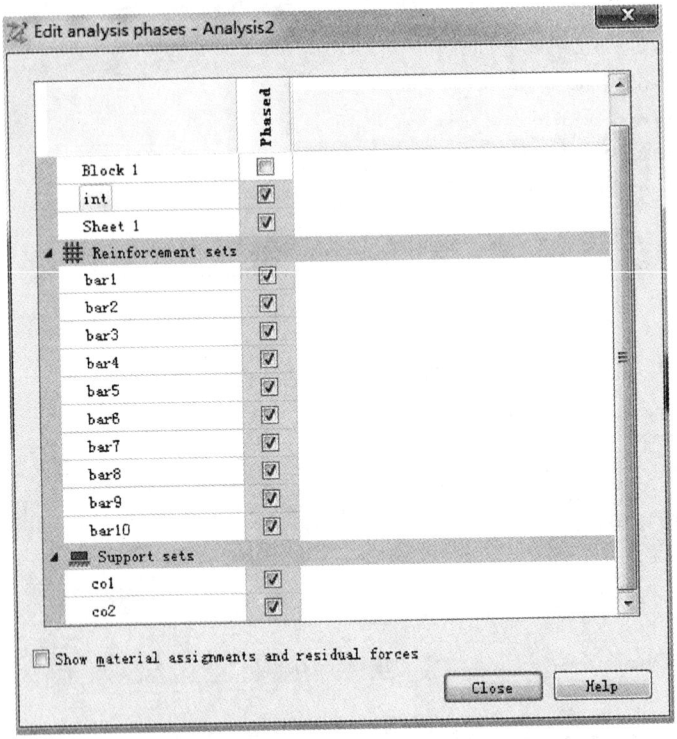

图 3.2-37　阶段性分析第一阶段激活的单元

点击 Analysis,右击选择结构非线性分析(Structural nonlinear),右击 Add→Execute steps→Load steps 创立荷载工况 1,将其命名为 load,下方的荷载定义中选择荷载组合 1,并且荷载步长因子设定为 1,如图 3.2-38 所示。

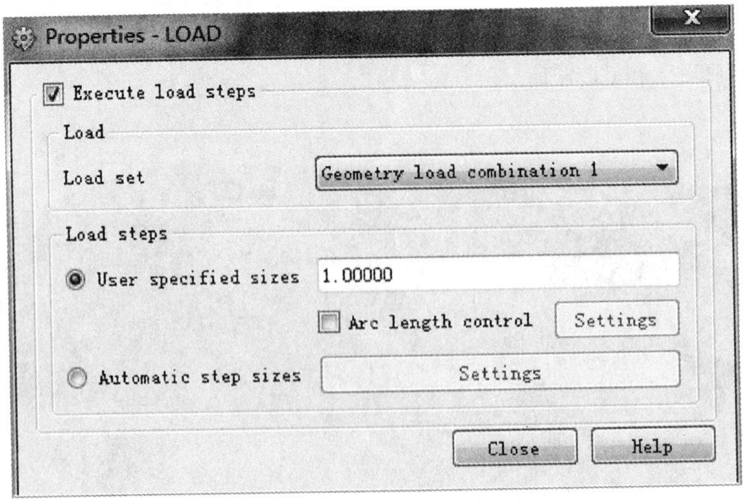

图 3.2-38　非线性分析荷载工况 1 定义界面

如图 3.2-39 所示,在下方的方程迭代(Equilibrium iterations)模块中设置最大迭代步数(Maximum number of iterations)为 20,收敛准则(Convergence norm)选择力和位移收敛(表示在迭代运算中只要力和位移二者有一个收敛该荷载步迭代计算结果就收敛)。在 settings 内设置收敛容差(Convergence tolerance)为 0.01,最大计算次数为 10000 次。如图 3.2-40 所示。

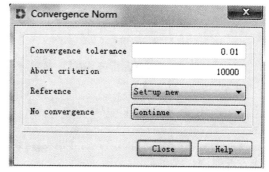

图 3.2-39　迭代方法和收敛准则设置　　　　图 3.2-40　收敛容差的设置

用上述方式选择 Time step 创建时间荷载工况,命名为 creep and shrinkage,时间荷载子步长设置为 1 天、1 年、10 年、50 年、100 年(对应的要施加的秒数分别为 86400 s, 3.14496e+07 s, 2.83824e+08 s, 1.26144e+09 s, 1.57680e+09 s),如图 3.2-41 所示。

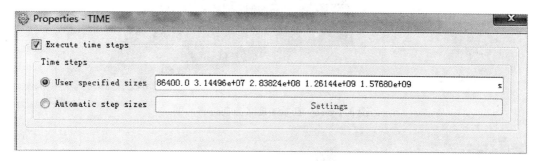

图 3.2-41　时间荷载步设置

时间荷载步计算最大迭代步数设置为 50 步,其他设置条件与上述荷载设置方式相同。

再用上述方法添加第二阶段,在该阶段中,所有单元均处于激活状态,命名为 phased1。如图 3.2-42 所示。同样创建第二阶段的非线性分析,这时应将荷载工况 1 作为初始荷载添加,迭代运算设置方式与之前相同。这时第二阶段作用时间设定为 100 年(3.1536e+09 s)之后。

点击 Run analysis,进行计算,待非线性计算完毕,查看 Z 方向的位移云图,如图 3.2-43、图 3.2-44、图 3.2-45 所示。

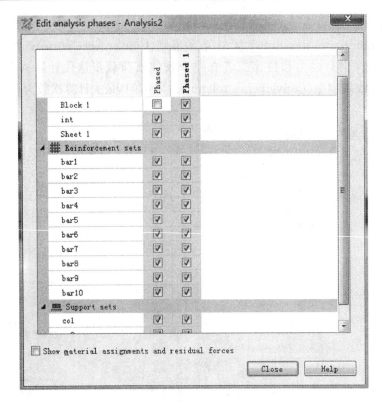

图 3.2－42　阶段性分析第二阶段激活的单元

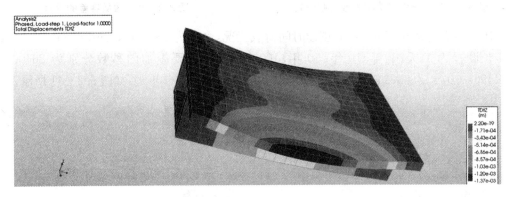

图 3.2－43　初始加载 Z 方向位移云图

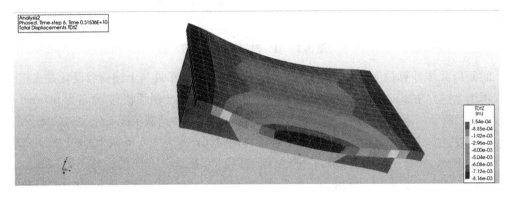

图 3.2－44　100 年后 Z 方向位移云图

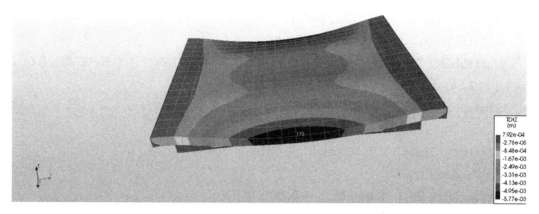

图 3.2‑45　启动钢板加固后的位移云图

为更好地说明问题,点击菜单栏 View→node selection,选择翼缘板部分的 110 号节点,右击 Output 下方 TDtZ,选择 Show table,读取徐变收缩等长期时变作用下该节点对应的最终 Z 方向位移为 8.158 mm,再点击启动钢板粘贴后的位移为 5.767 mm,恢复率为 29.31%,如图 3.2‑46、图 3.2‑47 所示。

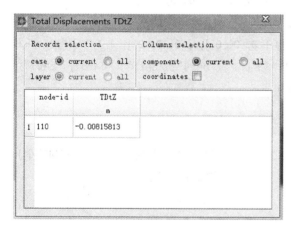

图 3.2‑46　钢板粘贴加固前

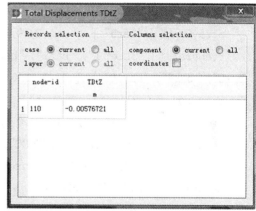

图 3.2‑47　钢板粘贴加固后

Python 语言如下:

```
###############################################################
# DianaIE 10.1 update 2017-04-25 13:38:53
# Python   3.3.4
# Session recorded at 2018-01-07 17:36:59
###############################################################
newProject( "Girder",100 )
setModelAnalysisAspects( [ "STRUCT" ] )
setModelDimension( "3D" )
```

```
setDefaultMeshOrder( "QUADRATIC" )
setDefaultMesherType( "HEXQUAD" )
setDefaultMidSideNodeLocation( "LINEAR" )
createSheet( "Sheet 1", [[ 0, 0, 0 ],[ 4, 0, 0 ],[ 4, 0, 2 ],[ 6, 0, 2 ],[ 6, 0, 2.4 ],[ -2, 0, 2.4 ],[ -2, 0, 2 ],[ 0, 0, 2 ]] )
createSheet( "Sheet 2", [[ 0.4, 0, 0.3 ],[ 3.6, 0, 0.3 ],[ 3.6, 0, 2 ],[ 0.4, 0, 2 ]] )
subtract( "Sheet 1", [ "Sheet 2" ], False, True )
saveProject( )
extrudeProfile( [ "Sheet 1" ], [ 0, 10, 0 ] )
createLine( "Line 1", [ 0.2, 0, 0.15 ], [ 0.2, 10, 0.15 ] )
renameShape( "Line 1", "bar1" )
arrayCopy( [ "bar1" ], [ 1.2, 0, 0 ], [ 0, 0, 0 ], [ 0, 0, 0 ], 3 )
saveProject( )
arrayCopy( [ "bar1" ], [ 0, 0, 2 ], [ 0, 0, 0 ], [ 0, 0, 0 ], 1 )
arrayCopy( [ "bar5" ], [ 1.2, 0, 0 ], [ 0, 0, 0 ], [ 0, 0, 0 ], 4 )
saveProject( )
arrayCopy( [ "bar5" ], [ -1.2, 0, 0 ], [ 0, 0, 0 ], [ 0, 0, 0 ], 1 )
translate( [ "Sheet 1", "bar1", "bar2", "bar3", "bar4", "bar5", "bar6", "bar7", "bar8", "bar9", "bar10" ], [ 0, 0, 0.01 ] )
saveProject( )
createBlock( "Block 1", [ 0, 0, 0 ], [ 4, 10, 0.01 ] )
addMaterial( "concrete", "CONCDC", "MC1990", [ "CREEP", "CRKIDX", "SHRINK" ] )
setParameter( "MATERIAL", "concrete", "MC90CO/GRADE", "C50" )
setUnit( "ANGLE", "DEGREE" )
setUnit( "TEMPER", "CELSIU" )
setParameter( "MATERIAL", "concrete", "MC90CO/RH", 60 )
setParameter( "MATERIAL", "concrete", "CONCDI/YOUNG", 3.8926e+10 )
setParameter( "MATERIAL", "concrete", "CONCDI/YOUN28", 3.15e+10 )
setParameter( "MATERIAL", "concrete", "CONCDI/POISON", 0.15 )
setParameter( "MATERIAL", "concrete", "CONCDI/THERMX", 1e-05 )
setParameter( "MATERIAL", "concrete", "CONCDI/DENSIT", 2500 )
setParameter( "MATERIAL", "concrete", "CONCDI/FCK28", 50000000 )
setParameter( "MATERIAL", "concrete", "CONCDI/FCM28", 58000000 )
setParameter( "MATERIAL", "concrete", "CONCCP/AGETYP", "AGING" )
setParameter( "MATERIAL", "concrete", "CONCCP/AGING", 2419200 )
setParameter( "MATERIAL", "concrete", "CONCSH/CURAGE", 86400 )
clearReinforcementAspects( [ "Sheet 1" ] )
```

```
    setElementClassType( "SHAPE", [ "Sheet 1" ], "STRSOL" )
    assignMaterial( "concrete", "SHAPE", [ "Sheet 1" ] )
    resetGeometry( "SHAPE", [ "Sheet 1" ] )
    resetElementData( "SHAPE", [ "Sheet 1" ] )
    hide( "SHAPE", [ "Sheet 1" ] )
    hide( "SHAPE", [ "bar1" ] )
    hide( "SHAPE", [ "bar2" ] )
    hide( "SHAPE", [ "bar3" ] )
    hide( "SHAPE", [ "bar4" ] )
    hide( "SHAPE", [ "bar5" ] )
    hide( "SHAPE", [ "bar6" ] )
    hide( "SHAPE", [ "bar7" ] )
    hide( "SHAPE", [ "bar8" ] )
    hide( "SHAPE", [ "bar9" ] )
    hide( "SHAPE", [ "bar10" ] )
    show( "SHAPE", [ "bar1" ] )
    show( "SHAPE", [ "bar2" ] )
    show( "SHAPE", [ "bar3" ] )
    show( "SHAPE", [ "bar4" ] )
    show( "SHAPE", [ "bar5" ] )
    show( "SHAPE", [ "bar6" ] )
    show( "SHAPE", [ "bar7" ] )
    show( "SHAPE", [ "bar8" ] )
    show( "SHAPE", [ "bar9" ] )
    show( "SHAPE", [ "bar10" ] )
    addMaterial( "bar", "MCSTEL", "ISOTRO", [] )
    setParameter( "MATERIAL", "bar", "LINEAR /ELASTI /YOUNG", 2.1e + 11 )
    setParameter( "MATERIAL", "bar", "LINEAR /ELASTI /POISON", 0.3 )
    setParameter( "MATERIAL", "bar", "LINEAR /MASS /DENSIT", 7800 )
    addGeometry( "Element geometry 1", "RELINE", "REBAR", [] )
    rename( "GEOMET", "Element geometry 1", "bar" )
    setParameter( "GEOMET", "bar", "REIEMB /CROSSE", 5.0265e - 05 )
    setReinforcementAspects( [ "bar1", "bar2", "bar3", "bar4", "bar5", "bar6",
"bar7", "bar8", "bar9", "bar10" ] )
    assignMaterial( "bar", "SHAPE", [ "bar1", "bar2", "bar3", "bar4", "bar5",
"bar6", "bar7", "bar8", "bar9", "bar10" ] )
    assignGeometry( "bar", "SHAPE", [ "bar1", "bar2", "bar3", "bar4", "bar5",
"bar6", "bar7", "bar8", "bar9", "bar10" ] )
```

```
resetElementData( "SHAPE", [ "bar1", "bar2", "bar3", "bar4", "bar5", "bar6", "bar7", "bar8", "bar9", "bar10" ] )
setReinforcementDiscretization( [ "bar1", "bar2", "bar3", "bar4", "bar5", "bar6", "bar7", "bar8", "bar9", "bar10" ], "SECTION" )
saveProject(  )
hide( "SHAPE", [ "bar1" ] )
hide( "SHAPE", [ "bar2" ] )
hide( "SHAPE", [ "bar3" ] )
hide( "SHAPE", [ "bar4" ] )
hide( "SHAPE", [ "bar5" ] )
hide( "SHAPE", [ "bar6" ] )
hide( "SHAPE", [ "bar7" ] )
hide( "SHAPE", [ "bar9" ] )
hide( "SHAPE", [ "bar8" ] )
hide( "SHAPE", [ "bar10" ] )
addMaterial( "int", "INTERF", "BONDSL", [] )
setParameter( "MATERIAL", "int", "LINEAR /ELAS6 /DSNZ", 3.65e+16 )
setParameter( "MATERIAL", "int", "LINEAR /ELAS6 /DSSX", 3.65e+12 )
setParameter( "MATERIAL", "int", "LINEAR /ELAS6 /DSSY", 3.65e+12 )
setParameter( "MATERIAL", "int", "BOSLIP /BONDSL", 3 )
setParameter( "MATERIAL", "int", "BOSLIP /BONDS3 /DISTAU", [] )
setParameter( "MATERIAL", "int", "BOSLIP /BONDS3 /DISTAU", [ 0, 0, 1, 1e+14, 10, 0, 100, 0 ] )
addGeometry( "Element geometry 2", "SHEET", "STRINT", [] )
remove( "GEOMET", "Element geometry 2" )
createSurfaceConnection( "int" )
setParameter( "GEOMETRYCONNECTION", "int", "CONTYP", "INTER" )
setParameter( "GEOMETRYCONNECTION", "int", "MODE", "AUTO" )
attachTo( "GEOMETRYCONNECTION", "int", "SOURCE", "Block 1", [[ 2.294292, 5.73573, 0.01 ]] )
setElementClassType( "GEOMETRYCONNECTION", "int", "STRINT" )
assignMaterial( "int", "GEOMETRYCONNECTION", "int" )
resetGeometry( "GEOMETRYCONNECTION", "int" )
resetElementData( "GEOMETRYCONNECTION", "int" )
saveProject(  )
show( "SHAPE", [ "Sheet 1" ] )
show( "SHAPE", [ "bar1" ] )
show( "SHAPE", [ "bar2" ] )
```

```
show( "SHAPE", [ "bar3" ] )
show( "SHAPE", [ "bar4" ] )
show( "SHAPE", [ "bar5" ] )
hide( "SHAPE", [ "bar1" ] )
hide( "SHAPE", [ "bar2" ] )
hide( "SHAPE", [ "bar3" ] )
hide( "SHAPE", [ "bar4" ] )
hide( "SHAPE", [ "bar5" ] )
hide( "SHAPE", [ "Sheet 1" ] )
addMaterial( "steel", "MCSTEL", "ISOTRO", [] )
setParameter( "MATERIAL", "steel", "LINEAR /ELASTI /YOUNG", 2.1e + 11 )
setParameter( "MATERIAL", "steel", "LINEAR /ELASTI /POISON", 0.3 )
setParameter( "MATERIAL", "steel", "LINEAR /MASS /DENSIT", 7800 )
clearReinforcementAspects( [ "Block 1" ] )
setElementClassType( "SHAPE", [ "Block 1" ], "STRSOL" )
assignMaterial( "steel", "SHAPE", [ "Block 1" ] )
resetGeometry( "SHAPE", [ "Block 1" ] )
resetElementData( "SHAPE", [ "Block 1" ] )
saveProject( )
show( "SHAPE", [ "Sheet 1" ] )
show( "SHAPE", [ "bar1" ] )
show( "SHAPE", [ "bar2" ] )
show( "SHAPE", [ "bar4" ] )
show( "SHAPE", [ "bar3" ] )
show( "SHAPE", [ "bar5" ] )
show( "SHAPE", [ "bar6" ] )
show( "SHAPE", [ "bar7" ] )
show( "SHAPE", [ "bar8" ] )
show( "SHAPE", [ "bar9" ] )
show( "SHAPE", [ "bar10" ] )
addSet( "GEOMETRYLOADSET", "gravity" )
createModelLoad( "gravity", "gravity" )
saveProject( )
addSet( "GEOMETRYLOADSET", "Geometry load case 2" )
rename( "GEOMETRYLOADSET", "Geometry load case 2", "load" )
createSurfaceLoad( "load", "load" )
setParameter( "GEOMETRYLOAD", "load", "FORCE /VALUE", - 50000 )
setParameter( "GEOMETRYLOAD", "load", "FORCE /DIRECT", 3 )
```

    attach("GEOMETRYLOAD", "load", "Sheet 1", [[2.588584, 5.73573, 2.41]])
    setDefaultGeometryLoadCombinations( )
    setGeometryLoadCombinationFactor("Geometry load combination 1", "gravity", 1)
    remove("GEOMETRYLOADCOMBINATION", "Geometry load combination 1")
    setGeometryLoadCombinationFactor("Geometry load combination 2", "gravity", 1)
    remove("GEOMETRYLOADCOMBINATION", "Geometry load combination 2")
    addGeometryLoadCombination("")
    setGeometryLoadCombinationFactor("Geometry load combination 1", "gravity", 1)
    setGeometryLoadCombinationFactor("Geometry load combination 1", "load", 1)
    setGeometryLoadCombinationFactor("Geometry load combination 1", "load", 1)
    setTimeDependentLoadFactors (" GEOMETRYLOADCOMBINATION ", " Geometry load combination 1", [0, 1.5768e+10], [1, 1])
    saveProject( )
    setTimeDependentLoadFactors (" GEOMETRYLOADCOMBINATION ", " Geometry load combination 1", [0, 1.5768e+10], [1, 1])
    addSet("GEOMETRYSUPPORTSET", "co1")
    createSurfaceSupport("co1", "co1")
    setParameter("GEOMETRYSUPPORT", "co1", "AXES", [1, 2])
    setParameter("GEOMETRYSUPPORT", "co1", "TRANSL", [1, 1, 1])
    setParameter("GEOMETRYSUPPORT", "co1", "ROTATI", [0, 0, 0])
    attach("GEOMETRYSUPPORT", "co1", "Sheet 1", [[0.1705708, -1.1202023e-16, 1.0349259]])
    saveProject( )
    addSet("GEOMETRYSUPPORTSET", "Geometry support set 2")
    rename("GEOMETRYSUPPORTSET", "Geometry support set 2", "co2")
    createSurfaceSupport("co2", "co2")
    setParameter("GEOMETRYSUPPORT", "co2", "AXES", [1, 2])
    setParameter("GEOMETRYSUPPORT", "co2", "TRANSL", [1, 1, 1])
    setParameter("GEOMETRYSUPPORT", "co2", "ROTATI", [0, 0, 0])
    attach("GEOMETRYSUPPORT", "co2", "Sheet 1", [[0.1705708, 10, 1.0349259]])
    saveProject( )
    saveProject( )
    setElementSize(["Sheet 1"], 0.5, -1, True)
    setMesherType(["Sheet 1"], "HEXQUAD")
    setMidSideNodeLocation(["Sheet 1"], "LINEAR")
    saveProject( )
    setElementSize(["Block 1"], 0.5, -1, True)
    setMesherType(["Block 1"], "HEXQUAD")

```
setMidSideNodeLocation( [ "Block 1" ], "LINEAR" )
saveProject( )
hide( "SHAPE", [ "Sheet 1" ] )
hide( "SHAPE", [ "bar1" ] )
hide( "SHAPE", [ "bar2" ] )
hide( "SHAPE", [ "bar3" ] )
hide( "SHAPE", [ "bar4" ] )
hide( "SHAPE", [ "bar5" ] )
hide( "SHAPE", [ "bar6" ] )
hide( "SHAPE", [ "bar8" ] )
hide( "SHAPE", [ "bar7" ] )
hide( "SHAPE", [ "bar9" ] )
hide( "SHAPE", [ "bar10" ] )
setElementSize( "Block 1", 2, [[ 2.294292, 5.73573, 0.01 ]], 0.5, 0.5, True )
saveProject( )
show( "SHAPE", [ "Sheet 1" ] )
show( "SHAPE", [ "bar1" ] )
show( "SHAPE", [ "bar2" ] )
show( "SHAPE", [ "bar3" ] )
show( "SHAPE", [ "bar4" ] )
show( "SHAPE", [ "bar5" ] )
show( "SHAPE", [ "bar6" ] )
show( "SHAPE", [ "bar7" ] )
show( "SHAPE", [ "bar8" ] )
show( "SHAPE", [ "bar9" ] )
show( "SHAPE", [ "bar10" ] )
generateMesh( [] )
hideView( "GEOM" )
showView( "MESH" )
addAnalysis( "Analysis2" )
addAnalysisCommand( "Analysis2", "PHASE", "Phased" )
renameAnalysis( "Analysis2", "Analysis2" )
setActivePhase( "Analysis2", "Phased" )
setActivePhase( "Analysis2", "Phased" )
setActiveInPhase( "Analysis2", "ELEMENTSET", [ "Block 1" ], [ "Phased" ], False )
addAnalysisCommand( "Analysis2", "NONLIN", "Structural nonlinear" )
setActivePhase( "Analysis2", "Phased" )
renameAnalysisCommandDetail( "Analysis2", "Structural nonlinear", "EXECUT
```

(1)", "laod")
    setAnalysisCommandDetail("Analysis2", "Structural nonlinear", "EXECUT(1) / ITERAT /MAXITE", 20)
    setAnalysisCommandDetail("Analysis2", "Structural nonlinear", "EXECUT(1) / ITERAT /MAXITE", 20)
    setAnalysisCommandDetail("Analysis2", "Structural nonlinear", "EXECUT(1) / ITERAT /MAXITE", 20)
    setAnalysisCommandDetail("Analysis2", "Structural nonlinear", "EXECUT(1) / ITERAT /CONVER /DISPLA /NOCONV", "CONTIN")
    setAnalysisCommandDetail("Analysis2", "Structural nonlinear", "EXECUT(1) / ITERAT /CONVER /FORCE /NOCONV", "CONTIN")
    setActivePhase("Analysis2", "Phased")
    setAnalysisCommandDetail ("Analysis2", "Structural nonlinear", "EXECUT / EXETYP", "TIME")
    renameAnalysisCommandDetail("Analysis2", "Structural nonlinear", "EXECUT (2)", "creep")
    renameAnalysisCommandDetail("Analysis2", "Structural nonlinear", "EXECUT (2)", "creep and shrinkage")
    setAnalysisCommandDetail("Analysis2", "Structural nonlinear", "EXECUT(2) / TIME /STEPS /EXPLIC /SIZES", "86400. 0 3. 14496e + 07 2. 83824e + 08 1. 26144e + 09 1.57680e + 09")
    setAnalysisCommandDetail("Analysis2", "Structural nonlinear", "EXECUT(2) / TIME /STEPS /EXPLIC /SIZES", "86400. 0 3. 14496e + 07 2. 83824e + 08 1. 26144e + 09 1.57680e + 09")
    setAnalysisCommandDetail("Analysis2", "Structural nonlinear", "EXECUT(2) / TIME /STEPS /EXPLIC /SIZES", "86400. 0 3. 14496e + 07 2. 83824e + 08 1. 26144e + 09 1.57680e + 09")
    setAnalysisCommandDetail("Analysis2", "Structural nonlinear", "EXECUT(2) / TIME /STEPS /EXPLIC /SIZES", "86400. 0 3. 14496e + 07 2. 83824e + 08 1. 26144e + 09 1.57680e + 09")
    setAnalysisCommandDetail("Analysis2", "Structural nonlinear", "EXECUT(2) / TIME /STEPS /EXPLIC /SIZES", "86400. 0 3. 14496e + 07 2. 83824e + 08 1. 26144e + 09 1.57680e + 09")
    setAnalysisCommandDetail("Analysis2", "Structural nonlinear", "EXECUT(2) / ITERAT /MAXITE", 50)
    setAnalysisCommandDetail("Analysis2", "Structural nonlinear", "EXECUT(2) / ITERAT /CONVER /DISPLA /NOCONV", "CONTIN")
    setAnalysisCommandDetail("Analysis2", "Structural nonlinear", "EXECUT(2) /

ITERAT /CONVER /FORCE /NOCONV", "CONTIN")
　　addAnalysisCommand("Analysis2", "PHASE", "Phased 1")
　　setActivePhase("Analysis2", "Phased 1")
　　addAnalysisCommand("Analysis2", "NONLIN", "Structural nonlinear 1")
　　removeAnalysisCommandDetail("Analysis2", "Structural nonlinear 1", "EXECUT
(1)")
　　setActivePhase("Analysis2", "Phased 1")
　　setAnalysisCommandDetail("Analysis2", "Structural nonlinear 1", "EXECUT /
EXETYP", "START")
　　renameAnalysisCommand("Analysis2", "Structural nonlinear 1", "Structural
nonlinear 1")
　　addAnalysisCommandDetail("Analysis2", "Structural nonlinear 1", "EXECUT(1) /
START /INITIA /STRESS")
　　setAnalysisCommandDetail("Analysis2", "Structural nonlinear 1", "EXECUT(1) /
START /INITIA /STRESS", True)
　　setAnalysisCommandDetail("Analysis2", "Structural nonlinear 1", "EXECUT(1) /
START /LOAD /PREVIO", False)
　　saveProject( )
　　setAnalysisCommandDetail("Analysis2", "Structural nonlinear 1", "EXECUT(1) /
ITERAT /MAXITE", 20)
　　setAnalysisCommandDetail("Analysis2", "Structural nonlinear 1", "EXECUT(1) /
ITERAT /CONVER /DISPLA /NOCONV", "CONTIN")
　　setAnalysisCommandDetail("Analysis2", "Structural nonlinear 1", "EXECUT(1) /
ITERAT /CONVER /FORCE /NOCONV", "CONTIN")
　　addAnalysisCommandDetail("Analysis2", "Structural nonlinear 1", "MODEL /
EVALUA /REINFO /INTERF")
　　setAnalysisCommandDetail("Analysis2", "Structural nonlinear 1", "MODEL /
EVALUA /REINFO /INTERF", True)
　　addAnalysisCommandDetail("Analysis2", "Structural nonlinear", "MODEL /
EVALUA /REINFO /INTERF")
　　setAnalysisCommandDetail("Analysis2", "Structural nonlinear", "MODEL /
EVALUA /REINFO /INTERF", True)
　　setAnalysisCommandDetail("Analysis2", "Structural nonlinear 1", "EXECUT(1) /
START /TIME", 3.1536e + 09)
　　runSolver("Analysis2")
　　showView("RESULT")
　　setResultPlot("contours", "Total Displacements /node", "TDtZ")
　　setResultCase(["Analysis2", "Output", "Phased, Time - step 2, Time 86400."])

setResultCase(["Analysis2", "Output", "Phased, Time – step 3, Time 0.31536E + 08"])
setResultCase(["Analysis2", "Output", "Phased, Time – step 4, Time 0.31536E + 09"])
setResultCase(["Analysis2", "Output", "Phased, Time – step 5, Time 0.15768E + 10"])
setResultCase(["Analysis2", "Output", "Phased, Time – step 6, Time 0.31536E + 10"])
setResultCase(["Analysis2", "Output", "Phased 1, Start – step 1, Load – factor 1.0000"])
setResultCase(["Analysis2", "Output", "Phased, Time – step 6, Time 0.31536E + 10"])
setResultCase(["Analysis2", "Output", "Phased, Time – step 5, Time 0.15768E + 10"])
setResultCase(["Analysis2", "Output", "Phased, Time – step 4, Time 0.31536E + 09"])
setResultCase(["Analysis2", "Output", "Phased, Time – step 3, Time 0.31536E + 08"])
setResultCase(["Analysis2", "Output", "Phased, Time – step 2, Time 86400."])
setResultCase(["Analysis2", "Output", "Phased, Load – step 1, Load – factor 1.0000"])
setResultCase(["Analysis2", "Output", "Phased, Time – step 2, Time 86400."])
setResultCase(["Analysis2", "Output", "Phased, Load – step 1, Load – factor 1.0000"])
setResultCase(["Analysis2", "Output", "Phased, Time – step 2, Time 86400."])
setResultCase(["Analysis2", "Output", "Phased, Time – step 3, Time 0.31536E + 08"])
setResultCase(["Analysis2", "Output", "Phased, Time – step 4, Time 0.31536E + 09"])
setResultCase(["Analysis2", "Output", "Phased, Time – step 5, Time 0.15768E + 10"])
setResultCase(["Analysis2", "Output", "Phased, Time – step 6, Time 0.31536E + 10"])
setResultCase(["Analysis2", "Output", "Phased 1, Start – step 1, Load – factor 1.0000"])
setResultCase(["Analysis2", "Output", "Phased, Time – step 6, Time 0.31536E + 10"])
setResultCase(["Analysis2", "Output", "Phased 1, Start – step 1, Load – factor

1.0000"])

　　showIds("NODE", [110])
　　setResultCase(["Analysis2", "Output", "Phased, Time‐step 6, Time 0.31536E+10"])
　　setResultCase(["Analysis2", "Output", "Phased 1, Start‐step 1, Load‐factor 1.0000"])

　提取上述.py文件，将文件中的混凝土抗压强度改为C40，对应的混凝土28天抗压强度标准值改为40 MPa，抗压强度平均值改为48 MPa，根据弹性模量与混凝土抗压强度值关系转换式$2.15 \times 10^4 \times \alpha_E \times \left(\dfrac{f_{cm}}{10}\right)^{\frac{1}{3}}$算得混凝土弹性模量为$3.6267 \times 10^{10}$ Pa。各参数在.py文件中命令流位置如图3.2-48所示。

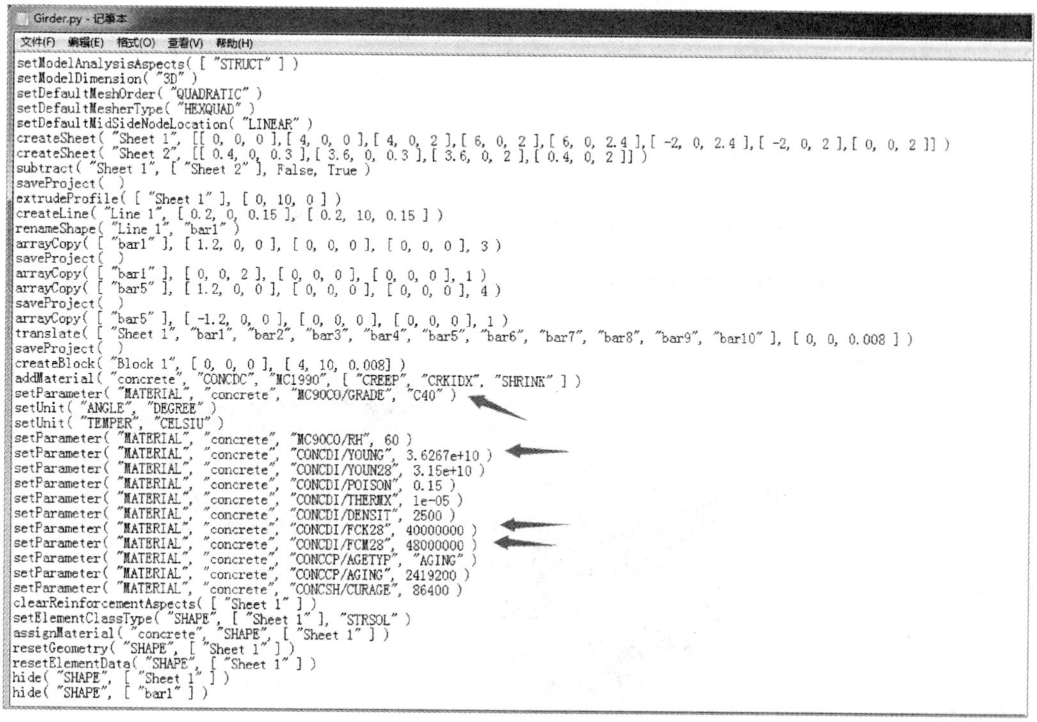

图 3.2-48　.py 文件中命令流语句中要修改的各参量位置

　点击保存，打开DianaIE界面，这里介绍两种在Diana中由.py文件生成.dpf文件的操作方式。

　方法一：点击DianaIE菜单栏下方的Run saved script，选择修改参量和保存后的Girder.py文件，Diana会自动计算并在GUI操作界面上生成位移云图，同时生成.dpf文件。

　方法二：打开DianaIE界面，复制修改参数后的.py文件中Python命令流语句，将其粘贴到Command consolo命令流语句栏，同样可以生成上述结果，粘贴位置如图3.2-49所示。

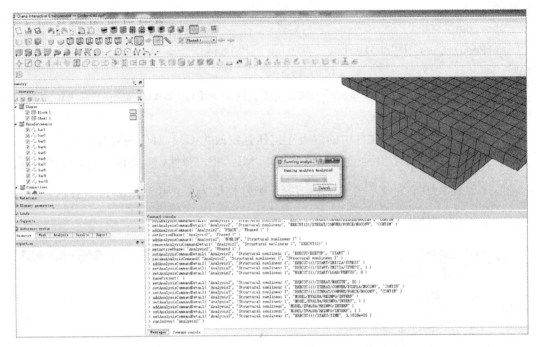

图 3.2‑49　Python 语句粘贴到命令流语句栏自动计算

100 年后的位移云图如图 3.2‑50 所示。

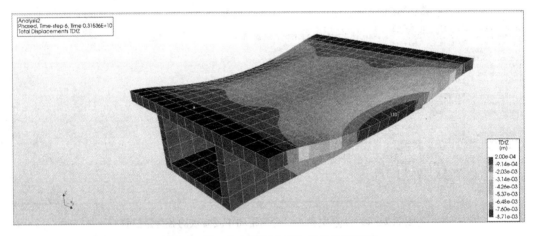

图 3.2‑50　C40 混凝土 100 年后位移云图

同样选择 110 号节点,提取计算结果,查看钢板粘贴前和粘贴后的该节点 Z 向位移,如图 3.2‑51 和图 3.2‑52 所示。

通过粘贴前后 110 节点的 Z 向位移可得,钢板粘贴前位移量为 8.710 76 mm,粘贴后位移量为 6.319 84 mm,即粘贴加固后位移恢复率为 27.45%。

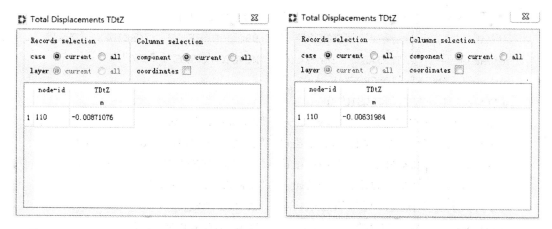

图 3.2-51　110 号节点钢板粘贴前 Z 向位移　　图 3.2-52　110 号节点钢板粘贴后 Z 向位移

经验总结：当 Diana 数值模型中几何参数相同时，用户可以使用上述两种方法将修改参数后的 .py 文件导入 DianaIE 中批量生成 .dpf 文件。这种建模方式节省大量的重复建模时间，提高建模效率。也便于对模型中出错之处进行检查和修正。

Python 命令流语句如下：

```
newProject( "Girder",100 )
setModelAnalysisAspects( [ "STRUCT" ] )
setModelDimension( "3D" )
setDefaultMeshOrder( "QUADRATIC" )
setDefaultMesherType( "HEXQUAD" )
setDefaultMidSideNodeLocation( "LINEAR" )
createSheet( "Sheet 1", [[ 0, 0, 0 ],[ 4, 0, 0 ],[ 4, 0, 2 ],[ 6, 0, 2 ],[ 6, 0, 2.4 ],[ -2, 0, 2.4 ],[ -2, 0, 2 ],[ 0, 0, 2 ]] )
createSheet( "Sheet 2", [[ 0.4, 0, 0.3 ],[ 3.6, 0, 0.3 ],[ 3.6, 0, 2 ],[ 0.4, 0, 2 ]] )
subtract( "Sheet 1", [ "Sheet 2" ], False, True )
saveProject(  )
extrudeProfile( [ "Sheet 1" ], [ 0, 10, 0 ] )
createLine( "Line 1", [ 0.2, 0, 0.15 ], [ 0.2, 10, 0.15 ] )
renameShape( "Line 1", "bar1" )
arrayCopy( [ "bar1" ], [ 1.2, 0, 0 ], [ 0, 0, 0 ], [ 0, 0, 0 ], 3 )
saveProject(  )
arrayCopy( [ "bar1" ], [ 0, 0, 2 ], [ 0, 0, 0 ], [ 0, 0, 0 ], 1 )
arrayCopy( [ "bar5" ], [ 1.2, 0, 0 ], [ 0, 0, 0 ], [ 0, 0, 0 ], 4 )
saveProject(  )
arrayCopy( [ "bar5" ], [ -1.2, 0, 0 ], [ 0, 0, 0 ], [ 0, 0, 0 ], 1 )
translate( [ "Sheet 1", "bar1", "bar2", "bar3", "bar4", "bar5", "bar6", "bar7", "bar8", "bar9", "bar10" ], [ 0, 0, 0.01 ] )
saveProject(  )
```

```
createBlock( "Block 1", [ 0, 0, 0 ], [ 4, 10, 0.01 ] )
addMaterial( "concrete", "CONCDC", "MC1990", [ "CREEP", "CRKIDX", "SHRINK" ] )
setParameter( "MATERIAL", "concrete", "MC90CO /GRADE", "C50" )
setUnit( "ANGLE", "DEGREE" )
setUnit( "TEMPER", "CELSIU" )
setParameter( "MATERIAL", "concrete", "MC90CO /RH", 60 )
setParameter( "MATERIAL", "concrete", "CONCDI /YOUNG", 3.8926e+10 )
setParameter( "MATERIAL", "concrete", "CONCDI /YOUN28", 3.15e+10 )
setParameter( "MATERIAL", "concrete", "CONCDI /POISON", 0.15 )
setParameter( "MATERIAL", "concrete", "CONCDI /THERMX", 1e-05 )
setParameter( "MATERIAL", "concrete", "CONCDI /DENSIT", 2500 )
setParameter( "MATERIAL", "concrete", "CONCDI /FCK28", 50000000 )
setParameter( "MATERIAL", "concrete", "CONCDI /FCM28", 58000000 )
setParameter( "MATERIAL", "concrete", "CONCCP /AGETYP", "AGING" )
setParameter( "MATERIAL", "concrete", "CONCCP /AGING", 2419200 )
setParameter( "MATERIAL", "concrete", "CONCSH /CURAGE", 86400 )
clearReinforcementAspects( [ "Sheet 1" ] )
setElementClassType( "SHAPE", [ "Sheet 1" ], "STRSOL" )
assignMaterial( "concrete", "SHAPE", [ "Sheet 1" ] )
resetGeometry( "SHAPE", [ "Sheet 1" ] )
resetElementData( "SHAPE", [ "Sheet 1" ] )
hide( "SHAPE", [ "Sheet 1" ] )
hide( "SHAPE", [ "bar1" ] )
hide( "SHAPE", [ "bar2" ] )
hide( "SHAPE", [ "bar3" ] )
hide( "SHAPE", [ "bar4" ] )
hide( "SHAPE", [ "bar5" ] )
hide( "SHAPE", [ "bar6" ] )
hide( "SHAPE", [ "bar7" ] )
hide( "SHAPE", [ "bar8" ] )
hide( "SHAPE", [ "bar9" ] )
hide( "SHAPE", [ "bar10" ] )
show( "SHAPE", [ "bar1" ] )
show( "SHAPE", [ "bar2" ] )
show( "SHAPE", [ "bar3" ] )
show( "SHAPE", [ "bar4" ] )
show( "SHAPE", [ "bar5" ] )
show( "SHAPE", [ "bar6" ] )
show( "SHAPE", [ "bar7" ] )
show( "SHAPE", [ "bar8" ] )
```

```
show( "SHAPE", [ "bar9" ] )
show( "SHAPE", [ "bar10" ] )
addMaterial( "bar", "MCSTEL", "ISOTRO", [] )
setParameter( "MATERIAL", "bar", "LINEAR /ELASTI /YOUNG", 2.1e+11 )
setParameter( "MATERIAL", "bar", "LINEAR /ELASTI /POISON", 0.3 )
setParameter( "MATERIAL", "bar", "LINEAR /MASS /DENSIT", 7800 )
addGeometry( "Element geometry 1", "RELINE", "REBAR", [] )
rename( "GEOMET", "Element geometry 1", "bar" )
setParameter( "GEOMET", "bar", "REIEMB /CROSSE", 5.0265e-05 )
setReinforcementAspects( [ "bar1", "bar2", "bar3", "bar4", "bar5", "bar6", "bar7", "bar8", "bar9", "bar10" ] )
assignMaterial( "bar", "SHAPE", [ "bar1", "bar2", "bar3", "bar4", "bar5", "bar6", "bar7", "bar8", "bar9", "bar10" ] )
assignGeometry( "bar", "SHAPE", [ "bar1", "bar2", "bar3", "bar4", "bar5", "bar6", "bar7", "bar8", "bar9", "bar10" ] )
resetElementData( "SHAPE", [ "bar1", "bar2", "bar3", "bar4", "bar5", "bar6", "bar7", "bar8", "bar9", "bar10" ] )
setReinforcementDiscretization( [ "bar1", "bar2", "bar3", "bar4", "bar5", "bar6", "bar7", "bar8", "bar9", "bar10" ], "SECTION" )
saveProject( )
hide( "SHAPE", [ "bar1" ] )
hide( "SHAPE", [ "bar2" ] )
hide( "SHAPE", [ "bar3" ] )
hide( "SHAPE", [ "bar4" ] )
hide( "SHAPE", [ "bar5" ] )
hide( "SHAPE", [ "bar6" ] )
hide( "SHAPE", [ "bar7" ] )
hide( "SHAPE", [ "bar9" ] )
hide( "SHAPE", [ "bar8" ] )
hide( "SHAPE", [ "bar10" ] )
addMaterial( "int", "INTERF", "BONDSL", [] )
setParameter( "MATERIAL", "int", "LINEAR /ELAS6 /DSNZ", 3.65e+16 )
setParameter( "MATERIAL", "int", "LINEAR /ELAS6 /DSSX", 3.65e+12 )
setParameter( "MATERIAL", "int", "LINEAR /ELAS6 /DSSY", 3.65e+12 )
setParameter( "MATERIAL", "int", "BOSLIP /BONDSL", 3 )
setParameter( "MATERIAL", "int", "BOSLIP /BONDS3 /DISTAU", [] )
setParameter( "MATERIAL", "int", "BOSLIP /BONDS3 /DISTAU", [ 0, 0, 1, 1e+14, 10, 0, 100, 0 ] )
addGeometry( "Element geometry 2", "SHEET", "STRINT", [] )
remove( "GEOMET", "Element geometry 2" )
```

```
createSurfaceConnection( "int" )
setParameter( "GEOMETRYCONNECTION", "int", "CONTYP", "INTER" )
setParameter( "GEOMETRYCONNECTION", "int", "MODE", "AUTO" )
attachTo( "GEOMETRYCONNECTION", "int", "SOURCE", "Block 1", [[ 2.294292, 5.73573, 0.01 ]] )
setElementClassType( "GEOMETRYCONNECTION", "int", "STRINT" )
assignMaterial( "int", "GEOMETRYCONNECTION", "int" )
resetGeometry( "GEOMETRYCONNECTION", "int" )
resetElementData( "GEOMETRYCONNECTION", "int" )
saveProject( )
show( "SHAPE", [ "Sheet 1" ] )
show( "SHAPE", [ "bar1" ] )
show( "SHAPE", [ "bar2" ] )
show( "SHAPE", [ "bar3" ] )
show( "SHAPE", [ "bar4" ] )
show( "SHAPE", [ "bar5" ] )
hide( "SHAPE", [ "bar1" ] )
hide( "SHAPE", [ "bar2" ] )
hide( "SHAPE", [ "bar3" ] )
hide( "SHAPE", [ "bar4" ] )
hide( "SHAPE", [ "bar5" ] )
hide( "SHAPE", [ "Sheet 1" ] )
addMaterial( "steel", "MCSTEL", "ISOTRO", [] )
setParameter( "MATERIAL", "steel", "LINEAR /ELASTI /YOUNG", 2.1e+11 )
setParameter( "MATERIAL", "steel", "LINEAR /ELASTI /POISON", 0.3 )
setParameter( "MATERIAL", "steel", "LINEAR /MASS /DENSIT", 7800 )
clearReinforcementAspects( [ "Block 1" ] )
setElementClassType( "SHAPE", [ "Block 1" ], "STRSOL" )
assignMaterial( "steel", "SHAPE", [ "Block 1" ] )
resetGeometry( "SHAPE", [ "Block 1" ] )
resetElementData( "SHAPE", [ "Block 1" ] )
saveProject( )
show( "SHAPE", [ "Sheet 1" ] )
show( "SHAPE", [ "bar1" ] )
show( "SHAPE", [ "bar2" ] )
show( "SHAPE", [ "bar4" ] )
show( "SHAPE", [ "bar3" ] )
show( "SHAPE", [ "bar5" ] )
show( "SHAPE", [ "bar6" ] )
show( "SHAPE", [ "bar7" ] )
```

```
show( "SHAPE", [ "bar8" ] )
show( "SHAPE", [ "bar9" ] )
show( "SHAPE", [ "bar10" ] )
addSet( "GEOMETRYLOADSET", "gravity" )
createModelLoad( "gravity", "gravity" )
saveProject( )
addSet( "GEOMETRYLOADSET", "Geometry load case 2" )
rename( "GEOMETRYLOADSET", "Geometry load case 2", "load" )
createSurfaceLoad( "load", "load" )
setParameter( "GEOMETRYLOAD", "load", "FORCE/VALUE", -50000 )
setParameter( "GEOMETRYLOAD", "load", "FORCE/DIRECT", 3 )
attach( "GEOMETRYLOAD", "load", "Sheet 1", [[ 2.588584, 5.73573, 2.41 ]] )
setDefaultGeometryLoadCombinations( )
setGeometryLoadCombinationFactor( "Geometry load combination 1", "gravity", 1 )
remove( "GEOMETRYLOADCOMBINATION", "Geometry load combination 1" )
setGeometryLoadCombinationFactor( "Geometry load combination 2", "gravity", 1 )
remove( "GEOMETRYLOADCOMBINATION", "Geometry load combination 2" )
addGeometryLoadCombination( "" )
setGeometryLoadCombinationFactor( "Geometry load combination 1", "gravity", 1 )
setGeometryLoadCombinationFactor( "Geometry load combination 1", "load", 1 )
setGeometryLoadCombinationFactor( "Geometry load combination 1", "load", 1 )
setTimeDependentLoadFactors ( " GEOMETRYLOADCOMBINATION ", " Geometry load combination 1", [ 0, 1.5768e+10 ], [ 1, 1 ] )
saveProject( )
setTimeDependentLoadFactors ( " GEOMETRYLOADCOMBINATION ", " Geometry load combination 1", [ 0, 1.5768e+10 ], [ 1, 1 ] )
addSet( "GEOMETRYSUPPORTSET", "co1" )
createSurfaceSupport( "co1", "co1" )
setParameter( "GEOMETRYSUPPORT", "co1", "AXES", [ 1, 2 ] )
setParameter( "GEOMETRYSUPPORT", "co1", "TRANSL", [ 1, 1, 1 ] )
setParameter( "GEOMETRYSUPPORT", "co1", "ROTATI", [ 0, 0, 0 ] )
attach( "GEOMETRYSUPPORT", "co1", "Sheet 1", [[ 0.1705708, -1.1202023e-16, 1.0349259 ]] )
saveProject( )
addSet( "GEOMETRYSUPPORTSET", "Geometry support set 2" )
rename( "GEOMETRYSUPPORTSET", "Geometry support set 2", "co2" )
createSurfaceSupport( "co2", "co2" )
setParameter( "GEOMETRYSUPPORT", "co2", "AXES", [ 1, 2 ] )
setParameter( "GEOMETRYSUPPORT", "co2", "TRANSL", [ 1, 1, 1 ] )
setParameter( "GEOMETRYSUPPORT", "co2", "ROTATI", [ 0, 0, 0 ] )
```

```
attach( "GEOMETRYSUPPORT", "co2", "Sheet 1", [[ 0.1705708, 10, 1.0349259 ]] )
setElementSize( [ "Sheet 1" ], 0.5, -1, True )
setMesherType( [ "Sheet 1" ], "HEXQUAD" )
setMidSideNodeLocation( [ "Sheet 1" ], "LINEAR" )
saveProject( )
setElementSize( [ "Block 1" ], 0.5, -1, True )
setMesherType( [ "Block 1" ], "HEXQUAD" )
setMidSideNodeLocation( [ "Block 1" ], "LINEAR" )
saveProject( )
hide( "SHAPE", [ "Sheet 1" ] )
hide( "SHAPE", [ "bar1" ] )
hide( "SHAPE", [ "bar2" ] )
hide( "SHAPE", [ "bar3" ] )
hide( "SHAPE", [ "bar4" ] )
hide( "SHAPE", [ "bar5" ] )
hide( "SHAPE", [ "bar6" ] )
hide( "SHAPE", [ "bar8" ] )
hide( "SHAPE", [ "bar7" ] )
hide( "SHAPE", [ "bar9" ] )
hide( "SHAPE", [ "bar10" ] )
setElementSize( "Block 1", 2, [[ 2.294292, 5.73573, 0.01 ]], 0.5, 0.5, True )
saveProject( )
show( "SHAPE", [ "Sheet 1" ] )
show( "SHAPE", [ "bar1" ] )
show( "SHAPE", [ "bar2" ] )
show( "SHAPE", [ "bar3" ] )
show( "SHAPE", [ "bar4" ] )
show( "SHAPE", [ "bar5" ] )
show( "SHAPE", [ "bar6" ] )
show( "SHAPE", [ "bar7" ] )
show( "SHAPE", [ "bar8" ] )
show( "SHAPE", [ "bar9" ] )
show( "SHAPE", [ "bar10" ] )
generateMesh( [] )
hideView( "GEOM" )
showView( "MESH" )
addAnalysis( "Analysis2" )
addAnalysisCommand( "Analysis2", "PHASE", "Phased" )
renameAnalysis( "Analysis2", "Analysis2" )
setActivePhase( "Analysis2", "Phased" )
```

```
setActivePhase( "Analysis2", "Phased" )
setActiveInPhase( "Analysis2", "ELEMENTSET", [ "Block 1" ], [ "Phased" ], False )
addAnalysisCommand( "Analysis2", "NONLIN", "Structural nonlinear" )
setActivePhase( "Analysis2", "Phased" )
renameAnalysisCommandDetail( "Analysis2", "Structural nonlinear", "EXECUT(1)", "laod" )
setAnalysisCommandDetail( "Analysis2", "Structural nonlinear", "EXECUT(1)/ITERAT/MAXITE", 20 )
setAnalysisCommandDetail( "Analysis2", "Structural nonlinear", "EXECUT(1)/ITERAT/MAXITE", 20 )
setAnalysisCommandDetail( "Analysis2", "Structural nonlinear", "EXECUT(1)/ITERAT/MAXITE", 20 )
setAnalysisCommandDetail( "Analysis2", "Structural nonlinear", "EXECUT(1)/ITERAT/CONVER/DISPLA/NOCONV", "CONTIN" )
setAnalysisCommandDetail( "Analysis2", "Structural nonlinear", "EXECUT(1)/ITERAT/CONVER/FORCE/NOCONV", "CONTIN" )
setActivePhase( "Analysis2", "Phased" )
setAnalysisCommandDetail( "Analysis2", "Structural nonlinear", "EXECUT/EXETYP", "TIME" )
renameAnalysisCommandDetail( "Analysis2", "Structural nonlinear", "EXECUT(2)", "creep" )
renameAnalysisCommandDetail( "Analysis2", "Structural nonlinear", "EXECUT(2)", "creep and shrinkage" )
setAnalysisCommandDetail( "Analysis2", "Structural nonlinear", "EXECUT(2)/TIME/STEPS/EXPLIC/SIZES", "86400.0 3.14496e+07 2.83824e+08 1.26144e+09 1.57680e+09" )
setAnalysisCommandDetail( "Analysis2", "Structural nonlinear", "EXECUT(2)/TIME/STEPS/EXPLIC/SIZES", "86400.0 3.14496e+07 2.83824e+08 1.26144e+09 1.57680e+09" )
setAnalysisCommandDetail( "Analysis2", "Structural nonlinear", "EXECUT(2)/TIME/STEPS/EXPLIC/SIZES", "86400.0 3.14496e+07 2.83824e+08 1.26144e+09 1.57680e+09" )
setAnalysisCommandDetail( "Analysis2", "Structural nonlinear", "EXECUT(2)/TIME/STEPS/EXPLIC/SIZES", "86400.0 3.14496e+07 2.83824e+08 1.26144e+09 1.57680e+09" )
setAnalysisCommandDetail( "Analysis2", "Structural nonlinear", "EXECUT(2)/TIME/STEPS/EXPLIC/SIZES", "86400.0 3.14496e+07 2.83824e+08 1.26144e+09 1.57680e+09" )
setAnalysisCommandDetail( "Analysis2", "Structural nonlinear", "EXECUT(2)/ITERAT/MAXITE", 50 )
```

```
    setAnalysisCommandDetail( "Analysis2", "Structural nonlinear", "EXECUT(2) /
ITERAT /CONVER /DISPLA /NOCONV", "CONTIN" )
    setAnalysisCommandDetail( "Analysis2", "Structural nonlinear", "EXECUT(2) /
ITERAT /CONVER /FORCE /NOCONV", "CONTIN" )
    addAnalysisCommand( "Analysis2", "PHASE", "Phased 1" )
    setActivePhase( "Analysis2", "Phased 1" )
    addAnalysisCommand( "Analysis2", "NONLIN", "Structural nonlinear 1" )
    removeAnalysisCommandDetail( "Analysis2", "Structural nonlinear 1", "EXECUT
(1)" )
    setActivePhase( "Analysis2", "Phased 1" )
    setAnalysisCommandDetail( "Analysis2", "Structural nonlinear 1", "EXECUT /
EXETYP", "START" )
    renameAnalysisCommand( "Analysis2", "Structural nonlinear 1", "Structural
nonlinear 1" )
    addAnalysisCommandDetail( "Analysis2", "Structural nonlinear 1", "EXECUT(1) /
START /INITIA /STRESS" )
    setAnalysisCommandDetail( "Analysis2", "Structural nonlinear 1", "EXECUT(1) /
START /INITIA /STRESS", True )
    setAnalysisCommandDetail( "Analysis2", "Structural nonlinear 1", "EXECUT(1) /
START /LOAD /PREVIO", False )
    saveProject( )
    setAnalysisCommandDetail( "Analysis2", "Structural nonlinear 1", "EXECUT(1) /
ITERAT /MAXITE", 20 )
    setAnalysisCommandDetail( "Analysis2", "Structural nonlinear 1", "EXECUT(1) /
ITERAT /CONVER /DISPLA /NOCONV", "CONTIN" )
    setAnalysisCommandDetail( "Analysis2", "Structural nonlinear 1", "EXECUT(1) /
ITERAT /CONVER /FORCE /NOCONV", "CONTIN" )
    addAnalysisCommandDetail ( "Analysis2", "Structural nonlinear 1", " MODEL /
EVALUA /REINFO /INTERF" )
    setAnalysisCommandDetail ( "Analysis2", "Structural nonlinear 1", " MODEL /
EVALUA /REINFO /INTERF", True )
    addAnalysisCommandDetail ( "Analysis2", "Structural nonlinear", " MODEL /
EVALUA /REINFO /INTERF" )
    setAnalysisCommandDetail ( "Analysis2", "Structural nonlinear", " MODEL /
EVALUA /REINFO /INTERF", True )
    setAnalysisCommandDetail( "Analysis2", "Structural nonlinear 1", "EXECUT(1) /
START /TIME", 3.1536e+09 )
    runSolver( "Analysis2" )
    showView( "RESULT" )
    setResultPlot( "contours", "Total Displacements /node", "TDtZ" )
```

setResultCase(["Analysis2", "Output", "Phased, Time-step 2, Time 86400."])
    setResultCase(["Analysis2", "Output", "Phased, Time-step 3, Time 0.31536E+08"])
    setResultCase(["Analysis2", "Output", "Phased, Time-step 4, Time 0.31536E+09"])
    setResultCase(["Analysis2", "Output", "Phased, Time-step 5, Time 0.15768E+10"])
    setResultCase(["Analysis2", "Output", "Phased, Time-step 6, Time 0.31536E+10"])
    setResultCase(["Analysis2", "Output", "Phased 1, Start-step 1, Load-factor 1.0000"])
    setResultCase(["Analysis2", "Output", "Phased, Time-step 6, Time 0.31536E+10"])
    setResultCase(["Analysis2", "Output", "Phased, Time-step 5, Time 0.15768E+10"])
    setResultCase(["Analysis2", "Output", "Phased, Time-step 4, Time 0.31536E+09"])
    setResultCase(["Analysis2", "Output", "Phased, Time-step 3, Time 0.31536E+08"])
    setResultCase(["Analysis2", "Output", "Phased, Time-step 2, Time 86400."])
    setResultCase(["Analysis2", "Output", "Phased, Load-step 1, Load-factor 1.0000"])
    setResultCase(["Analysis2", "Output", "Phased, Time-step 2, Time 86400."])
    setResultCase(["Analysis2", "Output", "Phased, Load-step 1, Load-factor 1.0000"])
    setResultCase(["Analysis2", "Output", "Phased, Time-step 2, Time 86400."])
    setResultCase(["Analysis2", "Output", "Phased, Time-step 3, Time 0.31536E+08"])
    setResultCase(["Analysis2", "Output", "Phased, Time-step 4, Time 0.31536E+09"])
    setResultCase(["Analysis2", "Output", "Phased, Time-step 5, Time 0.15768E+10"])
    setResultCase(["Analysis2", "Output", "Phased, Time-step 6, Time 0.31536E+10"])
    setResultCase(["Analysis2", "Output", "Phased 1, Start-step 1, Load-factor 1.0000"])
    setResultCase(["Analysis2", "Output", "Phased, Time-step 6, Time 0.31536E+10"])
    setResultCase(["Analysis2", "Output", "Phased 1, Start-step 1, Load-factor 1.0000"])

showIds("NODE", [110])

setResultCase(["Analysis2", "Output", "Phased, Time‐step 6, Time 0.31536E+10"])

setResultCase(["Analysis2", "Output", "Phased 1, Start‐step 1, Load‐factor 1.0000"])

3.3 案例三：预应力混凝土连续梁长期性能分析

本案例改编自 Diana 9.4 版(Concrete modeling and analysis Tutorials and experiences)中的曲线筋后张法混凝土梁 bat 语言命令流操作案例，这里采用 10.1 版本中的 DianaIE 模块界面可视化 GUI 操作重新进行建模并提供 Python 编辑语言。算例模型为一根后张法的两跨混凝土连续梁，预应力筋布筋形式采用曲线筋，预应筋屈服强度 1860 MPa，横截面积为 2886 mm^2。对整个梁模型施加重力荷载、预应力筋荷载以及外荷载。其中外荷载中的活荷载为均布荷载，大小为 15 kN/m，恒荷载为 10 kN/m 均布荷载。施加的初始预应力荷载为 4400 kN，预应力筋张拉方式为两端张拉(Both end)，预应力筋回缩长度为 0.01 m。梁的截面为工字型截面，尺寸信息如图 3.3-1 和图 3.3-2 所示。混凝土强度等级为 C55，采用 L7BEN 梁单元建模，收缩徐变模型为欧洲规范 CEB‐FIP 1990 模型。本案例采用非线性分析，对均布荷载下曲线预应力筋的混凝土梁收缩徐变效应进行分析。其中时间荷载步设定为 1 年(31536000 s)，以收缩徐变下的挠度和曲线预应力筋的预应力损失作为时变性能观测指标来演示采用 Diana10.1 模拟收缩徐变的变化规律。

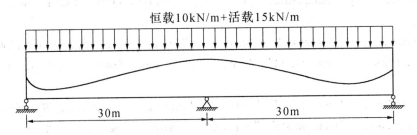

图 3.3-1 后张法混凝土连续梁预应力的位置图

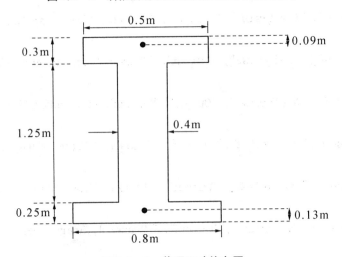

图 3.3-2 截面尺寸信息图

第三章　Diana 土木工程非线性建模案例分析

学习要义：
（1）掌握采用梁单元建立钢筋混凝土梁模型和工字型梁截面参数设置。
（2）掌握嵌入式预应力曲线筋正确建模方式。
（3）掌握 CEB-FIP90 混凝土徐变收缩模型和 Kelvin 链松弛模型的参数设置操作。
（4）掌握对荷载组合施加 Time-dependency 时变特征以及对时间荷载步的设置方式。
（5）掌握时变效应下非线性分析计算模块时间荷载步的设置方法。

首先打开 DianaIE 界面，点击上方菜单栏 File→New 创建一个新的 dpf 文件，因为采用 2D 直线梁单元，所以在弹出对话框默认划分类型中选择正六面体/矩形形状的单元，划分次数选择一阶。

注：作为 Diana10.1 算例例题，本书例题分析条件均为假定值，分析结果妥当性另当别论，望谅解。

打开 Diana 页面。点击 File→New，选择 D 盘为工作路径，在 D 盘创立新的文件夹，命名为例题，再创立新的 .dpf 文件，文件命名为 beam，选择 Structural 结构分析模块，模型的维数选择二维（Two dimensional），模型最大尺寸范围为 1 km，默认的网格形状为六面体/四边形形状（Hexa/Quad），默认网格划分阶数为线性（Linear）。如图 3.3-3 所示。

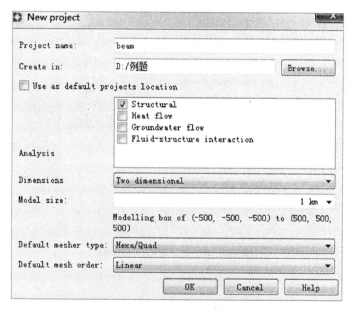

图 3.3-3　New project 操作界面

点击 OK 确认之后，进入 Diana 界面。单位设置如下：长度：米；质量：千克；力：牛顿；温度：摄氏度，角度：弧度，如图 3.3-4 所示。本书为了单位统一，计算方便，在长期性能的非线性计算中将时间基本单位设置成秒（s），用户也可以根据个人的喜好将时间设置成天（day）。只是将时间单位设置为天后，模型在非线性计算中的密度数值以及施加重力荷载时的重力加速度数值会相应发生改变。

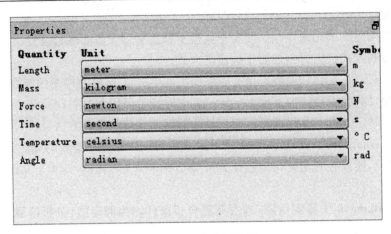

图 3.3-4　单位制设置

点击快捷工具栏图标 Add a polyline 创立三点直线,如图 3.3-5 所示。取消下方 Closed 闭合选项,输入三点的坐标,生成直线,如图 3.3-6 所示。

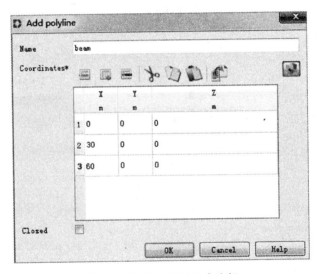

图 3.3-5　建立梁的三点坐标

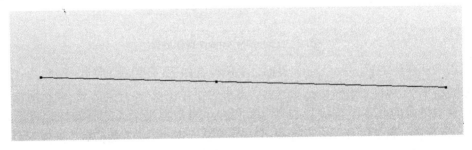

图 3.3-6

点击快捷工具栏图标 Add a curve 选择添加曲线筋几何坐标,如图 3.3-7 所示,输入 X 和 Y 轴的坐标,具体坐标见表 3-2。

表 3-2 曲线筋的几何坐标

1	(0,0)
2	(11.65,-0.5)
3	(29.1,0.6217)
4	(30,0.68)
5	(30.9,0.6217)
6	(47.45,-0.5)
7	(60,0)

图 3.3-7 Add a curve 快捷图标按钮

生成曲线筋后,命名为 reinfo,曲线筋形状如图 3.3-8 所示。

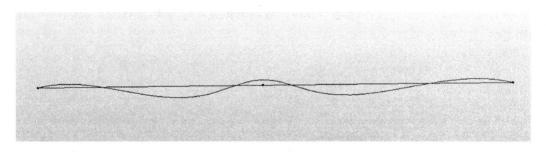

图 3.3-8 曲线筋形状几何模型

接下来赋予曲线预应力筋的材料属性,右击 Geometry 下方的 Reinforcement assignment,选中曲线筋,右击选择钢筋属性赋予为配筋和桩基础类型,如图 3.3-9 所示,命名为 reinfo,本案例中的预应力筋材料模型选择为线弹性模型(Linear elasticity),曲线筋的弹性模量为 $1.95 \times 10^{11} \text{N/m}^2$,屈服强度值(Yield stress)为 $1.86 \times 10^9 \text{N/m}^2$,如图 3.3-10 所示。

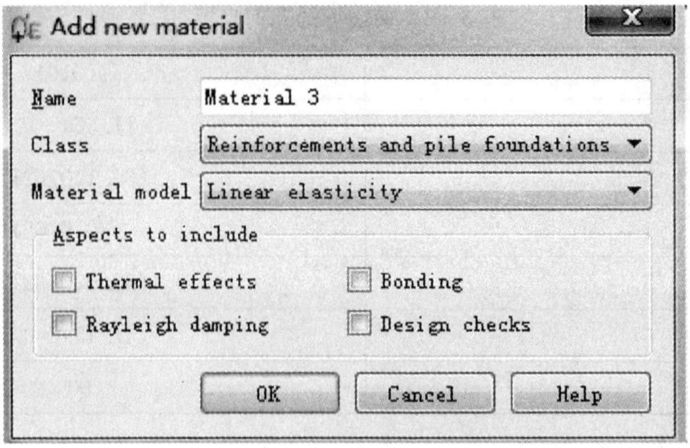

图 3.3-9　配筋及桩基础类型线弹性本构窗口

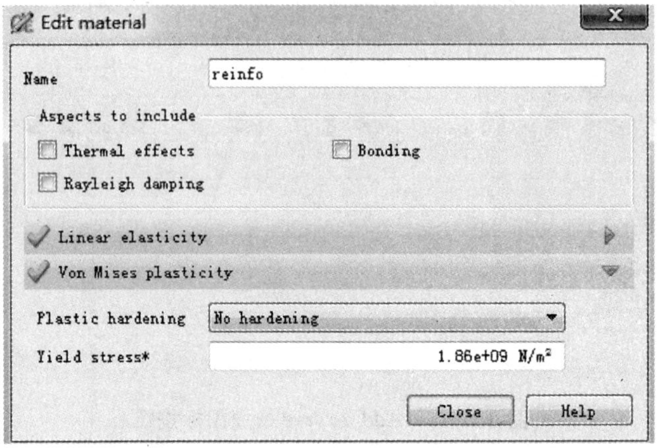

图 3.3-10　Ven Mises 塑性屈服强度设置

设置曲线筋的截面几何面积参数。其中曲线筋选择嵌入的方式,面积为 $2.886\times10^{-3}\,\mathrm{m}^2$,如图 3.3-11 所示。

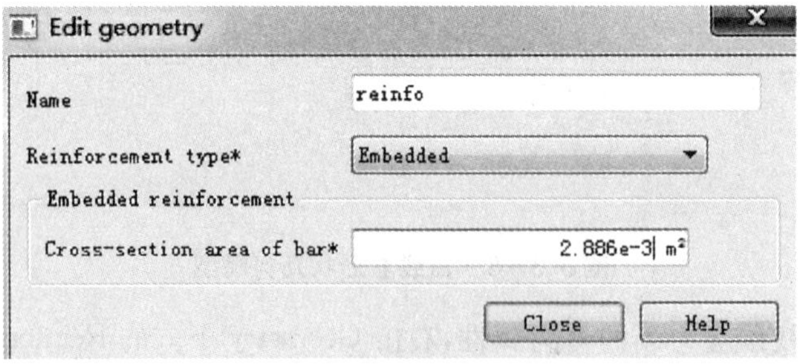

图 3.3-11　曲线预应力筋截面几何特性设置界面

设置混凝土材料特性。选中 beam,右击选择 Property assignment,单元类型选择 2D 的二类梁单元,如图 3.3-12 所示。命名为 concrete,如图 3.3-13 所示,采用欧洲 1990 规范,弹性模量根据计算公式 $E=2.15\times10^4\times\alpha_E\times\left(\dfrac{f_{cm}}{10}\right)^{\frac{1}{3}}$ 算得大小为 $3.975\times10^{10}\,\mathrm{N/m^2}$。混凝土 28 天弹性模量设定为 $3.5102\times10^{10}\,\mathrm{N/m^2}$,混凝土的密度为 $2500\,\mathrm{kg/m^3}$,如图 3.3-14 所示。

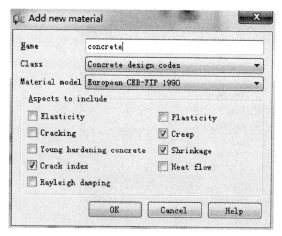

图 3.3‐12　混凝土本构 CEB‐FIP 1990 模型

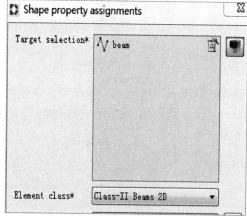

图 3.3‐13　第二类梁单元

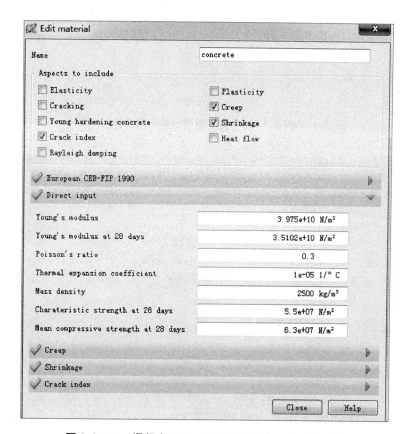

图 3.3‐14　混凝土 CEB‐FIP 1990 规范本构参数设置

在收缩徐变模块中，环境温度为 20 摄氏度，构件理论厚度（notational size）选择 0.58 m，由于混凝土的长期性能劣化主要与收缩徐变因素有关，选择以收缩徐变为主的混凝土老化理论的松弛模型 Kelvin 链，并且混凝土单元龄设置为 7 天（604800 s），收缩固化曲线结束时混凝土龄期为 1 天（86400 s）。分别如图 3.3‐15 和图 3.3‐16 所示。

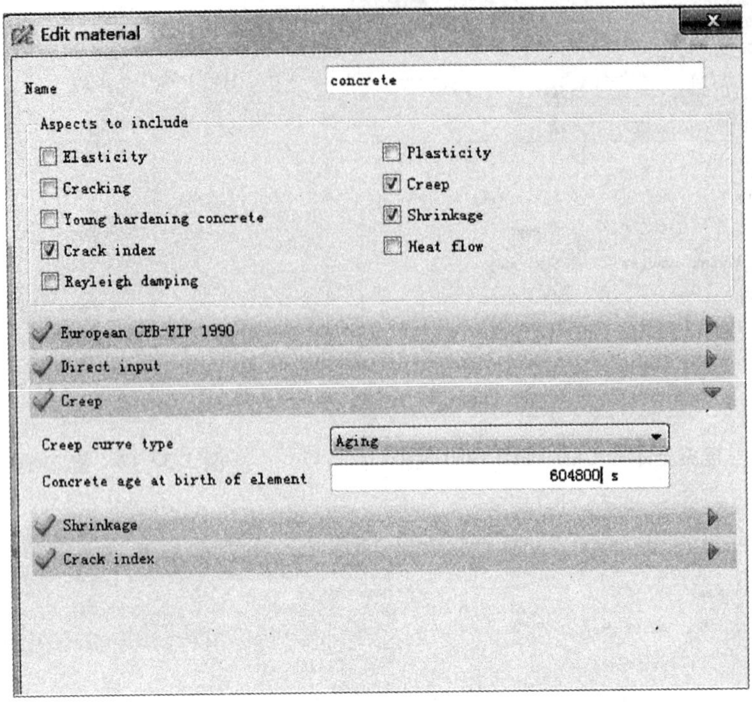

图 3.3-15 定义混装土单元龄参数

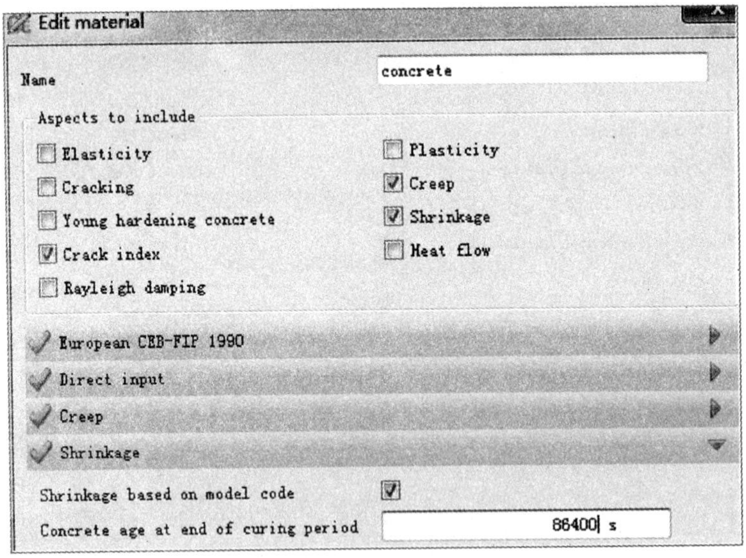

图 3.3-16 定义收缩固化养护时间

点击添加截面几何特性。命名为 concrete，选择工字形截面，输入整个工字型截面高度（Height）为 1.8 m，上翼缘宽度（Width of top flange）为 0.5 m，下翼缘宽度（Width of bottom flange）为 0.8 m。上翼缘厚度（Thickness of topflange）为 0.3 m，下翼缘厚度（Thickness of bottom flange）为 0.25 m，腹板厚度（Thickness of web）为 0.4 m。工字型几何截面的参数设置如图 3.3-17 所示。

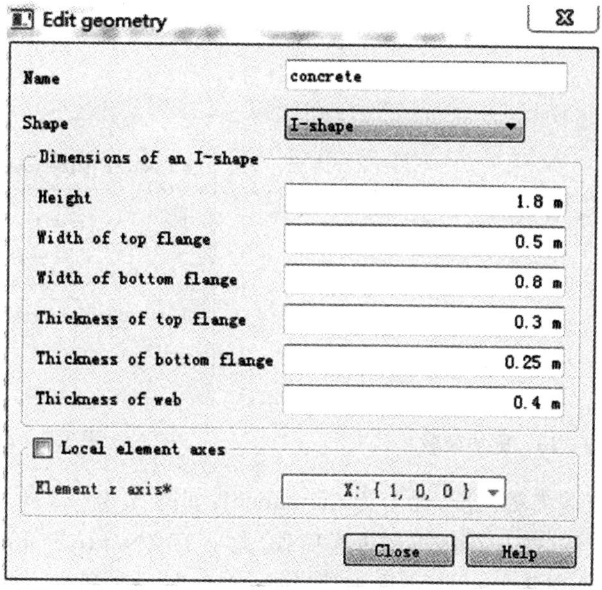

图 3.3 - 17 工字型截面几何特性参数定义界面

设置约束条件。根据整根梁的约束特点,在(0,0)点和(60,0)点施加 Y 向的位移约束,在中间节点(30,0)处施加 X 和 Y 方向的位移约束。约束施加操作分别如图 3.3-18 和图 3.3-19 所示。

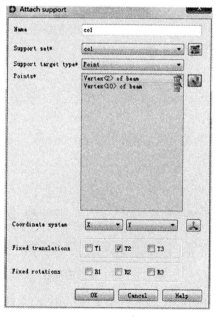

图 3.3 - 18 co1 约束施加

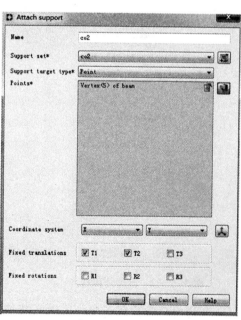

图 3.3 - 19 co2 约束施加

接下来施加荷载。首先点击 Add a new load,施加重力荷载工况,命名为 gravity,荷载类型选择恒载。如图 3.3-20、图 3.3-21 所示。

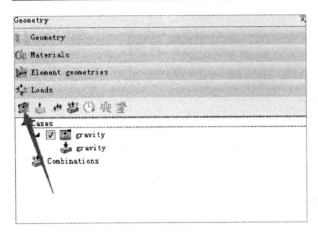
图 3.3-20 施加荷载

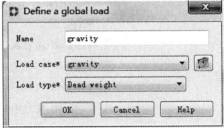
图 3.3-21 重力荷载

接下来施加外荷载荷载工况,命名为 permanent,如图 3.3-22 所示。框选中 beam,选择加载目标类型为直线,施加荷载类型为线荷载,大小 10 kN/m,方向为 Y 轴负向。用同样的方式施加活荷载,命名为 live,大小为 15 kN/m,方向为 Y 轴负向,如图 3.3-23 所示。

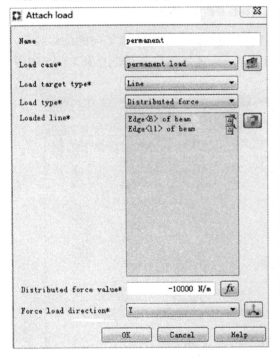
图 3.3-22 施加外荷载

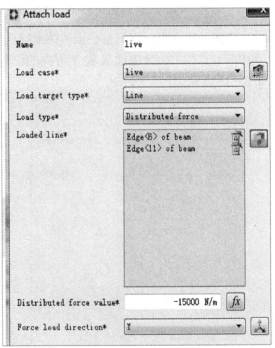
图 3.3-23 施加活荷载

添加预应力荷载荷载工况。采用两端张拉方式(Both ends),命名为 postte,在两端施加预应力荷载,大小均为 4.4×10^6 N,两端锚固段回缩长度为 0.01 m,后张法采用 CEB-FIP1990 欧洲规范,库仑摩擦系数为 0.22,握裹系数为 0.01/m。如图 3.3-24 所示。

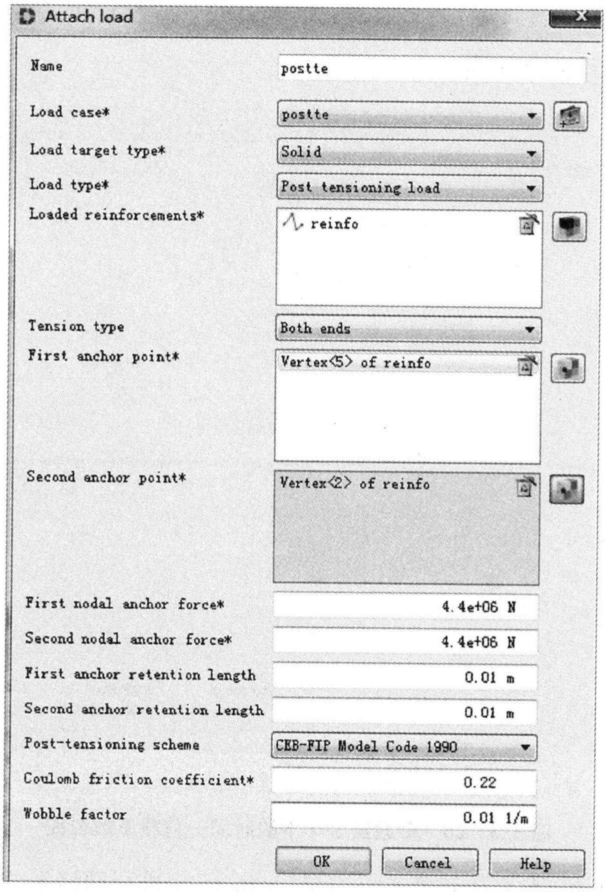

第三章 Diana 土木工程非线性建模案例分析

图 3.3 - 24 施加预应力荷载

创建荷载组合。将重力荷载和预应力荷载分成一组,将外荷载恒载和活载归为一组。如图 3.3 - 25 所示。

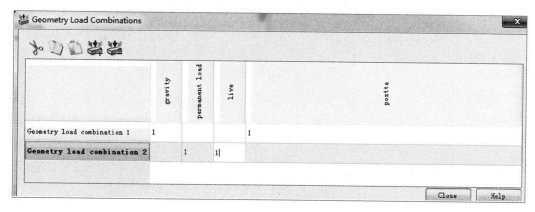

图 3.3 - 25 创建荷载组合

对每一个荷载组合均设置 Time dependency 时变特性参数,取时间(Time)为自变量,荷载系数(Factor)为因变量,取一年(31536000 s)作为总的收缩徐变时间上限,时间—荷载系数曲线坐标设置的时间点分别为开始、28 天、半年、一年这四个关键时间点。即(0,1),

(2419200,1),(15768000,1),(31536000,1),绘制时间—荷载系数关系曲线,其中荷载组合1下的时间—荷载系数曲线如图 3.3-26 所示。

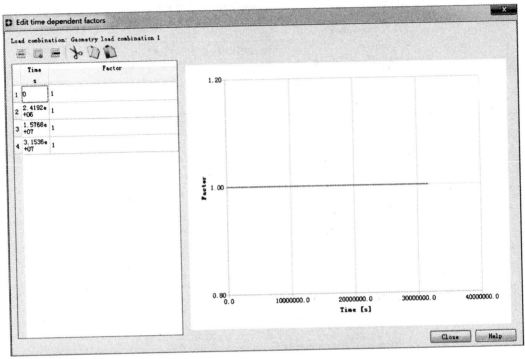

图 3.3-26　荷载组合 1 下的时间—荷载系数曲线

框选选中梁单元模型,右击选择 Set mesh properties 进行网格划分设置,采用 Divisions 份数划分方式进行划分,份数为 15 份,如图 3.3-27 所示。点击快捷键图标 Generate mesh of a shape 按钮,即可生成网格,如图 3.3-28 所示。

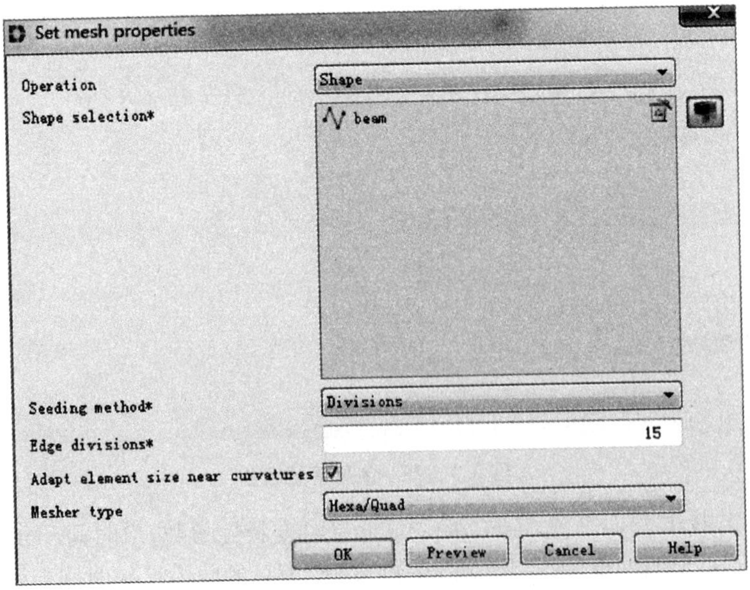

图 3.3-27　网格划分设置

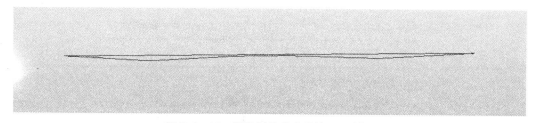

图 3.3-28　网格划分成功后的 GUI 图形

点击 mesh 下方的单元类型栏,查看并确认网格已经划分成需要的单元 L7BEN。

如图 3.3-29 所示,点击 Add an analysis 添加分析模块,生成计算分析模块。点击 Add command 添加计算控制命令文件,选择非线性分析,添加初始计算荷载模块,命名为 postte,并且添加物理非线性属性,嵌入的预应力筋单元采用完全粘结(Fully bonded)方式,如图 3.3-30 所示。

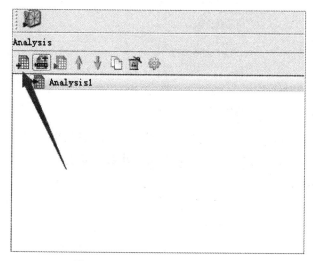

图 3.3-29　Add an analysis 图标

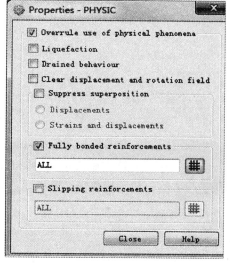

图 3.3-30　添加计算控制命令文件

初始计算荷载的荷载工况选择荷载组合 1,荷载步数设置为 1 步,荷载步长因子(User specified sizes)设置为 1。如图 3.3-31 所示。选择常规的牛顿—拉夫森迭代计算方法,最大迭代步数为 20 步,如图 3.3-31 和 3.3-32 所示。

图 3.3-31　初始计算荷载设置

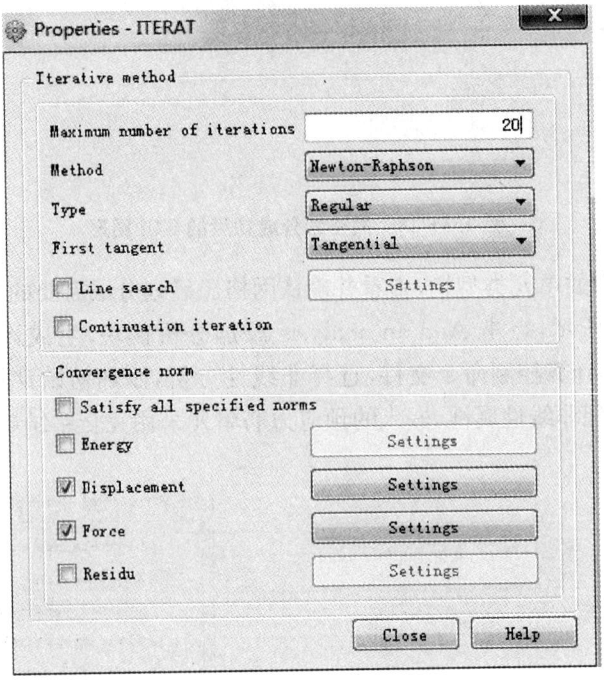

图 3.3-32 迭代计算设置

采用同样的方式添加荷载步,设置为荷载组合 2,荷载步设置为 1 步,荷载总步长因子设置为 1,表示一次性施加定义的所有荷载,迭代方程模块中最大迭代步数设定为 20 步,采用迭代方法与初始荷载步设置相同。如图 3.3-33 所示。

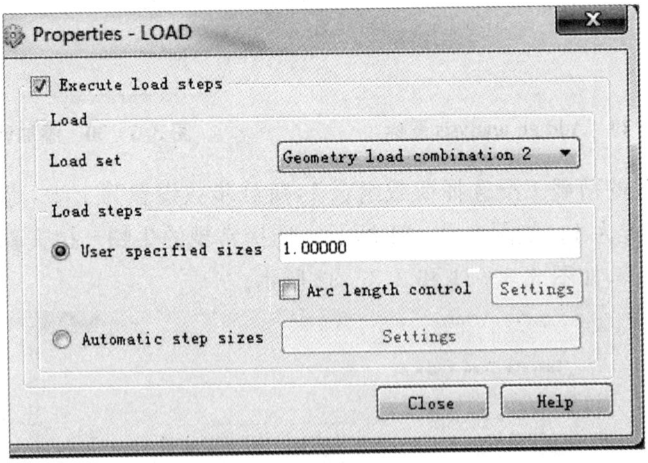

图 3.3-33 荷载组合 2 设置

添加时间荷载步,命名为 creep and shrinkage,最大迭代步数设定为 50 次。用户自定义时间荷载步对应的各时间点和时间子步长分别设置为第 28 天(2419200 s)、第 28 天到半年(13348000 s)、半年到一年(15768000 s)。如图 3.3-34 所示。

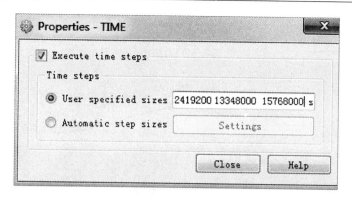

图 3.3-34 添加时间荷载步

点击 Run analysis 计算,生成应力云图,如图 3.3-35—图 3.3-38 所示。施加初始预应力筋,会出现明显反拱现象,随着长期荷载的施加,跨中挠度会有明显下降。

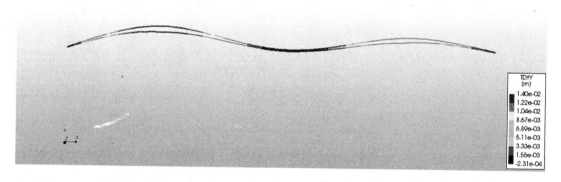

图 3.3-35 初始施加预应力荷载 Y 方向位移图,此时各跨出现反拱现象

由 28 天位移云图可以看出,在时变效应及荷载影响下挠度开始下降,反拱消失。

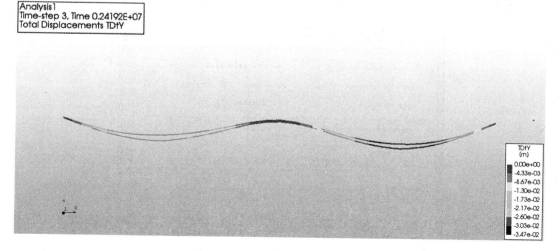

图 3.3-36 28 天后 Y 方向位移图

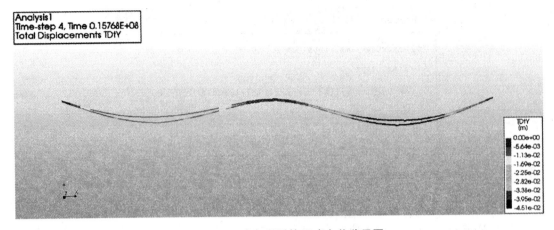

图 3.3-37　半年以后的 Y 方向位移云图

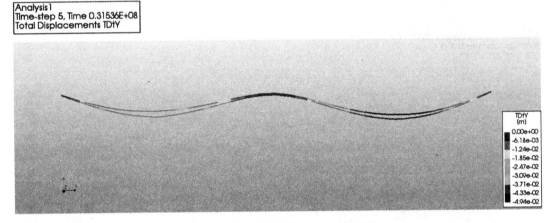

图 3.3-38　一年以后 Y 方向位移云图

为了更好地察看跨中位移,选择查看 23 号节点处的位移。点击快捷菜单栏 Viewer→选择 Node selection→选中节点,右击 Show id 显示器节点编号→点击 TDty 选择 Y 向的位移→右击选择 Show table,查看在收缩徐变等时变作用下该节点处的位移。如图 3.3-39—图 3.3-42 所示。

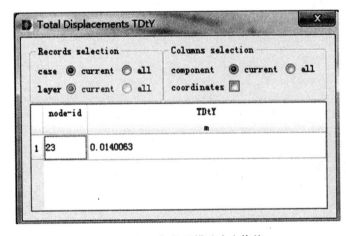

图 3.3-39　初始反拱跨中上挠值

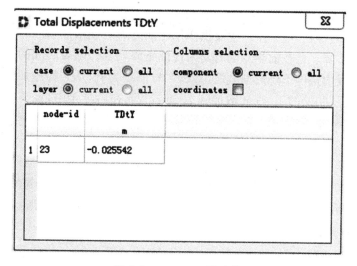

图 3.3-40 28 天挠度值

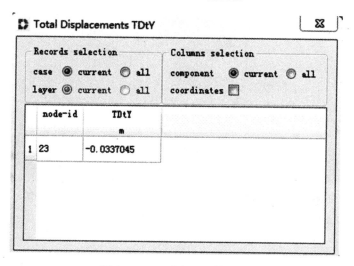

图 3.3-41 半年以后挠度值

图 3.3-42 一年以后挠度值

从上述 Diana 时变效应下 Y 方向位移值变化过程可以看出，在初始预应力筋张拉力作用下，由于预压力的存在，整个梁产生反拱作用。而随着时间增加，在收缩徐变的时变作用下整个梁的挠度下降。且在前半年的时间内挠度下降明显。随着时间的不断累积，在半年到一年这个时间段内挠度下降逐渐减缓。以 23 号节点为例，从开始施加预应力至第 28 天时变作用下，挠度从初始的 1.4 cm 增加到 2.55 cm，而在后半年中，该节点的挠度从 2.55 cm 增加至 3.37 cm 最终增加到 3.7 cm。Diana 整个数值模拟过程基本符合收缩徐变变化规律。

为了更进一步地验证和观察收缩徐变的时变效应，选取曲线预应力筋预应力损失作为研究对象，探究在时间效应为一年的时变作用下的预应力损失。其中施加初始预应力的云图如图 3.3-43 所示，半年及一年后的云图如图 3.3-44、图 3.3-45 所示。

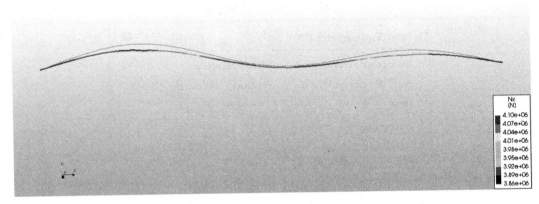

图 3.3-43　施加初始预应力下的曲线预应力筋预应力云图

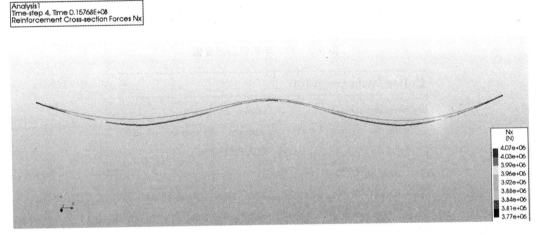

图 3.3-44　半年后的曲线预应力筋预应力云图

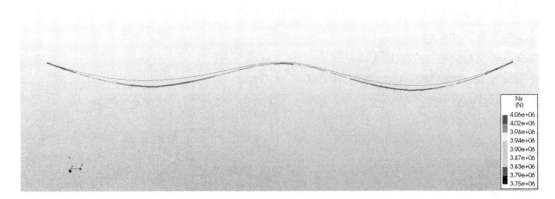

图 3.3-45　一年后的曲线预应力筋预应力云图

选取中部的 73 号节点作为研究对象。点击菜单栏 Viewer→Node selection,框选 73 号节点,再点击 Result 栏下方的 Output→Reinforcement Results→Reinforcement Cross-Section Force→N_x,右击 N_x 选择 Show table,提取各个时间节点(初始、28 天、半年、一年)对应的预应力值,采用 Origin 软件绘制预应力随时间衰减的曲线图,如图 3.3-46 所示。

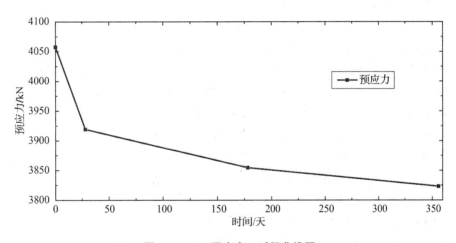

图 3.3-46　预应力—时间曲线图

由曲线图可以看出,在时变效应下,预应力筋的预应力损失速度在一开始会较快,随着时间增加,预应力筋预应力损失逐渐减缓,该趋势进一步验证了 Diana 可以有效模拟收缩徐变的机理和效果。

Python 文件生成的操作记录命令流如下:
＃＃

＃ DianaIE 10.1 update 2017-04-25 13:38:53

＃ Python 3.3.4

＃ Session recorded at 2017-06-20 16:36:37

＃＃

```
###########################
    newProject( "beam", 1000 )
    setModelAnalysisAspects( [ "STRUCT" ] )
    setModelDimension( "2D" )
    setDefaultMeshOrder( "LINEAR" )
    setDefaultMesherType( "HEXQUAD" )
    createPolyline( "beam", [[ 0, 0, 0 ],[ 30, 0, 0 ],[ 60, 0, 0 ]], False )
    saveProject( )
    createCurve( "concrete", [[ 0, 0, 0 ],[ 11.65, -0.5, 0 ],[ 29.1, 0.6217, 0 ],
[ 30, 0.68, 0 ],[ 30.9, 0.6217, 0 ],[ 47.45, -0.5, 0 ],[ 60, 0, 0 ]] )
    saveProject( )
    renameShape( "concrete", "reinfo" )
    addMaterial( "reinfo", "REINFO", "VMISES", [ ] )
    setParameter( "MATERIAL", "reinfo", "LINEAR /ELASTI /YOUNG", 1.95e+11 )
    setParameter( "MATERIAL", "reinfo", "PLASTI /HARDI1 /YLDSTR", 1 )
    setParameter( "MATERIAL", "reinfo", "PLASTI /HARDI1 /YLDSTR", 1 )
    remove( "MATERIAL", [ "reinfo" ] )
    addMaterial( "reinfo", "REINFO", "VMISES", [ ] )
    setParameter( "MATERIAL", "reinfo", "LINEAR /ELASTI /YOUNG", 1.95e+11 )
    setParameter( "MATERIAL", "reinfo", "PLASTI /HARDI1 /YLDSTR", 1.86e+09 )
    addGeometry( "Element geometry 1", "RELINE", "REBAR", [ ] )
    rename( "GEOMET", "Element geometry 1", "reinfo" )
    setParameter( "GEOMET", "reinfo", "REIEMB /CROSSE", 0.002886 )
    setReinforcementAspects( [ "reinfo" ] )
    assignMaterial( "reinfo", "SHAPE", [ "reinfo" ] )
    assignGeometry( "reinfo", "SHAPE", [ "reinfo" ] )
    resetElementData( "SHAPE", [ "reinfo" ] )
    setReinforcementDiscretization( [ "reinfo" ], "SECTION" )
    saveProject( )
    addMaterial( "concrete", "CONCDC", "MC1990", [ "CREEP", "CRKIDX", "SHRINK" ] )
    setParameter( "MATERIAL", "concrete", "MC90CO /GRADE", "C55" )
    setUnit( "TEMPER", "CELSIU" )
    saveProject( )
    setParameter( "MATERIAL", "concrete", "MC90CO /H", 0.58 )
    setParameter( "MATERIAL", "concrete", "MC90CO /RH", 69 )
    setParameter( "MATERIAL", "concrete", "CONCDI /YOUNG", 3.975e+10 )
    setParameter( "MATERIAL", "concrete", "CONCDI /YOUN28", 3.5102e+10 )
    setParameter( "MATERIAL", "concrete", "CONCDI /POISON", 0.3 )
    setParameter( "MATERIAL", "concrete", "CONCDI /THERMX", 1e-05 )
```

```
setParameter( "MATERIAL", "concrete", "CONCDI /DENSIT", 2500 )
setParameter( "MATERIAL", "concrete", "CONCDI /FCK28", 55000000 )
setParameter( "MATERIAL", "concrete", "CONCDI /FCM28", 63000000 )
setParameter( "MATERIAL", "concrete", "CONCCP /AGETYP", "AGING" )
setParameter( "MATERIAL", "concrete", "CONCCP /AGING", 604800 )
setParameter( "MATERIAL", "concrete", "CONCSH /CURAGE", 86400 )
addGeometry( "Element geometry 2", "LINE", "CLS2B2", [] )
rename( "GEOMET", "Element geometry 2", "concrete" )
setParameter( "GEOMET", "concrete", "SHAPE /BESHAP", "ISHAPE" )
setParameter( "GEOMET", "concrete", "SHAPE /ISHAPE", [ 1.8, 0.5, 0.8, 0.3,
0.25, 0.4 ] )
clearReinforcementAspects( [ "beam" ] )
setElementClassType( "SHAPE", [ "beam" ], "CLS2B2" )
assignMaterial( "concrete", "SHAPE", [ "beam" ] )
assignGeometry( "concrete", "SHAPE", [ "beam" ] )
resetElementData( "SHAPE", [ "beam" ] )
setParameter( "MATERIAL", "concrete", "MC90CO /CEMTYP", "RS" )
setParameter( "MATERIAL", "concrete", "MC90CO /CEMTYP", "NR" )
addSet( "GEOMETRYSUPPORTSET", "co1" )
createPointSupport( "co1", "co1" )
setParameter( "GEOMETRYSUPPORT", "co1", "AXES", [ 1, 2 ] )
setParameter( "GEOMETRYSUPPORT", "co1", "TRANSL", [ 0, 1, 0 ] )
setParameter( "GEOMETRYSUPPORT", "co1", "ROTATI", [ 0, 0, 0 ] )
attach( "GEOMETRYSUPPORT", "co1", "beam", [[ 0, 0, 0 ],[ 60, 0, 0 ]] )
createPointSupport( "co2", "co2" )
setParameter( "GEOMETRYSUPPORT", "co2", "AXES", [ 1, 2 ] )
setParameter( "GEOMETRYSUPPORT", "co2", "TRANSL", [ 1, 1, 0 ] )
setParameter( "GEOMETRYSUPPORT", "co2", "ROTATI", [ 0, 0, 0 ] )
attach( "GEOMETRYSUPPORT", "co2", "beam", [[ 30, 0, 0 ]] )
saveProject(   )
addSet( "GEOMETRYLOADSET", "Geometry load case 1" )
remove( "GEOMETRYLOADSET", [ "Geometry load case 1" ] )
addSet( "GEOMETRYLOADSET", "gravity" )
createModelLoad( "gravity", "gravity" )
saveProject(   )
addSet( "GEOMETRYLOADSET", "Geometry load case 2" )
rename( "GEOMETRYLOADSET", "Geometry load case 2", "permanent" )
createLineLoad( "permanent", "permanent" )
setParameter( "GEOMETRYLOAD", "permanent", "FORCE /VALUE", -10000 )
```

```
setParameter( "GEOMETRYLOAD", "permanent", "FORCE/DIRECT", 2 )
attach( "GEOMETRYLOAD", "permanent", "beam", [[ 15, 0, 0 ],[ 45, 0, 0 ]] )
saveProject( )
addSet( "GEOMETRYLOADSET", "Geometry load case 3" )
rename( "GEOMETRYLOADSET", "Geometry load case 3", "live" )
createLineLoad( "live", "live" )
setParameter( "GEOMETRYLOAD", "live", "FORCE/VALUE", -15000 )
setParameter( "GEOMETRYLOAD", "live", "FORCE/DIRECT", 2 )
attach( "GEOMETRYLOAD", "live", "beam", [[ 15, 0, 0 ],[ 45, 0, 0 ]] )
saveProject( )
addSet( "GEOMETRYLOADSET", "Geometry load case 4" )
rename( "GEOMETRYLOADSET", "Geometry load case 4", "postte" )
createBodyLoad( "postte", "postte" )
setParameter( "GEOMETRYLOAD", "postte", "LODTYP", "POSTEN" )
setParameter( "GEOMETRYLOAD", "postte", "POSTEN/BOTHEN/FORCE1", 4400000 )
setParameter( "GEOMETRYLOAD", "postte", "POSTEN/BOTHEN/FORCE2", 4400000 )
setParameter( "GEOMETRYLOAD", "postte", "POSTEN/BOTHEN/RETLE1", 0.01 )
setParameter( "GEOMETRYLOAD", "postte", "POSTEN/BOTHEN/RETLE2", 0.01 )
setParameter( "GEOMETRYLOAD", "postte", "POSTEN/SHEAR", 0.22 )
setParameter( "GEOMETRYLOAD", "postte", "POSTEN/WOBBLE", 0.01 )
attach( "GEOMETRYLOAD", "postte", [ "reinfo" ] )
attachTo( "GEOMETRYLOAD", "postte", "POSTEN/BOTHEN/PNTS1", "reinfo", [[ 0, 0, 0 ]] )
attachTo( "GEOMETRYLOAD", "postte", "POSTEN/BOTHEN/PNTS2", "reinfo", [[ 60, 0, 0 ]] )
setDefaultGeometryLoadCombinations( )
setGeometryLoadCombinationFactor( "Geometry load combination 1", "gravity", 1 )
remove( "GEOMETRYLOADCOMBINATION", "Geometry load combination 1" )
remove( "GEOMETRYLOADCOMBINATION", "Geometry load combination 2" )
remove( "GEOMETRYLOADCOMBINATION", "Geometry load combination 3" )
remove( "GEOMETRYLOADCOMBINATION", "Geometry load combination 4" )
addGeometryLoadCombination( "" )
setGeometryLoadCombinationFactor( "Geometry load combination 1", "gravity", 1 )
setGeometryLoadCombinationFactor( "Geometry load combination 1", "postte", 1 )
addGeometryLoadCombination( "" )
setGeometryLoadCombinationFactor( "Geometry load combination 2", "permanent", 1 )
setGeometryLoadCombinationFactor( "Geometry load combination 2", "live", 1 )
setTimeDependentLoadFactors ( " GEOMETRYLOADCOMBINATION ", " Geometry load combination 2", [ 0, 2419200, 15768000, 31536000 ], [ 1, 1, 1, 1 ] )
```

```
    setTimeDependentLoadFactors( " GEOMETRYLOADCOMBINATION", " Geometry load
combination 1", [ 0, 2419200, 15768000, 31536000 ], [ 1, 1, 1, 1 ] )
    generateMesh( [] )
    hide( "SHAPE", [ "reinfo" ] )
    setEdgeMeshSeed( [ "beam" ], 15 )
    setMesherType( [ "beam" ], "HEXQUAD" )
    saveProject(  )
    generateMesh( [] )
    hideView( "GEOM" )
    showView( "MESH" )
    addAnalysis( "Analysis1" )
    addAnalysisCommand( "Analysis1", "NONLIN", "Structural nonlinear" )
    renameAnalysis( "Analysis1", "Analysis1" )
    renameAnalysisCommand ( " Analysis1", " Structural nonlinear", " Structural
nonlinear" )
    addAnalysisCommandDetail ( " Analysis1", " Structural nonlinear", " MODEL /
EVALUA /REINFO /INTERF" )
    setAnalysisCommandDetail ( " Analysis1", " Structural nonlinear", " MODEL /
EVALUA /REINFO /INTERF", True )
    removeAnalysisCommandDetail( "Analysis1", "Structural nonlinear", "EXECUT
(1)" )
    setAnalysisCommandDetail ( " Analysis1", " Structural nonlinear", " EXECUT /
EXETYP", "START" )
    renameAnalysisCommandDetail( "Analysis1", "Structural nonlinear", "EXECUT
(1)", "postte" )
    addAnalysisCommandDetail( "Analysis1", "Structural nonlinear", "EXECUT(1) /
START /INITIA /STRESS" )
    setAnalysisCommandDetail( "Analysis1", "Structural nonlinear", "EXECUT(1) /
START /INITIA /STRESS", True )
    addAnalysisCommandDetail( "Analysis1", "Structural nonlinear", "EXECUT(1) /
PHYSIC" )
    setAnalysisCommandDetail( "Analysis1", "Structural nonlinear", "EXECUT(1) /
PHYSIC /LIQUEF", False )
    setAnalysisCommandDetail( "Analysis1", "Structural nonlinear", "EXECUT(1) /
PHYSIC /BOND", True )
    setAnalysisCommandDetail( "Analysis1", "Structural nonlinear", "EXECUT(1) /
ITERAT /MAXITE", 20 )
    setAnalysisCommandDetail ( " Analysis1", " Structural nonlinear", " EXECUT /
EXETYP", "LOAD" )
```

```
    renameAnalysisCommandDetail( "Analysis1", "Structural nonlinear", "EXECUT
(2)", "load" )
    setAnalysisCommandDetail( "Analysis1", "Structural nonlinear", "EXECUT(2) /
LOAD /LOADNR", 1 )
    setAnalysisCommandDetail( "Analysis1", "Structural nonlinear", "EXECUT(2) /
ITERAT /MAXITE", 20 )
    setAnalysisCommandDetail( "Analysis1", "Structural nonlinear", "EXECUT /
EXETYP", "TIME" )
    renameAnalysisCommandDetail( "Analysis1", "Structural nonlinear", "EXECUT
(3)", "creep and shrinkage" )
    setAnalysisCommandDetail( "Analysis1", "Structural nonlinear", "EXECUT(3) /
TIME /STEPS /EXPLIC /SIZES", "2419200 13348800 15768000" )
    setAnalysisCommandDetail( "Analysis1", "Structural nonlinear", "EXECUT(3) /
ITERAT /MAXITE", 50 )
    saveProject(  )
    runSolver( "Analysis1" )
    showView( "RESULT" )
```

3.4 案例四：平面单元钢筋混凝土梁的裂缝分析

本节案例模型为作者正在修改的《Probabilistic finite element reliability analysis based on the Least Angle Regression algorithm》一文中的Diana9.4版建模的钢筋混凝土简支梁可靠度计算算例。这里采用二次平面应力单元来模拟混凝土,用嵌入式的钢筋单元来模拟钢筋。由于混凝土的单元类型选择平面应力单元,因此箍筋和纵筋采用单根 Bar 钢筋单元来建模。采用弥散裂缝模型中的总应变裂缝模型模拟混凝土梁的开裂。整根梁长为4 m,高度为0.45 m,承受对称荷载作用,因此考虑采用如图3.4-1所示半结构模型,施加向下的竖向集中荷载总计为48 kN,且加载方式为逐级加载。其中箍筋间距均为150 mm,受压区和受拉区纵筋距混凝土边缘距离均为0.05 m。纵筋采用3Φ8类型,箍筋为Φ6@150。

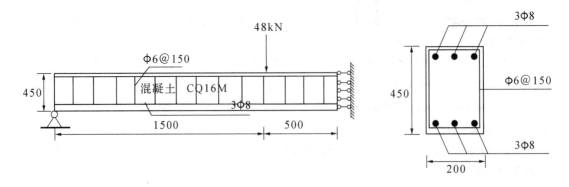

图 3.4-1 钢筋混凝土梁半结构及截面尺寸特征

混凝土及钢筋的参数设置见表 3-3。

表 3-3　混凝土及钢筋的参数设置

	混凝土	钢筋
单元类型	二次平面应力单元	Bar 单元
材料参数	泊松比 0.15 密度 2500 kg/m³ 弹性模量 $3×10^{10}$ N/m² 裂缝面剪力滞留系数 $\beta=0.01$ 拉伸软化模型：指数模型 开裂模型：混合裂缝模型 抗拉强度 $2.5×10^{6}$ N/m² 抗压强度 30 MPa	弹性模量 $2.1×10^{11}$ N/m² 屈服强度 $4.4×10^{8}$ N/m²
几何属性	4 m×0.45 m×0.2 m	等效纵筋横截面积 $1.51×10^{-4}$ m² 等效箍筋横截面积 $5.6×10^{-5}$ m²
荷载	48 kN，竖直向下	

学习要义：

（1）掌握平面应力单元的定义方式。

（2）学习总应变裂缝模型下的混凝土本构参数设置的界面操作方法。

（3）采用单根钢筋的建模方式来模拟纵筋和箍筋以及平面类单元中钢筋等效面积计算方法。

（4）掌握印刻投影（Imprint projection）施加点荷载的正确操作方法。

（5）掌握平面应力单元半对称结构约束施加方式。

（6）学习在后处理中查看各向裂缝分布云图、裂缝宽度云图和某一节点位移。

首先，启动 DianaIE，弹出 New project 对话框，选择 F 盘工作路径，在 F 盘创立一个文件夹，命名为例题，建立工作目录，将模型命名为 Quabeam，选择结构分析类型，2 维单元，模型尺寸范围为 10 m，单位设置如图 3.4-2、图 3.4-3 所示。

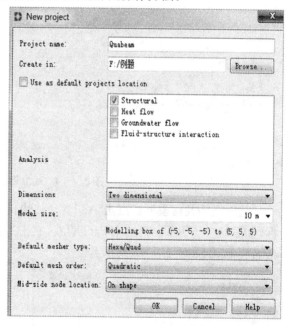

图 3.4-2　模型属性设置

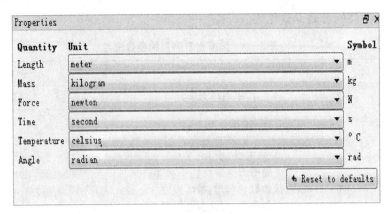

图 3.4-3 基本单位设置

建立混凝土梁几何模型。点击菜单栏上的快捷键 Add a sheet 创立平面,首先输入四个点坐标值(0,0,0),(2,0,0),(2,0.45,0),(0,0.45,0)(**注**:在 Diana10.1 的输入坐标对话框中,即使是二维坐标,每个坐标点表征平面外方向的数值 0 也不可缺少,本次模型由于是 2 维单元,并且设定在 XOY 面内建立模型,所以 Z 方向一律输入值为零。读者可根据自己喜好对构成二维平面的坐标轴进行调整)。点击 OK,生成如图 3.4-4 所示混凝土梁几何模型。

图 3.4-4 建立混凝土梁几何模型

建立钢筋几何模型。点击 Add a line 快捷键创立直线,命名为 bar1,生成直线模型坐标,如图 3.4-5 所示。再选中 bar1 直线,右击选择 Array copy,向上平移 0.35 m,生成直线,命名为 bar2,如图 3.4-6 所示。

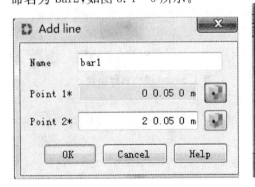

图 3.4-5 钢筋几何模型坐标

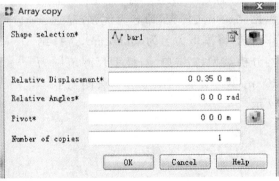

图 3.4-6 Array copy 创建 bar2

生成的模型如图 3.4-7 所示。

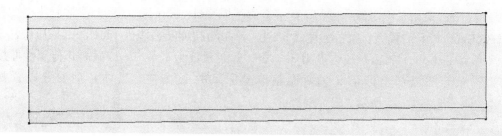

图 3.4-7　生成钢筋模型

创立箍筋,如图 3.4-8、图 3.4-9 所示,采用先创建单根直线再 Array copy 的复制平移方式创立箍筋。间距为 0.15 m,复制数量为 13 次。生成结果如图 3.4-10 所示。

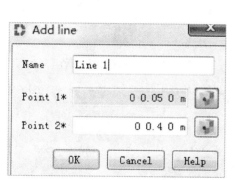

图 3.4-8　创建单根直线

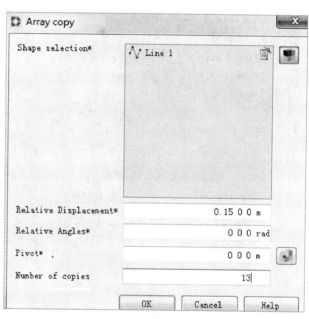

图 3.4-9　Array copy 创建箍筋

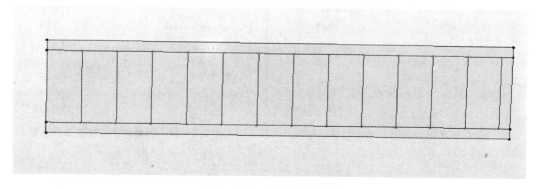

图 3.4-10　生成箍筋模型

赋予材料属性。点击选中面 sheet1,右击选择 Property assignments,赋予材料属性。

单元类型选择普通平面应力单元(Regular Plane stress),材料属性命名为concrete,混凝土选择Concrete and mansory选项,弹性模量$3\times10^{10}\,\mathrm{N/m^2}$,泊松比0.15,混凝土的密度为$2500\,\mathrm{kg/m^3}$,采用总应变裂缝模型下的混合裂缝模型(Rotating to fixed)拉伸软化模型采用指数(Exponential)模型。抗拉强度$2.5\times10^{6}\,\mathrm{N/m^2}$,拉伸断裂能为150 N/m,抗压特性采用抛物线(Parabolic)特性,抗压强度为30 MPa,剪滞系数为常数0.01,截面几何属性厚度为0.2 m,单元的局部坐标系x轴对应整体坐标系的X方向。具体各项参数设置如图3.4-11—图3.4-15所示。

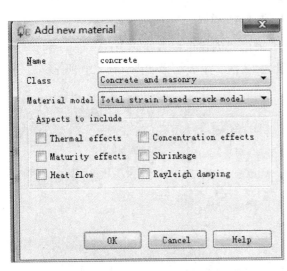

图3.4-11 添加concrete材料属性

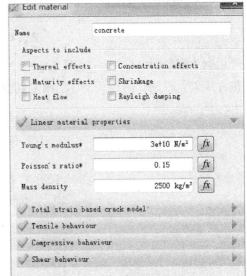

图3.4-12 编辑材料属性

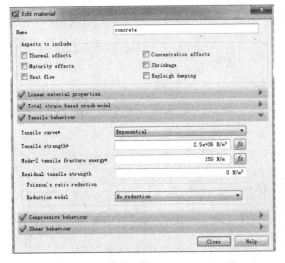

图3.4-13 拉伸软化指数模型设置

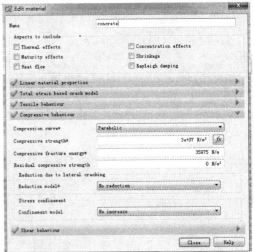

图3.4-14 抛物线抗压模型参数设置

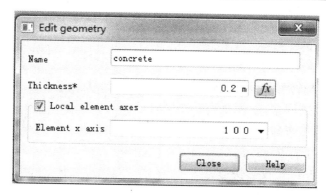

图 3.4‑15　截面几何特性编辑

赋予钢筋的材料属性和截面几何特性。选择配筋和桩基础选项,材料类型为冯·米塞斯(Von Mises)塑性屈服准则,钢筋弹性模量为 $2\times10^{11}\,\text{N/m}^2$。屈服应力为 $4.4\times10^8\,\text{N/m}^2$,由于模型中需要将三根纵筋面积折算在一起,因此纵筋的面积为 $\frac{\pi}{4}\times8^2\times3\approx151\,\text{mm}^2$。如图 3.4‑16—图 3.4‑19 所示。用同样的方式定义箍筋。命名为 Grid,面积按两层叠加,为 $\frac{\pi}{4}\times6^2\times2=56.5487\,\text{mm}^2$,由于箍筋采用的本构模型和定义方式相同,这里不再赘述。

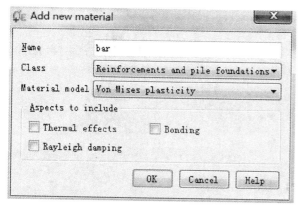

图 3.4‑16　添加钢筋类型

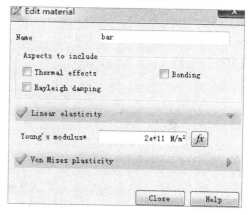

图 3.4‑17　编辑材料属性

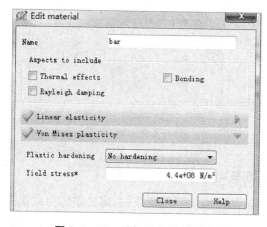

图 3.4‑18　编辑钢筋材料属性

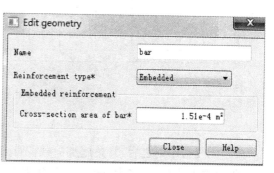

图 3.4‑19　编辑钢筋几何属性单位的几何点

由于要施加的集中荷载的坐标点并不在之前创立面的那些点上,直接创立单个几何点会在之后的单元网格划分中产生无法分配材料和几何属性的错误从而无法生成网格,故添加荷载前首先考虑采用投影印刻的方式。首先创立一个点,坐标为(1.5,0.5,0),再点击上方的 Project 投影快捷方式,弹出快捷键,操作类型选择边缘,

图 3.4-20　Project 投影图标

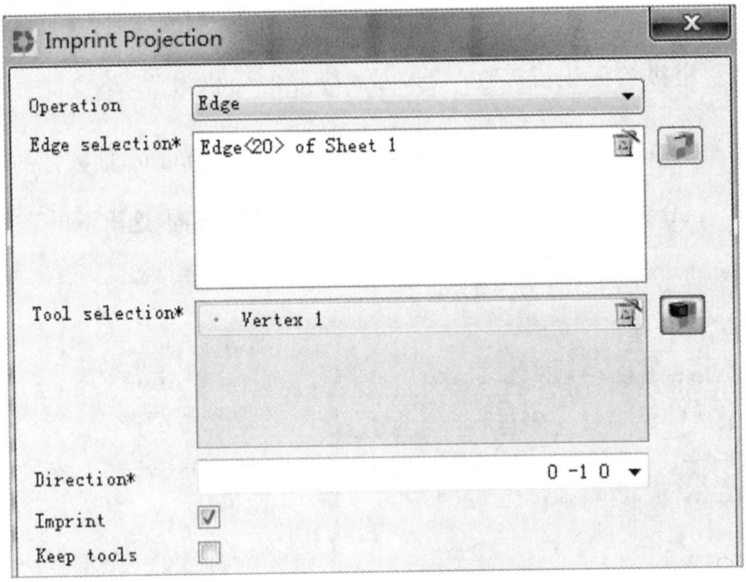

并且将集中荷载所要作用的那条线选做边缘,将创立的点选做工具,方向选择整体坐标系下 Y 轴负向,勾选印刻。这时候发现点已经投印在了面上。如图 3.4-20—图 3.4-22 所示。

图 3.4-21　操作类型设置

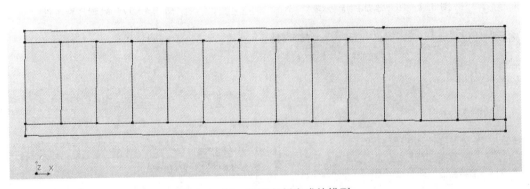

图 3.4-22　投影印刻完成的模型

完成投影印刻之后,开始施加荷载。首先点击几何模块下的荷载栏,定义整体荷载快捷键,创立重力荷载,命名为 gravity,荷载类型选择恒载,如图 3.4-23 所示。其次在已经投影印刻的点上施加集中荷载 10 kN 作为初始基准值(注意:后面的计算模块针对外荷载会

设置多个荷载步以10 kN为一级逐级加载,所以这里先施加10 kN荷载)。点击load下方的Add a new loadcase,生成新的荷载工况,右击选择Attach load,添加集中荷载,如图3.4－24、图3.4－25所示。

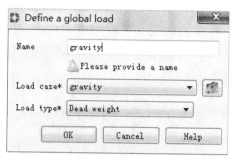

图3.4－23 创立重力荷载

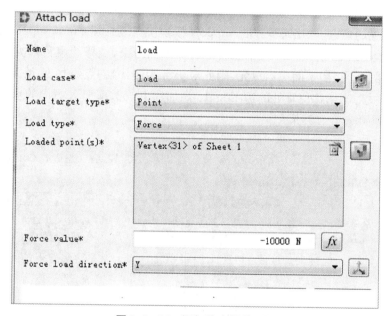

图3.4－24 添加集中荷载工况

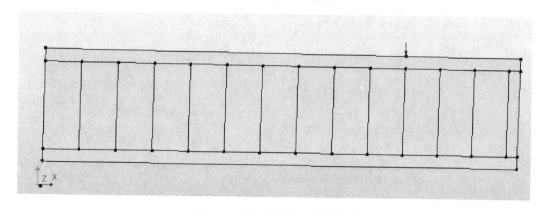

图3.4－25 荷载添加后的模型

添加荷载组合。将重力荷载设置为荷载组合1,将外荷载设置成荷载组合2,如图3.4－26所示。

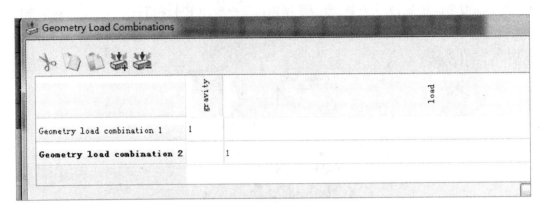

图 3.4-26 荷载组合设置

约束采用正对称半结构约束。点击快捷菜单栏 Gemetny→Analysis→Add a new support set,施加约束,点击几何下方的 Support 模块快捷键 Add a new support set,建立新的约束,命名为 co1,如图 3.4-27 所示。右击选择添加约束,选择约束目标类型为点约束。由于本节案例混凝土采用平面应力单元,因此对坐标为(0,0)的点添加 X 及 Y 方向位移约束,用同样的方式对半结构位置处的边 Edge16 添加约束 co2,施加 X 方向的水平位移约束,生成约束信息如图 3.4-28 所示。

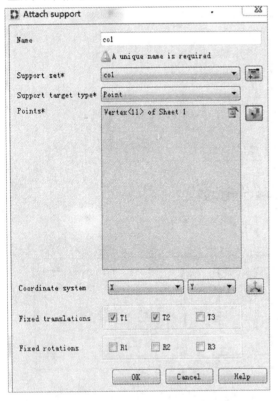

图 3.4-27 约束 1 施加方式

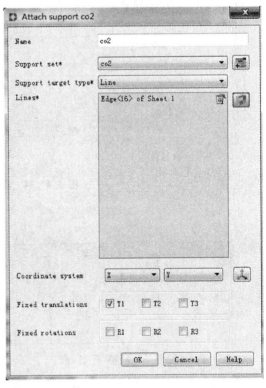

图 3.4-28 约束 2 施加方式

点击 OK,完成约束施加步骤。查看正对称半结构生成的约束情况,约束施加结果如图 3.4-29 所示。

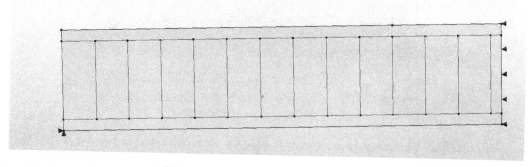

图 3.4-29　正对称半结构约束施加方式

网格划分采用单元尺寸划分，单元的大小选择为 0.1 m，网格类型为六面体/正方形，中间位置处节点采用线性插值方式，如图 3.4-30 所示。点击 OK 后，点击快捷键图标 Generate mesh of a shape 按钮，如图 3.4-31 所示。

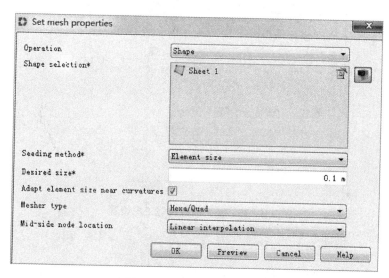

图 3.4-30　划分网络设置

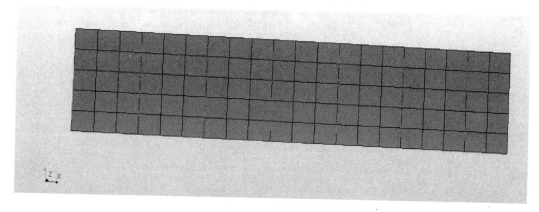

图 3.4-31　生成网格

网格划分结束后，检查网格类型（CQ16M），确认是需要的平面应力单元类型，如图 3.4-32所示。

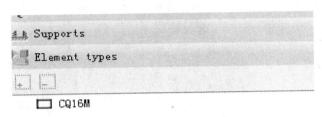

图 3.4-32 确认 CQ16M 网格类型

添加计算模块 Add an analysis,生成分析模块 1。右击选择 Add command 添加计算控制指令,在控制指令中选择结构分析类型为非线性分析,在结构非线性分析中选择添加荷载步。创立重力荷载的计算荷载步和集中外荷载的计算荷载步,其中重力荷载选择荷载组合1,设置计算步数为 1,荷载总步长为 1,选择最大迭代次数 20 次,采用位移或力收敛的收敛准则。如图 3.4-33 所示。

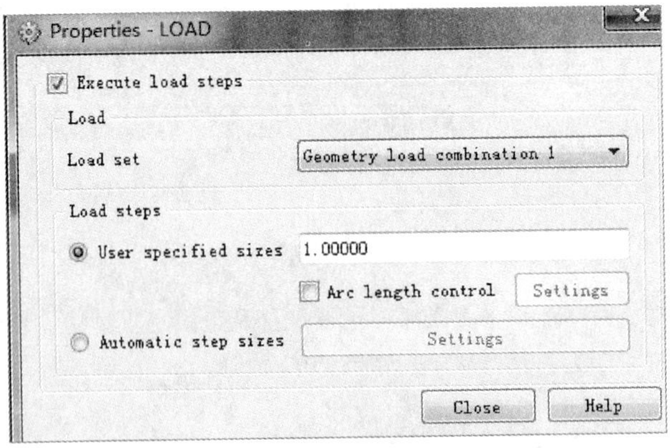

图 3.4-33 重力荷载荷载工况设置

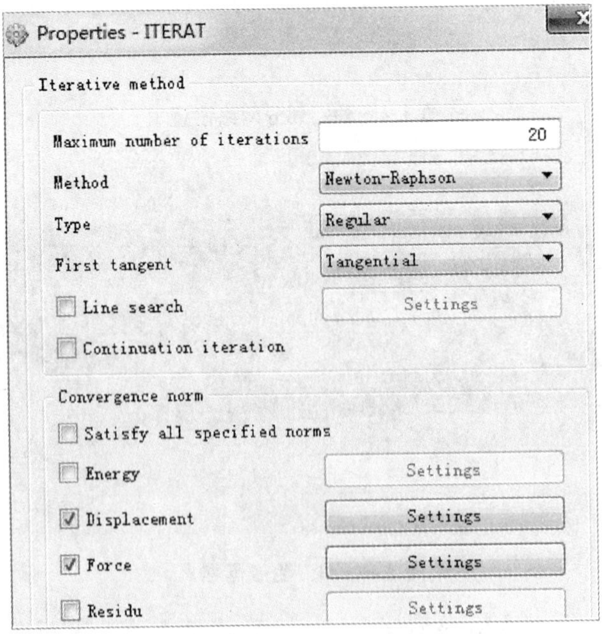

图 3.4-34 重力荷载计算模块设置

集中荷载选择荷载组合 2,考虑到非线性计算中力与位移特征,计算荷载步设置为 1.00000(4) 0.200000(4),同样选择最大迭代次数 20 次,采用位移或力收敛的收敛准则。如图 3.4-35、图 3.4-36 所示。

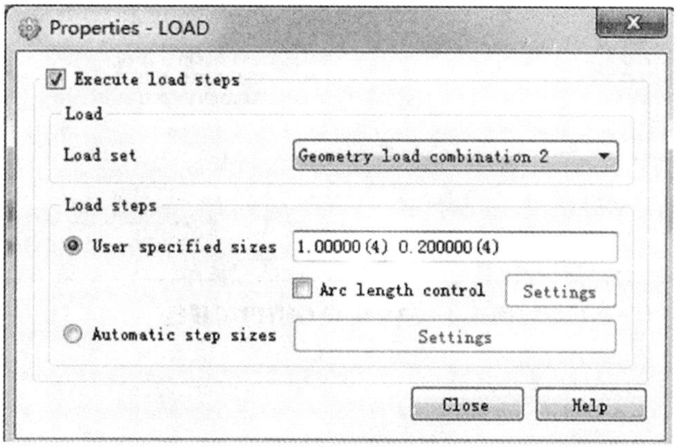

图 3.4-35 集中荷载设置

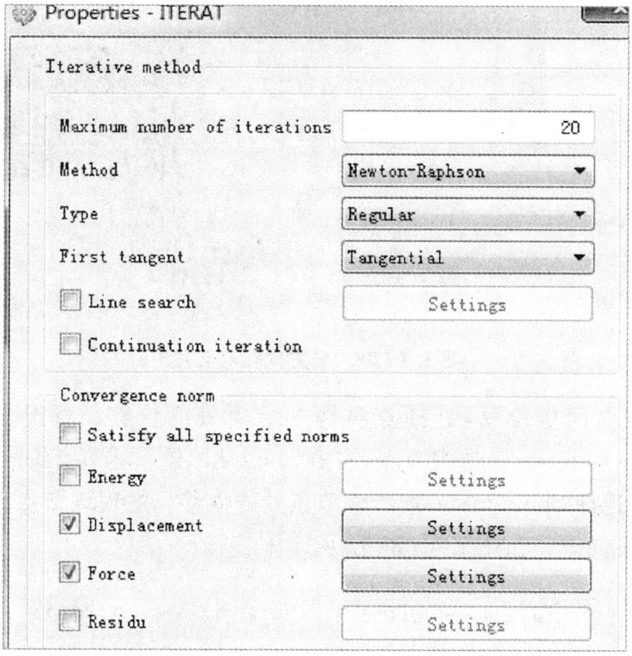

图 3.4-36 集中荷载计算模块设置

右击选择编辑 OUTPUT 属性,计算输出结果选择用户选项(User selecttion)方式,选择的各输出结果选项如图 3.4-37 所示。

计算完成后,查看后处理计算结果,选择 OUTPUT,点击整体位移下 Y 向的跨中位移(TDtY),如图 3.4-38 所示。

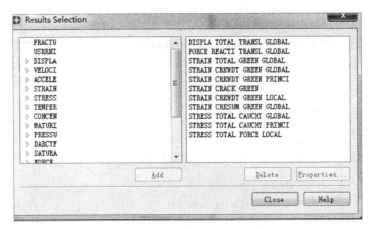

图 3.4-37　编辑 OUTPUT 属性

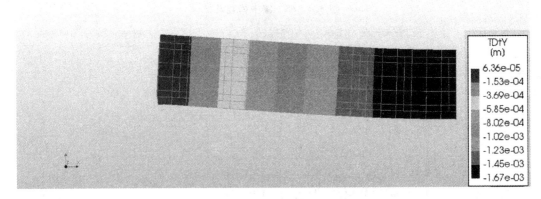

图 3.4-38　Y 方向位移云图

为了更好地查看跨中位移,选择查看跨中节点处的位移。点击上方的快捷菜单栏 Viewer→选择 node selection→选中节点,右击 Show ids 显示器节点编号→点击 TDty 选择 Y 向的位移→右击选择 Show table,查看该节点处的位移。发现该节点编号为 3,Y 方向位移为 1.66 mm,方向向下。如图 3.4-39 所示。

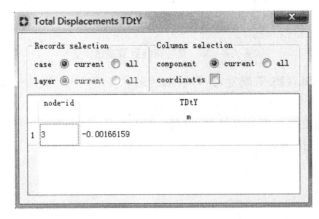

图 3.4-39　3 号节点 Y 方向位移

查看裂缝宽度和裂缝分布云图。点击荷载步到第 5 荷载步时 Output 下方 Element reults 模块,出现 Crack-widths,Crack Strains 和 Summed Crack Strains 三个与裂缝有关的后处理查看新选项,首先点击 Crack-widths 下方 EcwXX 查看局部坐标系下 x 方向单元裂缝宽度云图,如图 3.4-40 所示。

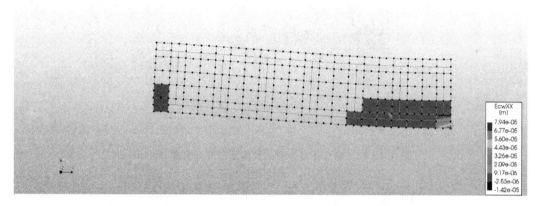

图 3.4-40　第 5 荷载步时整体坐标系下裂缝宽度云图(EcwXX)

点击荷载步至最大荷载步第 9 荷载步,查看 x 方向的单元裂缝宽度分布云图。这时候可以看到裂缝正在扩展,并且在初始开裂位置的上方产生新的裂缝,且从云图中不难看出,裂缝最大宽度位于梁的跨中部位受拉区。如图 3.4-41 所示。

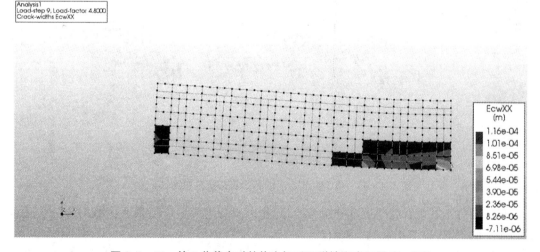

图 3.4-41　第 9 荷载步时整体坐标系下裂缝宽度云图(EcwXX)

选择第 5 荷载步,点击 Crack Strains 下方的 Eknn 查看裂缝刚产生时的累积应变分布位置图,可以看到在梁的支座处和跨中位置处的下方出现裂缝,如图 3.4-42 所示。

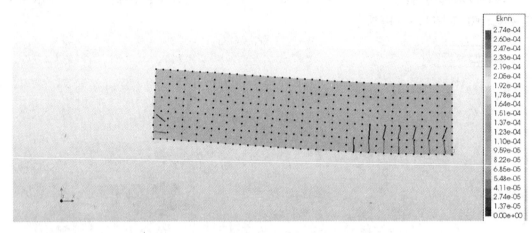

图 3.4-42　第 5 荷载步时的裂缝正应变分布图（Eknn）

选择第 7 荷载步时，这时荷载系数 load factor＝4.4，裂缝分布如图 3.4-43 所示。

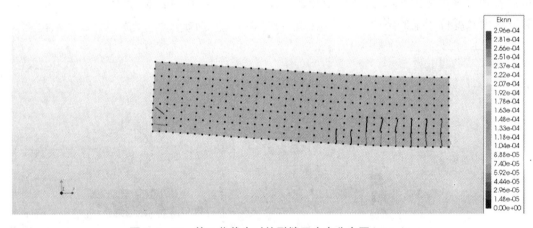

图 3.4-43　第 7 荷载步时的裂缝正应变分布图（Eknn）

最终荷载步下查看裂缝位置分布云图，此时可以看出裂缝应变最大值处对应的位置也是位于梁的下部，如图 3.4-44 所示。

有兴趣的读者可以采用增加荷载步数或增大子步长因子的方式增加荷载值，探究和查看在承载力极限状态下的简支梁开裂特征。

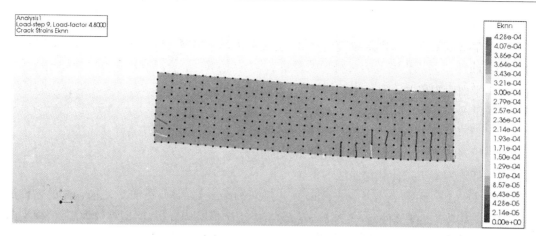

图 3.4-44 第 9 荷载步时的裂缝正应变分布图（Eknn）

3.5 案例五：门式框架

本节案例几何尺寸改编自 Diana9.4 版 Concrete modeling and analysis Tutorials and experiences 混凝土建模分析教程门式框架案例。本书另辟蹊径，在10.1版的GUI可视化界面中依次采用多向固定裂缝模型、总应变旋转裂缝模型、纤维混凝土和总应变固定裂缝模型建模，待各模型计算完毕后在加载大小相同的条件下绘制荷载—位移曲线以供读者对比。门式框架最大长度1.62 m，最大高度为 0.87 m，配筋一律采用 2Φ6 规格，框架的厚度为 0.07 m，采用 2D 平面应力单元建模。混凝土弹性模量 3.15×10^{10} N/m^2。泊松比 0.2，应变极限值（Ultimate strain）0.0011，剪滞系数 0.2。设定单个荷载步大小为 0.2 kN，荷载步数为 40 步。由于对称施加等步长节点荷载，因此在建模中考虑采用半结构进行建模，其半结构尺寸如图 3.5-1 所示。

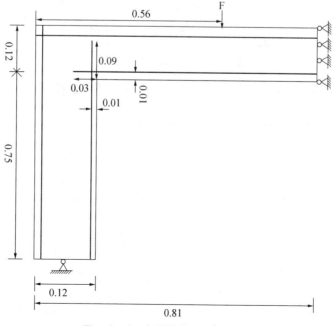

图 3.5-1 半结构模型（单位：m）

学习要义：

(1) 学习混凝土多向固定裂缝模型、总应变旋转裂缝模型、纤维混凝土和总应变固定裂缝模型各线性拉伸软化模型的材料参数设置。

(2) 平面应力单元正对称半结构施加方式。

(3) 学习投影印刻功能的使用。

(4) 学习采用 Divisions 按边长比例划分网格方法。

(5) 学习计算输出结果选项中开裂结果的查看。

(6) 纤维混凝土拉伸软化曲线的设置。

打开 DianaIE 界面，点击 New→File，弹出 New project 对话框，在 D 盘之前创立的名称为例题文件夹中创立 dpf 文件，命名为 Frame，分析类型选择结构分析，模型维数为二维，模型的尺寸范围选择 10 m，默认划分类型六面体/四边形单元，划分次数为二次，如图 3.5-2 所示。

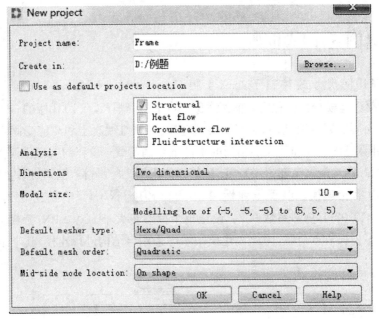

图 3.5-2　创建新模型窗口

设置国际单位(长度：米，质量：千克，力：牛顿，时间：秒，温度：摄氏度，角度：角)，如图 3.5-3 所示。

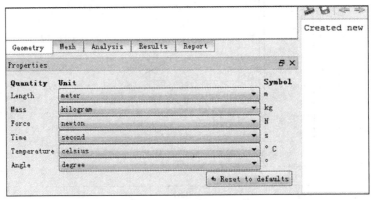

图 3.5-3　模型单位设置

建立几何模型。输入各点坐标(见表 3-4)。点击 OK,生成半结构几何模型,如图 3.5-4 所示。

表 3-4　几何模型各点坐标

P1	(0,0)
P2	(0.06,0)
P3	(0.12,0)
P4	(0.12 0.75)
P5	(0.81,0.75)
P6	(0.81,0.87)
P7	(0,0.87)

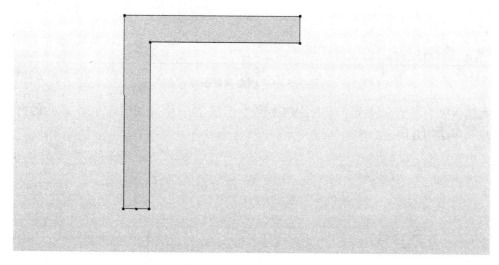

图 3.5-4　半结构几何模型

建立钢筋单元。本例中采用嵌入式钢筋 Bar 单元建模,如图 3.5-5 所示。点击快捷工具栏选择 Add a line 创立钢筋,各钢筋坐标位置见表 3-5。以 bar1 为例,生成的几何模型如图 3.5-6 所示。

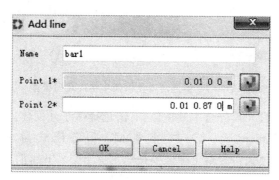

图 3.5-5　建立钢筋单元

表 3-5　各钢筋坐标位置

bar1	(0.01, 0), (0.87, 0)
bar2	(0.11, 0), (0.11, 0.84)
bar3	(0.09, 0.76), (0.81, 0.76)
bar4	(0, 0.86), (0.81, 0.86)

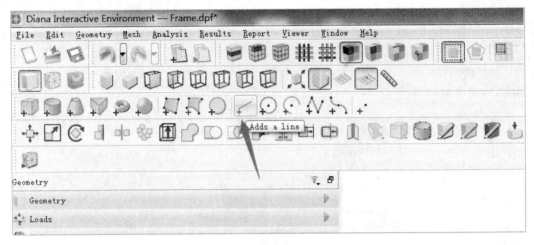

图 3.5-6　创立钢筋图标

每输入完一次坐标后,点击 OK,四根钢筋全部建立后,生成如图 3.5-7 所示的钢筋混凝土框架半结构几何模型。

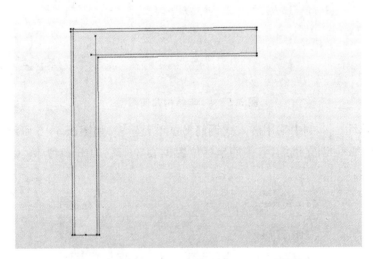

图 3.5-7　钢筋混凝土框架半结构几何模型

赋予几何体材料属性。首先选中几何体,右击选择赋予几何体属性,材料类型为混凝土和砌体(Concrete and mansory)结构材料类型。裂缝模式为多向固定裂缝(Multi-directional fixed crack)模式。混凝土弹性模量 $3.15 \times 10^{10} \mathrm{N/m^2}$,泊松比 0.2,多向固定裂缝模式下的拉伸截断(Tension cut-off)为常数,拉伸软化曲线选择基于极限应变下的线性拉伸软化模型(Linear based on ultimate strain)，抗拉强度 2.2 MPa，应变极限值(Ultimate strain)为 0.0011，剪滞系数为 0.2。如图 3.5-8—图 3.5-10 所示。

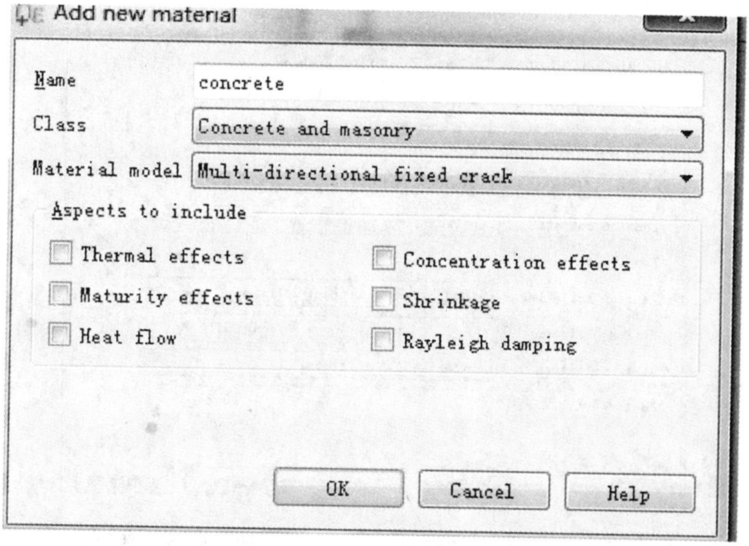

图 3.5-8　多向固定裂缝本构模型

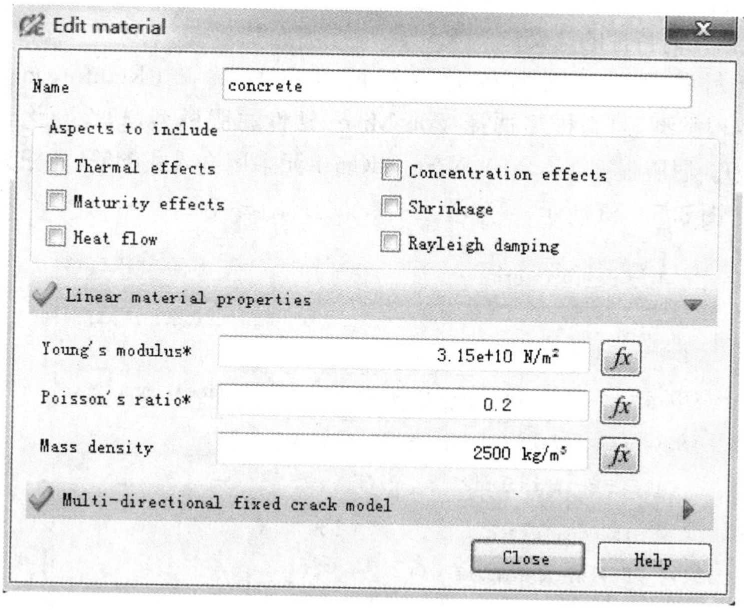

图 3.5-9　多向固定裂缝模型下混凝土基本参数设置

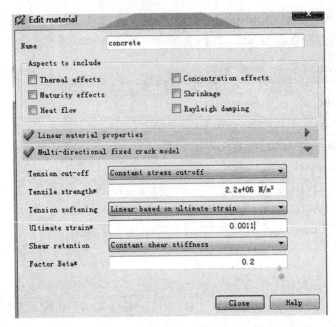

图 3.5-10　多向固定裂缝模型线性拉伸软化曲线参数设置

点击图标 ↔，弹出截面几何属性定义菜单对话框，输入平面应力单元厚度 0.07 m。点击 OK，完成截面几何特性的定义。

赋予钢材材料属性。钢筋单元选择配筋和桩基础(Reinforcements and pile foundations)本构类型，材料模型选择 Von Mises 塑性屈服模型，塑性硬化类型为非硬化(No hardening)。屈服强度 4.5×10^8 N/m²，钢筋单元采用嵌入式钢筋，面积为 56.55 mm²。如图 3.5-11—图 3.5-13 所示。

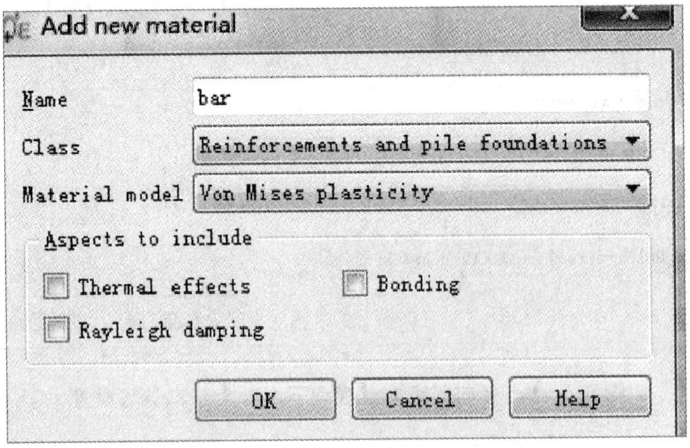

图 3.5-11　钢材塑性本构模型

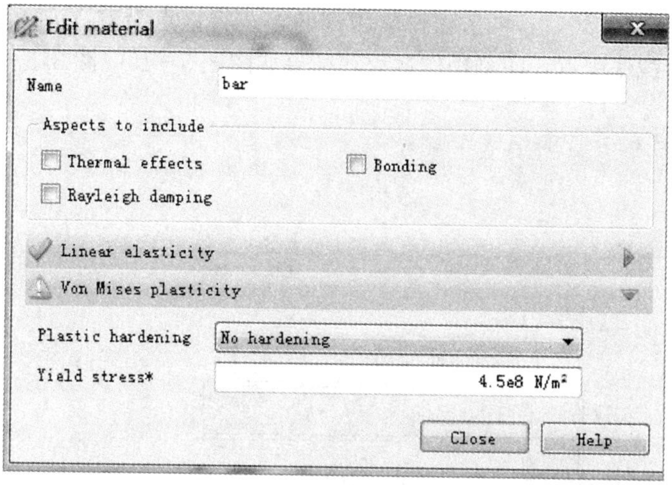

图 3.5‑12　钢材塑性本构模型塑性硬化方式和屈服强度设置

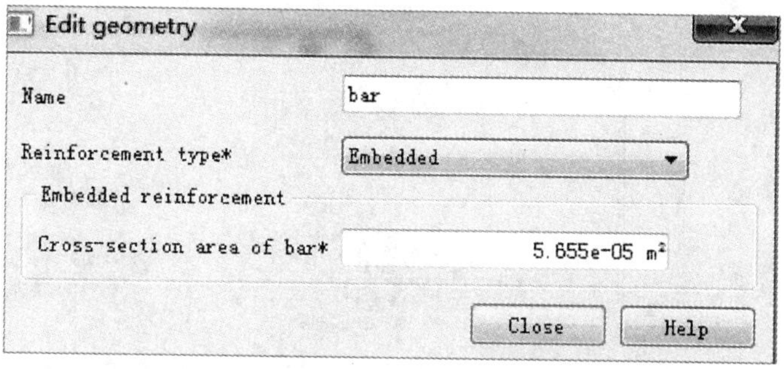

图 3.5‑13　钢材塑性本构模型截面几何特性设置

施加约束。对(0.06,0)点施加 X 和 Y 方向的线位移约束，对正对称半结构的对称截面施加 X 方向位移约束，其中 co1 约束施加如图 3.5‑14 所示。

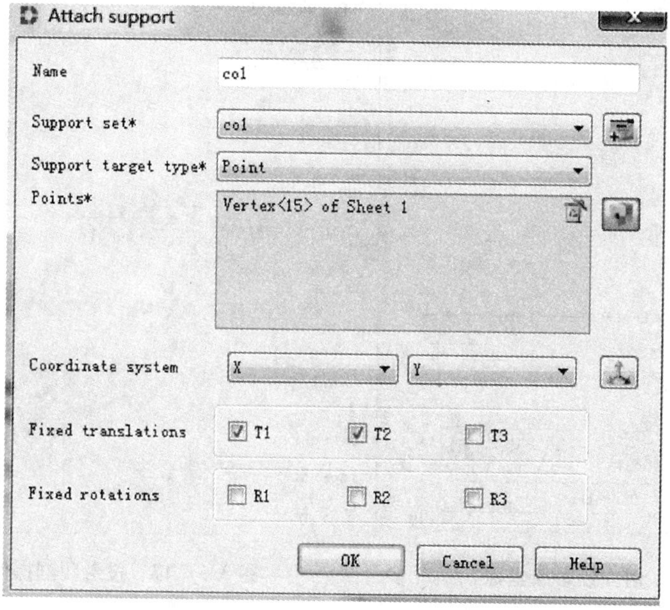

图 3.5‑14　co1 约束施加图示

创立点,坐标为(0.56,1),再采用投影印刻的方式将该点沿着 Y 轴负向投影印刻在结构上,作为集中荷载作用点。印刻操作过程和结果如图 3.5-15—图 3.5-18 所示。

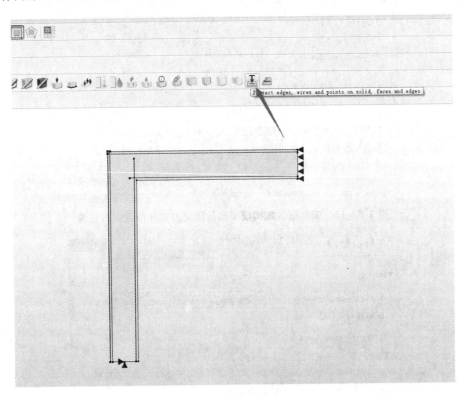

图 3.5-15　投影印刻快捷工具栏位置

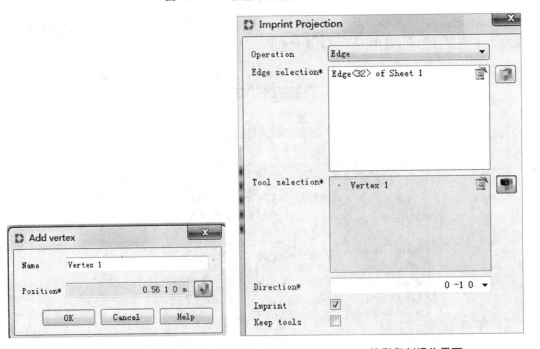

图 3.5-16　创立投影点坐标　　　　图 3.5-17　投影印刻操作界面

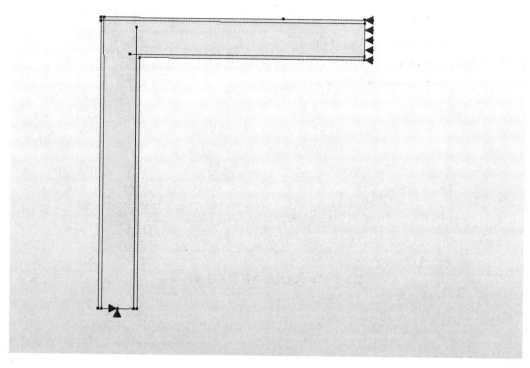

图 3.5-18　投影印刻操作完毕后在模型上生成投影点

施加重力荷载工况,命名为 gravity,荷载方式选择恒载。如图 3.5-19 所示。

施加集中荷载工况。荷载作用目标类型(Load target type)为点荷载,作用的荷载类型为集中荷载(Force)。荷载施加大小为 0.2 kN (200 N),作用方向为 Y 轴负向。在后续非线性计算模块荷载工况分析中再施加倍数荷载步。这里需要说明的是,Diana10.1 中荷载施加数值的方式很多,根据作者经验至少有以下三种:① 在Attach load 模块的 Force value 中直接施加荷载,大小与所要施加的荷载大小相同;② 在 Attach load 模块的 Force value 中输入一个初

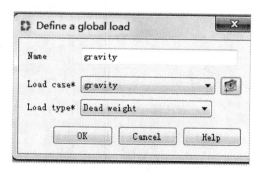

图 3.5-19　施加重力荷载工况

始值,这个初始值通常为要施加的荷载值的一部分,在荷载工况组合(load combinations)中定义荷载因子(load factor)时输入倍数;③ 输入基准值,定义荷载因子(load factor)为 1,在后面 Analysis 模块定义荷载步(load steps)时输入该荷载因子的倍数。本节案例采用第③种荷载定义方式,施加荷载的 GUI 操作界面如图 3.5-20 所示,生成集中荷载如图 3.5-21 所示。

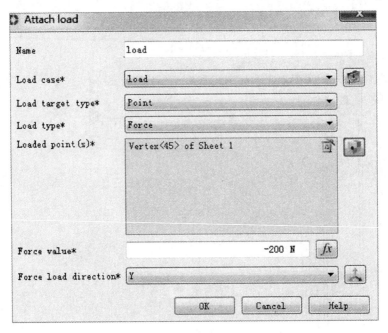

图 3.5-20　施加集中荷载的操作界面

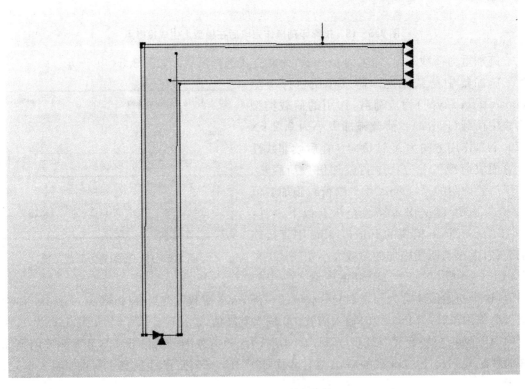

图 3.5-21　生成集中荷载

接下来开始网格划分。与之前的尺寸划分方式有所不同,本案例的网格划分方式采用份数划分(Divisions),操作类型为对边(Edge)进行网格划分。由于模型中存在长边和短边,按照各边长度的不同,分别选择不同的划分份数(所有较长边划分为8份,较短边划分为4份)。用鼠标左键拾取长边,选择Divisions的网格划分方式,网格划分份数选择为8份,如图3.5-22所示。点击OK,完成对模型中较长边的网格划分。

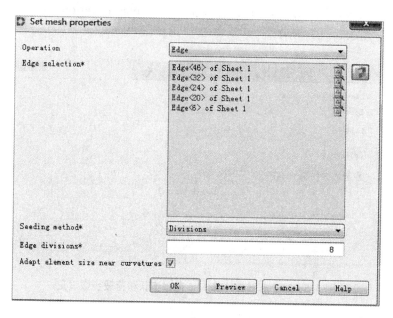

图3.5-22 对较长边的网格划分

用同样的方法将短边划分为4份。点击快捷键图标Generate mesh of a shape按钮,即可生成网格。

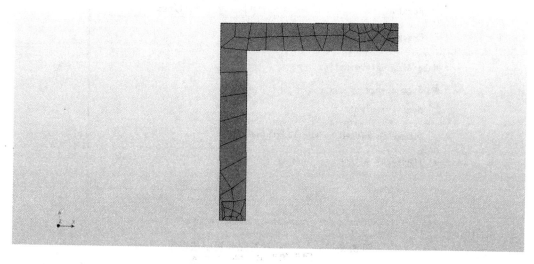

图3.5-23 生成单元网格划分

设置非线性计算选项模块。对于load荷载步工况,各荷载步子步长因子为1.0,设置计算步数为 40 步,采用弧长控制法(Arc length control),自动转化为弧长法控制(Automatic switch to arc-length method)的刚度参数(Stiffness parameter)为 0.25,在非线性计算的迭代属性设置中,最大迭代步数(Maximum number of iterations)设置成 20 步,采用牛顿-拉夫森(Newton-Raphon)迭代法,收敛方式为同时勾选力和位移收敛,即在非线性计算中同时针对力和位移进行迭代计算,只要有一个达到收敛准则即可保证收敛。迭代残差许可值(Convergence tolerance)为 0.05,迭代计算终止上限(Abort criterion)为 10000。上述操作如图 3.5-24—图 3.5-27 所示。

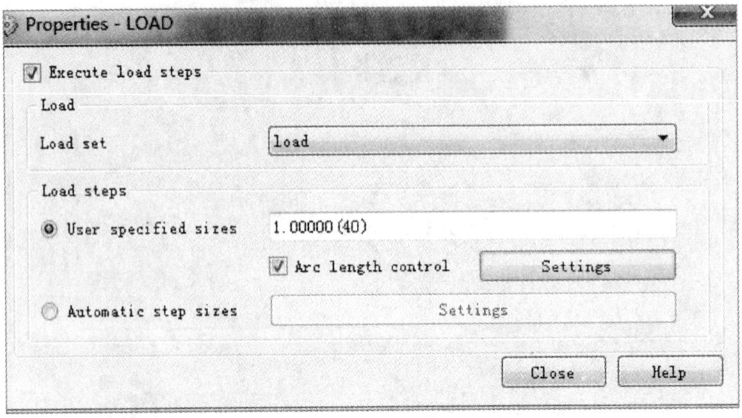

图 3.5-24　非线性计算 load 荷载步长和荷载步数设置

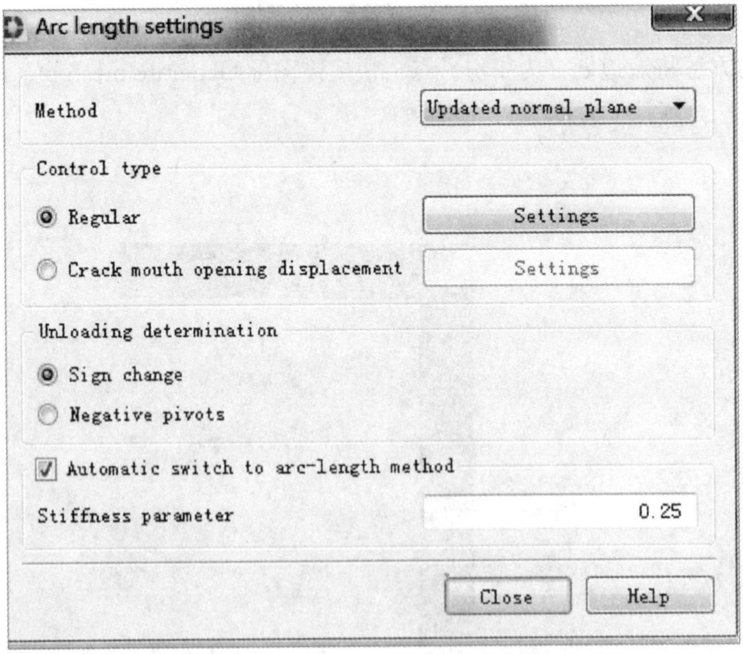

图 3.5-25　非线性计算弧长控制法设置

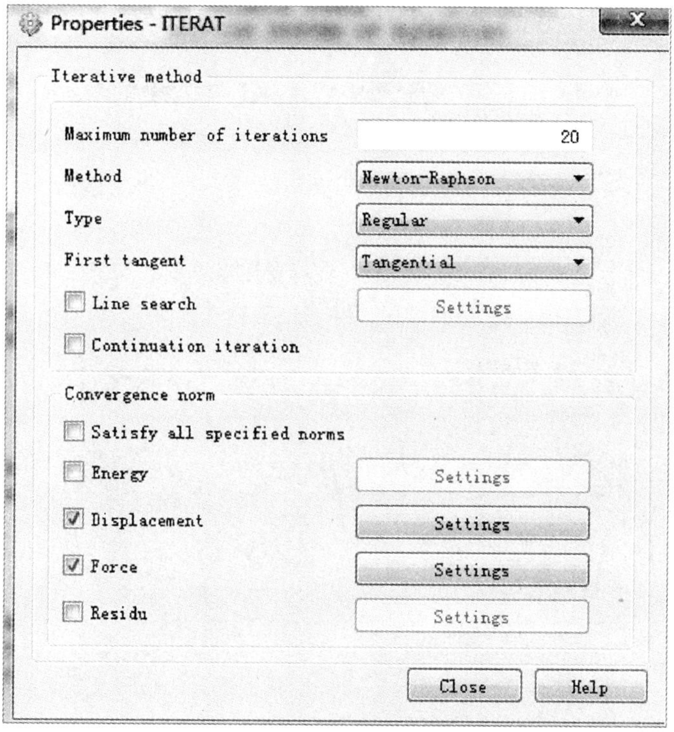

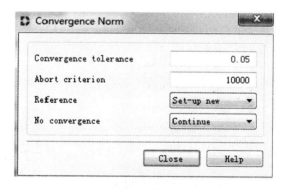

图 3.5-26 非线性计算迭代方法的设置

图 3.5-27 非线性计算收敛准则的设置

在 OUTPUT 输出选项中选择输出格式类型为原始 Diana 结果类型(Diana native),结果选项中选择用户选项(User selection)表示由用户自行确定要输出和查看的结果。其中选择整体坐标系下所有方向位移(DISPLA TOTAL TRANSL GLOBAL)、整体坐标系下所有外力(FORCE ENTERN TRANSL GLOBAL)、整体坐标系下塑性应变(STRAIN PLASTI GREEN GLOBAL)、开裂应变(STRAIN CRACK GREEN)、整体坐标系下的所有方向柯西应力(STRESS TOTAL CAUCHY GLOBAL),如图 3.5-28 和图 3.5-29 所示。

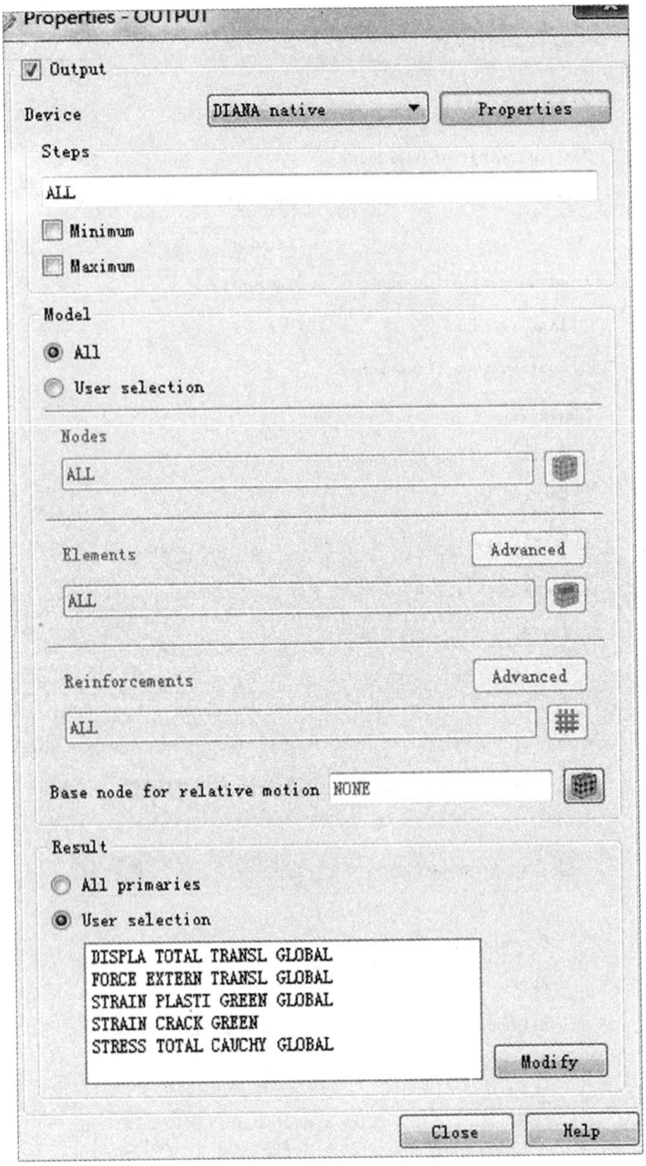

图 3.5-28　OUTPUT 输出结果定义界面

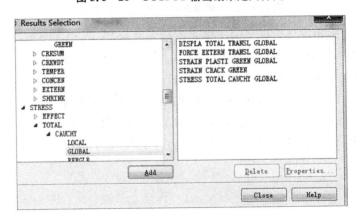

图 3.5-29　OUTPUT 用户选项设置模块

查看计算结果。选择最终荷载步计算结果作为后处理查看对象,查看 Y 方向的位移云图(TDtY)、整体裂缝宽度分布图(Eknn)和 X 方向主应力云图(SXX),如图 3.5-30—图 3.5-32所示。

图 3.5-30　Y 方向的位移云图

图 3.5-31　整体裂缝应变 Eknn 分布云图

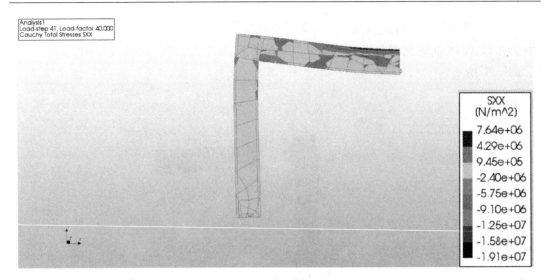

图 3.5‑32　X 方向主应力云图

点击上方菜单栏 Viewer Node Selection→点击 Results 结果栏下方的 TDtY→点击鼠标右键,选择 show ids,选中跨中位置的节点(3 号节点),再右击 TDtY 选择 Show table,查看计算结束后的 3 号节点最终 Y 方向位移值为 3.43 mm。

待上述计算完毕后,删去 Geometry 栏下方的材料栏和几何栏中的原 concrete 选项,添加混凝土材料本构模型为总应变裂缝模型中的旋转裂缝模型,混凝土的线性材料模型和拉伸软化模型下的各参数与多向固定裂缝模型中相同。其中裂缝走向(Crack orientation)选择旋转(Rotating),混凝土受压曲线模型(Compression curve)选择抛物线(Parabolic),抗压强度为 23.4 MPa,压缩断裂能为 4500 N/m,其抗压强度参数设置如图 3.5‑33 所示。

图 3.5‑33　总应变旋转裂缝模型受压模型参数设置

点击 Generate mesh of a shape 快捷菜单按钮,重新生成单元网格后点击 Run analysis 按钮 开始计算,待计算结束后,查看同一节点处的 Y 向位移值为 3.27 mm。

再次采用相同的操作方式,定义纤维混凝土在总应变旋转裂缝模型下的本构参数,即在拉伸软化曲线中选择纤维混凝土(fib fiber reinforced concrete)拉伸软化模型,纤维混凝土应力—应变曲线类型选择 CMOD 模式,其中混凝土抗拉强度值和裂纹口张开极限值与多向固定裂缝模型下极限应变值相同,根据纤维混凝土的应力—应变曲线特征,表征其应力—应变曲线的各参数设置如图 3.5-34 所示。

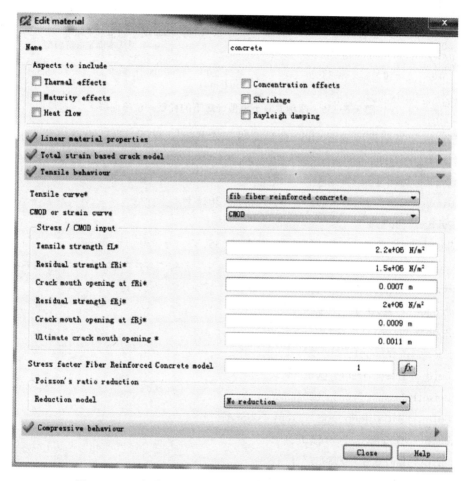

图 3.5-34　纤维混凝土 CMOD 模式下拉伸行为各参数设置图

重新网格划分后,点击 Run analysis 开始计算。待整个计算结束后,点击第 41 荷载步,查看荷载施加完毕后同一节点位置(3 号节点)处的 Y 向位移值,为 2.23 mm。

采用相同的方法计算总应变固定裂缝模型下的 Y 方向位移为 2.84 mm。计算完毕后,分别提取上述各模型在第 1、第 6、第 11、第 16、第 21、第 26、第 31、第 36、第 41 荷载步下的计算结果,绘制 Origin 曲线,比较在相同荷载作用下的 3 号节点 Y 向荷载—位移曲线,如图 3.5-35 所示。

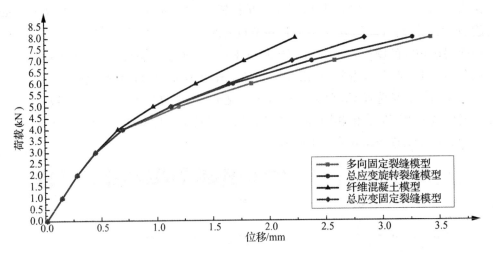

图 3.5-35　四种混凝土模型计算下的荷载—位移曲线

取多向固定裂缝模型的 Python 命令流语言如下：

```
##################################################
################################
# DianaIE 10.1 update 2017-04-25 13:38:53
# Python   3.3.4
# Session recorded at 2017-06-20 16:36:37
##################################################
############################
    newProject( "frame1", 10 )
    setModelAnalysisAspects( [ "STRUCT" ] )
    setModelDimension( "2D" )
    setDefaultMeshOrder( "QUADRATIC" )
    setDefaultMesherType( "HEXQUAD" )
    setDefaultMidSideNodeLocation( "ONSHAP" )
    setUnit( "TEMPER", "CELSIU" )
    setUnit( "ANGLE", "DEGREE" )
    saveProject( )
    createSheet( "Sheet 1", [[ 0, 0, 0 ],[ 0.06, 0, 0 ],[ 0.12, 0, 0 ],[ 0.12, 0.75,
0 ],[ 0.81, 0.75, 0 ],[ 0.81, 0.87, 0 ],[ 0, 0.87, 0 ]] )
    createLine( "bar1", [ 0.01, 0, 0 ], [ 0.01, 0.87, 0 ] )
    saveProject( )
    createLine( "bar2", [ 0.11, 0, 0 ], [ 0.11, 0.84, 0 ] )
    saveProject( )
    createLine( "bar3", [ 0.09, 0.76, 0 ], [ 0.09, 0.81, 0 ] )
    removeShape( [ "bar3" ] )
    createLine( "bar3", [ 0.09, 0.76, 0 ], [ 0.81, 0.76, 0 ] )
```

```
saveProject( )
createLine( "bar4", [ 0, 0.86, 0 ], [ 0.81, 0.86, 0 ] )
saveProject( )
addMaterial( "concrete", "CONCR", "MDFC", [] )
setParameter( "MATERIAL", "concrete", "LINEAR /ELASTI /YOUNG", 3.15e+10 )
setParameter( "MATERIAL", "concrete", "LINEAR /ELASTI /POISON", 0.2 )
setParameter( "MATERIAL", "concrete", "LINEAR /MASS /DENSIT", 2500 )
setParameter( "MATERIAL", "concrete", "MDFIXC /CRKVA1 /CRKVAL", 2200000 )
setParameter( "MATERIAL", "concrete", "MDFIXC /TAUCRI", 1 )
setParameter( "MATERIAL", "concrete", "MDFIXC /TAUCR1 /BETA", 0.2 )
setParameter( "MATERIAL", "concrete", "MDFIXC /TENSIO", 1 )
setParameter( "MATERIAL", "concrete", "MDFIXC /TENSI1 /TENVAL", 0.0011 )
setParameter( "MATERIAL", "concrete", "MDFIXC /TENSI1 /TENVAL", 0.0011 )
setParameter( "MATERIAL", "concrete", "MDFIXC /TENSI1 /TENVAL", 0.0011 )
setParameter( "MATERIAL", "concrete", "MDFIXC /TENSI1 /TENVAL", 0.0011 )
addGeometry( "Element geometry 1", "SHEET", "MEMBRA", [] )
setParameter( "GEOMET", "Element geometry 1", "THICK", 0.07 )
setParameter( "GEOMET", "Element geometry 1", "LOCAXS", True )
setParameter( "GEOMET", "Element geometry 1", "LOCAXS", False )
rename( "GEOMET", "Element geometry 1", "concrete" )
setElementClassType( "SHAPE", [ "Sheet 1" ], "MEMBRA" )
resetElementData( "SHAPE", [ "Sheet 1" ] )
assignMaterial( "concrete", "SHAPE", [ "Sheet 1" ] )
assignGeometry( "concrete", "SHAPE", [ "Sheet 1" ] )
addMaterial( "bar", "REINFO", "VMISES", [] )
setParameter( "MATERIAL", "bar", "LINEAR /ELASTI /YOUNG", 2.1e+11 )
setParameter( "MATERIAL", "bar", "PLASTI /HARDI1 /YLDSTR", 4.5e+08 )
setParameter( "MATERIAL", "bar", "PLASTI /HARDI1 /YLDSTR", 4.5e+08 )
setParameter( "MATERIAL", "bar", "PLASTI /HARDI1 /YLDSTR", 4.5e+08 )
addGeometry( "Element geometry 2", "RELINE", "REBAR", [] )
rename( "GEOMET", "Element geometry 2", "bar" )
setParameter( "GEOMET", "bar", "REIEMB /CROSSE", 5.655e-05 )
setReinforcementAspects( [ "bar1", "bar2", "bar3", "bar4" ] )
assignMaterial( "bar", "SHAPE", [ "bar1", "bar2", "bar3", "bar4" ] )
assignGeometry( "bar", "SHAPE", [ "bar1", "bar2", "bar3", "bar4" ] )
resetElementData( "SHAPE", [ "bar1", "bar2", "bar3", "bar4" ] )
setReinforcementDiscretization ( [ " bar1", " bar2", " bar3", " bar4" ], "ELEMENT" )
saveProject( )
```

```
addSet( "GEOMETRYSUPPORTSET", "co1" )
createPointSupport( "co1", "co1" )
setParameter( "GEOMETRYSUPPORT", "co1", "AXES", [ 1, 2 ] )
setParameter( "GEOMETRYSUPPORT", "co1", "TRANSL", [ 1, 1, 0 ] )
setParameter( "GEOMETRYSUPPORT", "co1", "ROTATI", [ 0, 0, 0 ] )
attach( " GEOMETRYSUPPORT", "co1", " Sheet 1", [[ 0.06, 1.8378494e - 34, -5.2871687e - 18 ]] )
addSet( "GEOMETRYSUPPORTSET", "Geometry support set 2" )
rename( "GEOMETRYSUPPORTSET", "Geometry support set 2", "co2" )
createLineSupport( "co2", "co2" )
setParameter( "GEOMETRYSUPPORT", "co2", "AXES", [ 1, 2 ] )
setParameter( "GEOMETRYSUPPORT", "co2", "TRANSL", [ 1, 0, 0 ] )
setParameter( "GEOMETRYSUPPORT", "co2", "ROTATI", [ 0, 0, 0 ] )
attach( "GEOMETRYSUPPORT", "co2", "Sheet 1", [[ 0.81, 0.81, -1.4938131e - 17 ]] )
saveProject( )
createVertex( "Vertex 1", [ 0.56, 1, 0 ] )
projection( " SHAPEEDGE", " Sheet 1", [[ 0.405, 0.87, 7.5632907e - 18 ]], [ "Vertex 1" ], [ 0, -1, 0 ], True )
removeShape( [ "Vertex 1" ] )
addSet( "GEOMETRYLOADSET", "gravity" )
createModelLoad( "gravity", "gravity" )
addSet( "GEOMETRYLOADSET", "Geometry load case 2" )
rename( "GEOMETRYLOADSET", "Geometry load case 2", "load" )
addSet( "GEOMETRYLOADSET", "Geometry load case 3" )
remove( "GEOMETRYLOADSET", [ "Geometry load case 3" ] )
createPointLoad( "load", "load" )
setParameter( "GEOMETRYLOAD", "load", "FORCE /VALUE", -200 )
setParameter( "GEOMETRYLOAD", "load", "FORCE /DIRECT", 2 )
attach( "GEOMETRYLOAD", "load", "Sheet 1", [[ 0.56, 0.87, -2.5015918e - 19 ]] )
setEdgeMeshSeed( "Sheet 1", [[ 0.81, 0.81, -1.4938131e - 17 ],[ 0.09, 2.3635266e - 34, -6.7994493e - 18 ],[ 0.03, 1.3121722e - 34, -3.7748881e - 18 ]], 4 )
saveProject( )
setEdgeMeshSeed( "Sheet 1", [[ 0.28, 0.87, 1.386446e - 17 ],[ 6.4817536e - 34, 0.435, 1.2858236e - 17 ],[ 0.12, 0.375, 4.7234798e - 18 ],[ 0.465, 0.75, 3.6746234e - 19 ],[ 0.685, 0.87, -6.5513284e - 18 ]], 4 )
setEdgeMeshSeed( " Sheet 1", [[ 6.4817536e - 34, 0.435, 1.2858236e - 17 ], [ 0.28, 0.87, 1.386446e - 17 ], [ 0.685, 0.87, -6.5513284e - 18 ], [ 0.465, 0.75,
```

```
3.6746234e - 19 ],[ 0.12, 0.375, 4.7234798e - 18 ]], 8 )
    saveProject( )
    generateMesh( [ ] )
    hideView( "GEOM" )
    showView( "MESH" )
    addAnalysis( "Analysis1" )
    addAnalysisCommand( "Analysis1", "NONLIN", "Structural nonlinear" )
    setAnalysisCommandDetail( "Analysis1", "Structural nonlinear", "EXECUT(1) /
LOAD /STEPS /EXPLIC /SIZES", "1.00000" )
    setAnalysisCommandDetail ( "Analysis1", "Structural nonlinear", "EXECUT /
EXETYP", "LOAD" )
    renameAnalysisCommandDetail( "Analysis1", "Structural nonlinear", "EXECUT
(2)", "new execute block 2" )
    setAnalysisCommandDetail( "Analysis1", "Structural nonlinear", "EXECUT(2) /
LOAD /STEPS /EXPLIC /SIZES", "1.00000" )
    setAnalysisCommandDetail( "Analysis1", "Structural nonlinear", "EXECUT(2) /
LOAD /LOADNR", 2 )
    saveProject( )
    setAnalysisCommandDetail( "Analysis1", "Structural nonlinear", "EXECUT(2) /
LOAD /STEPS /EXPLIC /SIZES", "1.00000(40)" )
    setAnalysisCommandDetail( "Analysis1", "Structural nonlinear", "EXECUT(2) /
LOAD /STEPS /EXPLIC /SIZES", "1.00000(40)" )
    addAnalysisCommandDetail( "Analysis1", "Structural nonlinear", "EXECUT(2) /
LOAD /STEPS /EXPLIC /ARCLEN" )
    setAnalysisCommandDetail( "Analysis1", "Structural nonlinear", "EXECUT(2) /
LOAD /STEPS /EXPLIC /ARCLEN", True )
    addAnalysisCommandDetail( "Analysis1", "Structural nonlinear", "EXECUT(2) /
LOAD /STEPS /EXPLIC /ARCLEN /AUTARC" )
    setAnalysisCommandDetail( "Analysis1", "Structural nonlinear", "EXECUT(2) /
LOAD /STEPS /EXPLIC /ARCLEN /AUTARC", True )
    setAnalysisCommandDetail( "Analysis1", "Structural nonlinear", "EXECUT(2) /
LOAD /STEPS /EXPLIC /ARCLEN /AUTARC", False )
    setAnalysisCommandDetail( "Analysis1", "Structural nonlinear", "EXECUT(2) /
LOAD /STEPS /EXPLIC /ARCLEN /METHOD", "SPHERI" )
    setAnalysisCommandDetail( "Analysis1", "Structural nonlinear", "EXECUT(2) /
LOAD /STEPS /EXPLIC /ARCLEN /METHOD", "UPDATE" )
    setAnalysisCommandDetail( "Analysis1", "Structural nonlinear", "EXECUT(2) /
LOAD /STEPS /EXPLIC /ARCLEN /AUTARC", True )
    setAnalysisCommandDetail( "Analysis1", "Structural nonlinear", "EXECUT(2) /
```

```
LOAD /STEPS /EXPLIC /ARCLEN /AUTARC", False )
    setAnalysisCommandDetail( "Analysis1", "Structural nonlinear", "EXECUT(2) /
LOAD /STEPS /EXPLIC /ARCLEN /AUTARC", True )
    setAnalysisCommandDetail( "Analysis1", "Structural nonlinear", "EXECUT(2) /
LOAD /STEPS /EXPLIC /ARCLEN /AUTARC /STFPAR", 0.25 )
    saveProject( )
    setAnalysisCommandDetail( "Analysis1", "Structural nonlinear", "OUTPUT(1) /
SELTYP", "USER" )
    addAnalysisCommandDetail( "Analysis1", "Structural nonlinear", "OUTPUT(1) /
USER" )
    addAnalysisCommandDetail( "Analysis1", "Structural nonlinear", "OUTPUT(1) /
USER /DISPLA(1) /TOTAL /TRANSL /GLOBAL" )
    addAnalysisCommandDetail( "Analysis1", "Structural nonlinear", "OUTPUT(1) /
USER /FORCE(1) /EXTERN /TRANSL /GLOBAL" )
    addAnalysisCommandDetail( "Analysis1", "Structural nonlinear", "OUTPUT(1) /
USER /STRAIN(2) /CRACK /GREEN" )
    addAnalysisCommandDetail( "Analysis1", "Structural nonlinear", "OUTPUT(1) /
USER /STRESS(1) /TOTAL /CAUCHY /GLOBAL" )
    setAnalysisCommandDetail( "Analysis1", "Structural nonlinear", "EXECUT(2) /
LOAD /STEPS /EXPLIC /ARCLEN /AUTARC /STFPAR", 0.001 )
    saveProject( )
    setAnalysisCommandDetail( "Analysis1", "Structural nonlinear", "EXECUT(2) /
ITERAT /MAXITE", 20 )
    addAnalysisCommandDetail( "Analysis1", "Structural nonlinear", "EXECUT(2) /
ITERAT /LINESE" )
    setAnalysisCommandDetail( "Analysis1", "Structural nonlinear", "EXECUT(2) /
ITERAT /LINESE", True )
    setAnalysisCommandDetail( "Analysis1", "Structural nonlinear", "EXECUT(2) /
ITERAT /LINESE", False )
    setAnalysisCommandDetail( "Analysis1", "Structural nonlinear", "EXECUT(2) /
ITERAT /CONTIN", True )
    setAnalysisCommandDetail( "Analysis1", "Structural nonlinear", "EXECUT(2) /
ITERAT /CONTIN", False )
    setAnalysisCommandDetail( "Analysis1", "Structural nonlinear", "EXECUT(2) /
ITERAT /MAXITE", 20 )
    setAnalysisCommandDetail( "Analysis1", "Structural nonlinear", "EXECUT(2) /
ITERAT /MAXITE", 20 )
    setAnalysisCommandDetail( "Analysis1", "Structural nonlinear", "EXECUT(2) /
ITERAT /METHOD /NEWTON /TYPNAM", "REGULA" )
```

```
    setAnalysisCommandDetail( "Analysis1", "Structural nonlinear", "EXECUT(2)/
ITERAT/MAXITE", 20 )
    setAnalysisCommandDetail( "Analysis1", "Structural nonlinear", "EXECUT(2)/
ITERAT/CONVER/DISPLA/TOLCON", 0.05 )
    setAnalysisCommandDetail( "Analysis1", "Structural nonlinear", "EXECUT(2)/
ITERAT/CONVER/DISPLA/NOCONV", "CONTIN" )
    setAnalysisCommandDetail( "Analysis1", "Structural nonlinear", "EXECUT(2)/
ITERAT/CONVER/FORCE/TOLCON", 0.05 )
    setAnalysisCommandDetail( "Analysis1", "Structural nonlinear", "EXECUT(2)/
ITERAT/CONVER/FORCE/NOCONV", "CONTIN" )
    runSolver( "Analysis1" )
    showView( "RESULT" )
    setResultCase( [ "Analysis1", "Output", "Load-step 41, Load-factor 40.000" ]
)
    setResultPlot( "contours", "Total Displacements/node", "TDtZ" )
    setResultPlot( "contours", "Total Displacements/node", "TDtY" )
    setResultPlot( "cracks", "Crack Strains/mappedcrack", "Eknn" )
    setResultPlot( "cracks", "Crack Strains/mappedcrack", "Gknt" )
    setResultPlot( "contours", "Cauchy Total Stresses/node", "SYY" )
    setResultPlot( "contours", "Cauchy Total Stresses/node", "SXX" )
    setViewSettingValue( "result view setting", "CRACK/AUTRNG", "AUTVIS" )
    setResultPlot( "contours", "Total Displacements/node", "TDtY" )
    showIds( "NODE", [ 3 ] )
    setResultCase( [ "Analysis1", "Output", "Load-step 41, Load-factor 40.000" ]
)
```

3.6 案例六：剪力墙滞回分析

本节案例参考自敦楼剪力墙滞回分析案例，整个模型分为上部加载梁、中间的剪力墙和下部的支撑梁三部分。其中加载梁长度为 2 m，高度为 0.4 m，厚度为 0.6 m。内部配有一根 CFRP 筋，上方承受 30 kN/m 的竖向均布荷载，下部支撑梁长度 2.6 m，高度 0.4 m，厚度为 0.6 m，剪力墙长为 1.8 m，高为 2 m，厚度大小为 0.4 m，其平面尺寸如图 3.6-1 所示。模型内配有钢筋网片和箍筋，钢筋均采用 Von Mises 塑性模型。整个模型采用混凝土平面应力单元建模，拉伸软化模型采用 Hordijk 模型，受压模型采用前川滞回模型。钢筋网片的单根钢筋折算面积均为 157 mm^2，两层箍筋面积折算为 100.53 mm^2，各部分的材料参数值见表(3-6)。

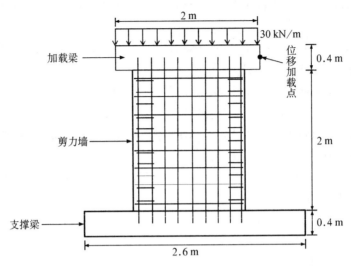

图 3.6-1 剪力墙平面尺寸图

表 3-6 各部分材料参数

剪力墙	混凝土弹性模量	$3.7 \times 10^{10} \, \text{N/m}^2$
	混凝土抗拉强度	$2.8 \times 10^6 \, \text{N/m}^2$
	混凝土抗压强度	$4.5 \times 10^7 \, \text{N/m}^2$
	剪滞系数	0.1
加载梁和支撑梁	混凝土弹性模量	$3.25 \times 10^{10} \, \text{N/m}^2$
FRP	弹性模量	$1.36 \times 10^{11} \, \text{N/m}^2$
	抗拉强度	$1.2 \times 10^9 \, \text{N/m}^2$
钢筋	弹性模量	$2.1 \times 10^{11} \, \text{N/m}^2$
	屈服强度	$4.4 \times 10^8 \, \text{N/m}^2$

注：作为 Diana10.1 算例例题，本书例题分析条件均为假定值，关于分析结果妥当性另当别论，望谅解。

学习要义：

(1) 总应变固定裂缝模型 Hordijk 拉伸软化模型参数设置

(2) 总应变固定裂缝模型下的前川模型受压参数设置

(3) FRP 本构参数模拟

(4) 点位移间接约束施加方法

(5) 分析模块中循环荷载位移施加方法

首先，启动 Diana10.1 界面，弹出如图 3.6-2 所示的操作界面。在 G 盘例题文件夹中创立 Shear wall 的 .dpf 文件，选择 2D 结构建模，模型文件的尺寸为 10 m，单元默认的划分类型为六面体/矩形单元，默认的划分阶数为 2 次，单元中间节点位置确定方式为线性插值。

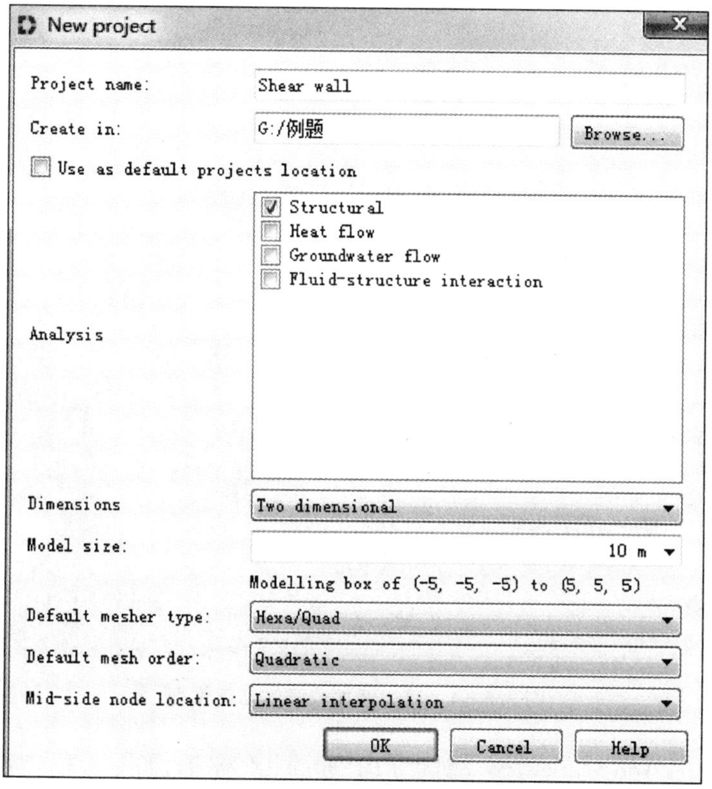

图 3.6-2 创建新模型的操作界面

点击 Add a sheet 快捷键图标  ，输入各点坐标，创立底部支撑梁的几何模型，命名为 support，输入各点坐标如图 3.6-3 所示，点击 OK 生成底部支撑梁的几何模型。

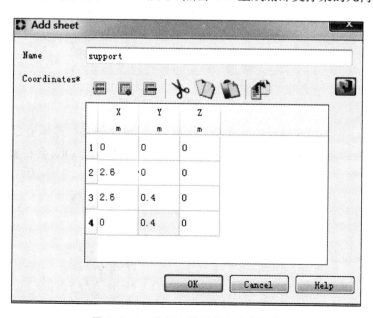

图 3.6-3 底部支撑梁的各点坐标值

再依次输入坐标值(0.4,0.4),(2.2,0.4),(2.2,2.6)和(0.4,2.6)生成剪力墙几何模

型。再采用相同方式依次输入坐标值(0.3,2.6),(2.3,2.6),(2.3,3)和(0.3,3),点击 OK 生成上部加载梁的几何模型。整体几何模型如图 3.6-4 所示。

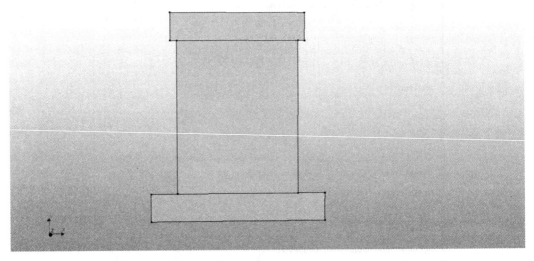

图 3.6-4 剪力墙加载整体几何模型

点击快捷图标 Add a line，创立 FRP 筋的几何坐标，如图 3.6-5 所示。

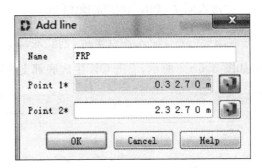

图 3.6-5 FRP 筋几何坐标

采用单根 Bar 钢筋单元建模的方式创立钢筋网片，首先创立与整体坐标系 Y 向一致的第一根竖向钢筋，命名为 bar1，输入坐标如图 3.6-6 所示。点击 OK，会在整个模型中间部分生成第一根钢筋。

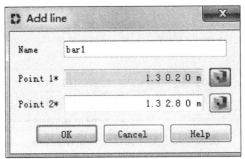

图 3.6-6 bar1 几何坐标

点击拾取 Geometry 栏下的 bar1,右击选择 Select 选项,再在界面中右击选择 Array copy 复制平移功能,选择平移距离 0.2m,方向 X 轴负向,复制份数为 4 份,生成左半部分的 Y 方向钢筋。操作界面如图 3.6-7 所示。

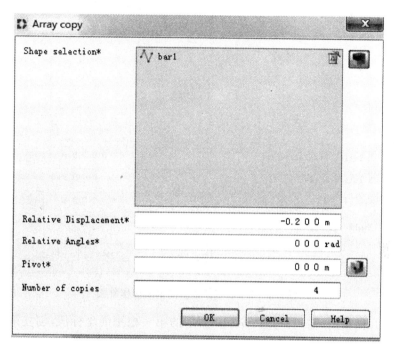

图 3.6-7 复制平移竖向左半部分钢筋操作界面

再以同样的操作方式和同样的平移距离向右平移 0.2 m,复制份数为 4 份,生成右半部分 Y 向分布钢筋。

创立沿整体坐标系 X 方向的第一根横向钢筋,如图 3.6-8 所示。命名为 barx1。

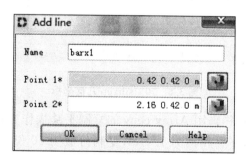

图 3.6-8 barx1 几何坐标

同样采用 Array copy 复制平移功能生成沿着 X 方向的分布钢筋,平移距离为 0.3 m,方向为沿着 Y 方向正向,复制份数 7 份,如图 3.6-9 所示。

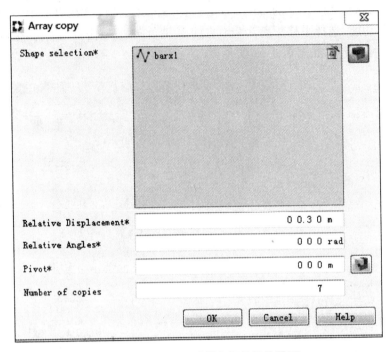

图 3.6-9　复制平移横向钢筋操作界面

建立箍筋几何模型,输入如图 3.6-10 所示的第一根箍筋坐标值,创立第一根箍筋,命名为 stirrup。

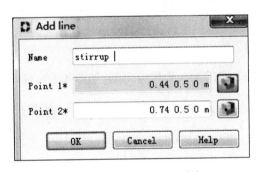

图 3.6-10　stirrup 几何坐标

如图 3.6-11 所示,同样采用 Array copy 的复制平移操作方式生成左半部分箍筋,平移方向为沿着 X 轴正向,平移距离为 1.4 m,复制份数为 10 份,点击 OK,生成左半部分箍筋。

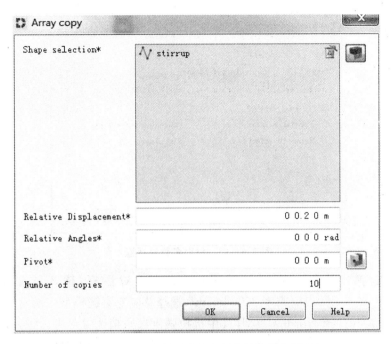

图 3.6-11　复制平移左半部分箍筋操作界面

采用同样方式生成右半部分箍筋,操作界面如图 3.6-12 所示。

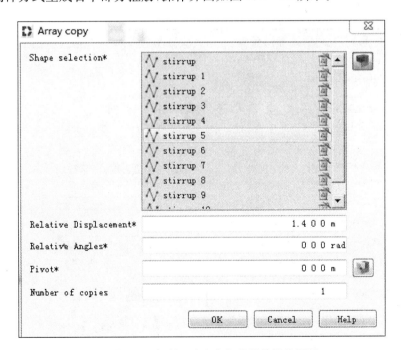

图 3.6-12　复制平移右半部分箍筋操作界面

在 GUI 图形可视化界面区用鼠标左键拾取 wall 几何模型,右击选择 Property assignments,弹出材料属性对话框,材料模型命名为 wall,材料类型选择混凝土和砌体结构(Concrete and mansory),材料模型为总应变裂缝模型(Total strain based crack model)。如图 3.6-13 所示。

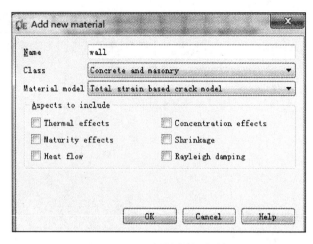

图 3.6-13　材料名称对话框

点击 OK，弹出编辑材料属性的 Edit material 对话框，输入混凝土弹性模量 $3.7×10^{10}$ N/m²，泊松比 0.15，密度 2500 kg/m³，选择总应变裂缝模型中的总应变固定裂缝模型，拉伸软化曲线选择 Hordijk 模型，其中抗拉强度极限值为 $2.8×10^6$ N/m²，拉伸断裂能为 200 N/m，残余强度为 100 N/m²，设定泊松比不衰减，混凝土抗压曲线为前川开裂混凝土曲线（Maekawa Cracked Concrete curves），抗压强度值为 45 MPa，剪力滞留系数为 0.1，如图 3.6-14—图 3.6-16 所示。

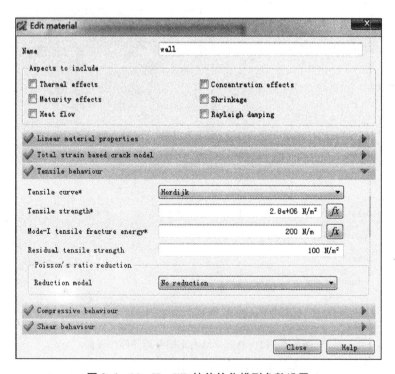

图 3.6-14　Hordijk 拉伸软化模型参数设置

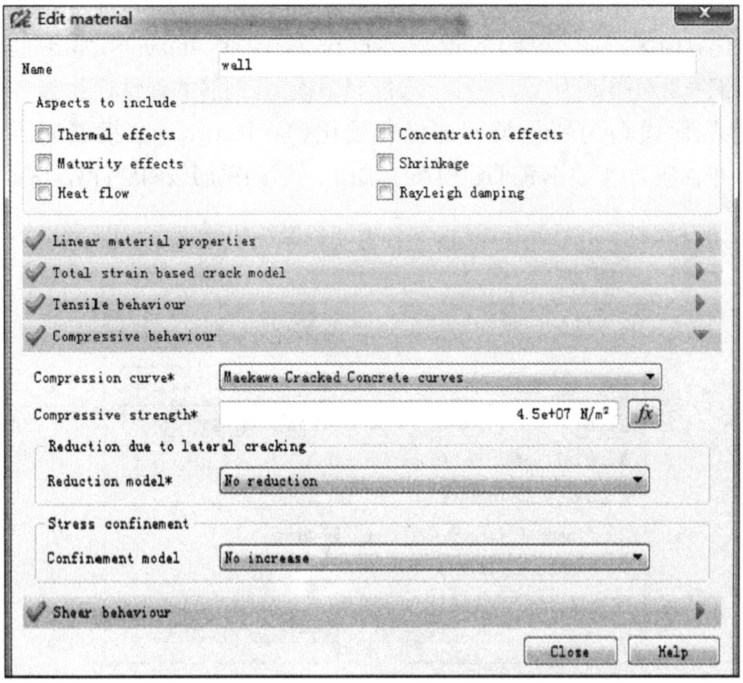

图 3.6‑15　混凝土受压前川模型参数设置

图 3.6‑16　剪滞系数设置

打开 Edit geometry 对话框,编辑截面几何特性,命名为 wall,输入剪力墙厚度值 0.4 m。如图 3.6‑17 所示。

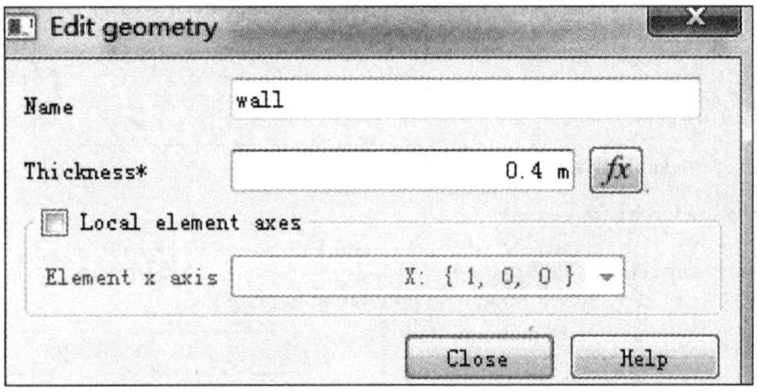

图 3.6‑17　剪力墙截面几何特性编辑框

定义竖向钢筋的材料属性,在 Geometry 栏下方选中 bar1~bar9,右击选择 Select 选项,再右击 GUI 可视化界面区,右击选择 Reinforcement property assignments,弹出配筋材料属性编辑框,命名为 bar,钢筋材料类型为 Von Mises 塑性屈服模型,钢筋的弹性模量为 2.1×10^{11} N/m^2,Von Mises 塑性屈服模型的硬化类型选择非硬化(No hardening),钢筋屈服应力为 4.4×10^8 N/m^2。由于平面应力单元中,钢筋建模在中轴面上,竖向钢筋按两层计算,因此在截面几何属性中定义截面面积为 $2\times\frac{1}{4}\pi\times10^2\approx157$ mm^2。上述各操作如图 3.6-18—图 3.6-20 所示。

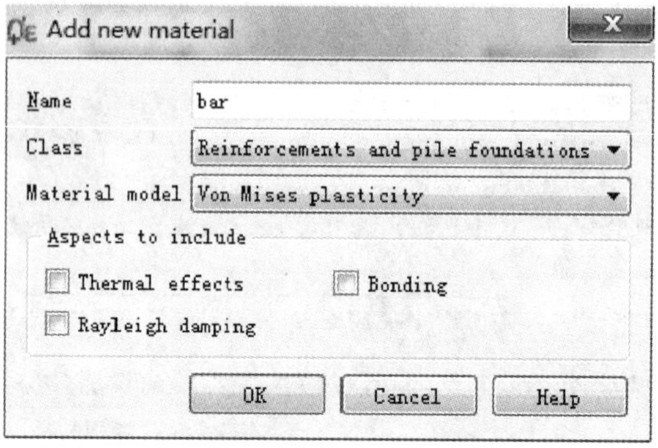

图 3.6-18 钢材 Von Mises 塑性屈服模型定义

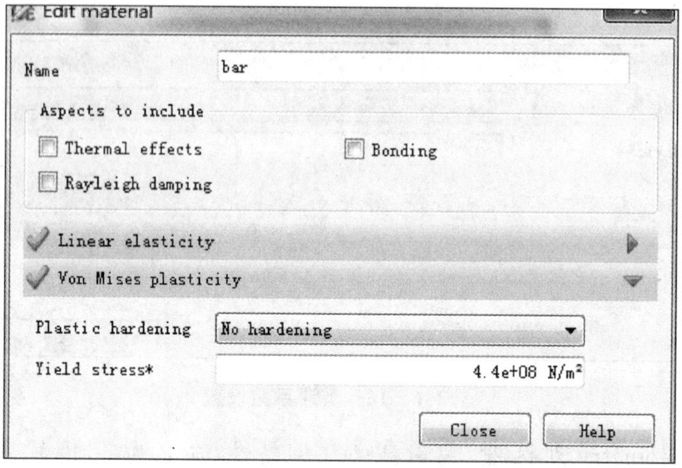

图 3.6-19 塑性屈服模型本构参数设置

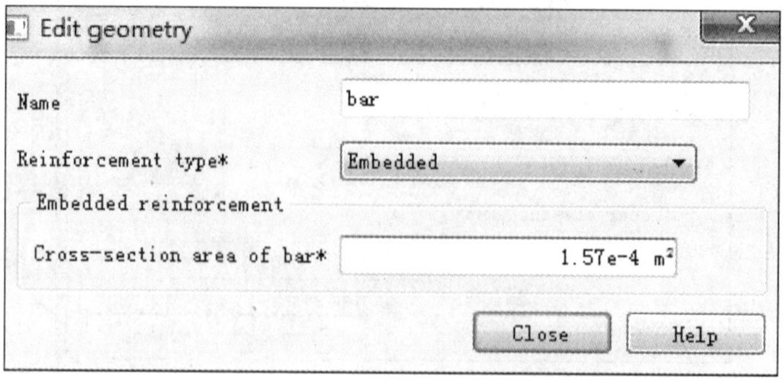

图 3.6-20 bar 截面几何特性参数设置

采用同样的方式定义横向钢筋 barx~barx8 的材料属性和截面几何属性。由于各参数值相同，这里不再赘述。

定义箍筋的材料属性与截面几何特性。命名为 stirrup。材料属性与 bar 和 barx 相同。截面几何特性定义如图 3.6-21 所示。箍筋截面面积折算后为 100.53 mm²。

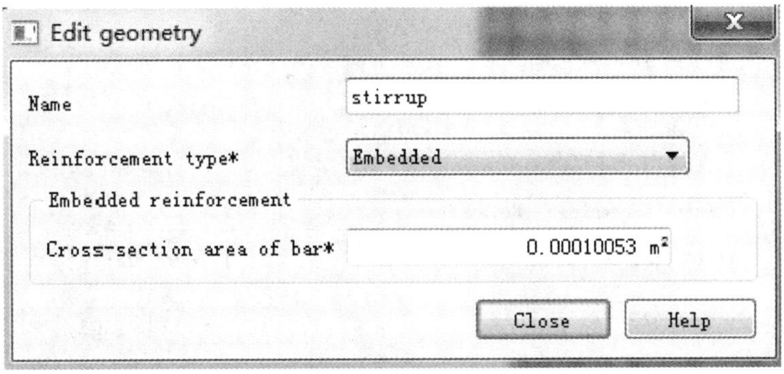

图 3.6-21　箍筋截面几何特性定义

定义 FRP 几何特性。由于 Diana10.1 中并没有专门用于定义 FRP 材料特性的模块，因此采用配筋和桩基础模块定义 FRP 材料特性。通常情况下 FRP 的弹性模量为普通钢筋的 25%~75%，抗拉强度为普通钢筋的 2~10 倍，因此本例采用弹性模量值为 1.36×10^{11} N/m²，抗拉强度为 1.2×10^{9} N/m²，塑性硬化类型选择非硬化（No hardening）。如图 3.6-22 和图 3.6-23 所示。

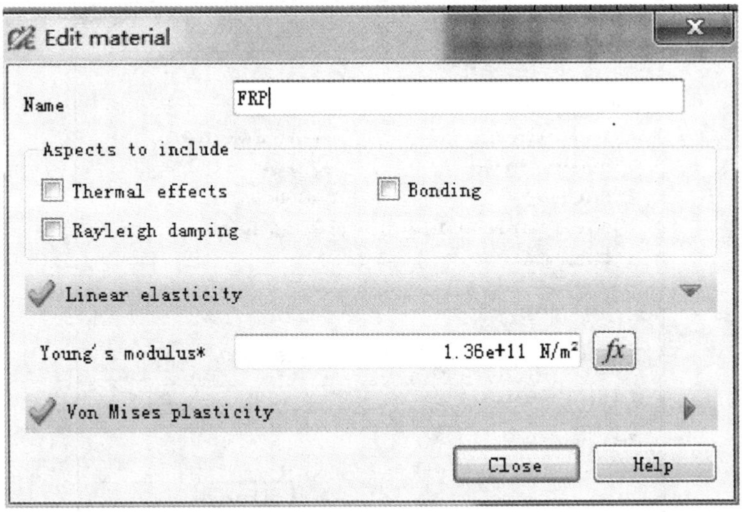

图 3.6-22　FRP 筋弹性模量定义

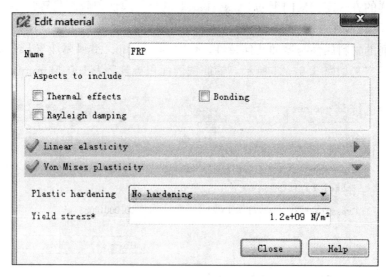

图 3.6‑23 FRP 筋屈服强度设置

设置上部加载梁和底部支撑梁的本构模型。为了减少上下梁自身破坏对墙体性能产生的影响,因此上下梁采用的材料模型为弹性模型,不考虑开裂或者塑性行为的影响。设置上部加载梁和底部支撑梁的混凝土弹性模量均为 3.25×10^{10} N/m², 泊松比 0.15,密度 2500 kg/m³,厚度均为 0.6 m。其中上部加载梁混凝土本构如图 3.6‑24 和图 3.6‑25 所示。

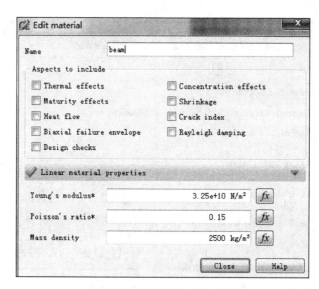

图 3.6‑24 上部加载梁混凝土本构设置

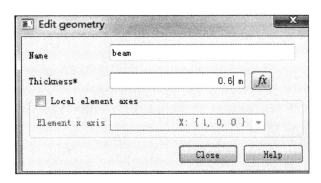

图 3.6-25　上部加载梁几何界面属性厚度设置

接下来设置滞回分析中位移加载的加载点。首先需要印刻一个加载点,以方便在后续操作中将水平位移荷载作用于该点。和之前各例题投影印刻的操作方式相同,创立点 Vertex 1,输入坐标后进行印刻投影。印刻投影方向沿着 X 轴负向(−1,0,0)。具体操作如图 3.6-26 和图 3.6-27 所示。

图 3.6-26　投影点坐标

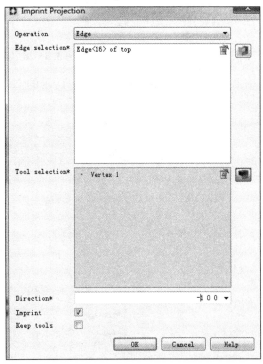

图 3.6-27　印刻投影操作界面

对底部支撑梁底边施加约束 X 和 Y 方向的平动位移线约束。命名为 co1,由于位移荷载必须作用在约束点上,因此在施加 co1 约束的同时对位移加载点施加水平 X 方向的点约束,命名为 co2。约束施加信息如图 3.6-28(a)和图 3.6-28(b)所示。

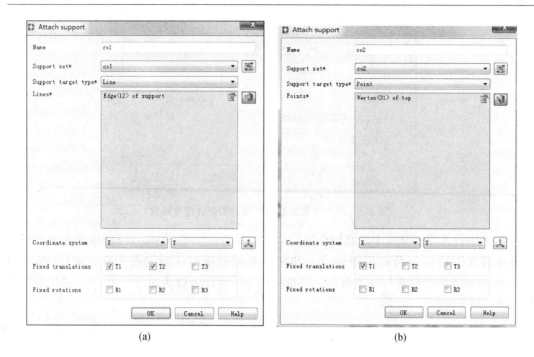

图 3.6-28　co1 和 co2 约束信息

点击 OK，生成如图 3.6-29 所示的 GUI 图形界面约束施加信息。

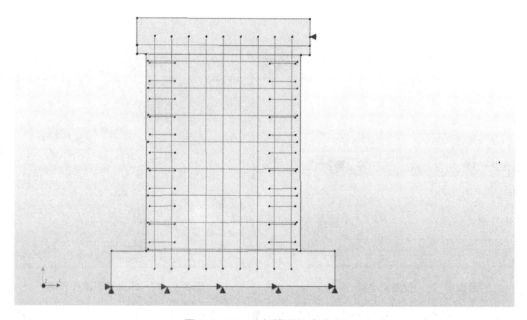

图 3.6-29　几何模型约束信息

点击 Geometry 栏下方 Load 栏中的图标 ，施加重力荷载，再对加载梁上方的边施加线荷载，线荷载大小为 30 kN/m，方向 Y 轴向下，命名为 load。如图 3.6-30 所示。

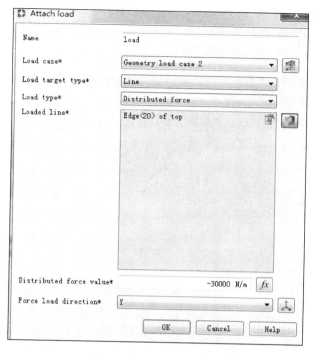

图 3.6-30 施加线荷载

接下来定义位移加载荷载,对已经施加 co2 水平约束的点施加点位移荷载。命名为 displacement。加载目标类型(Load target type)选择点(Point),加载类型(Load type)为 Prescribed deformation。位移量为 0.5 mm。方向沿着 X 轴负向。如图 3.6-31 所示。

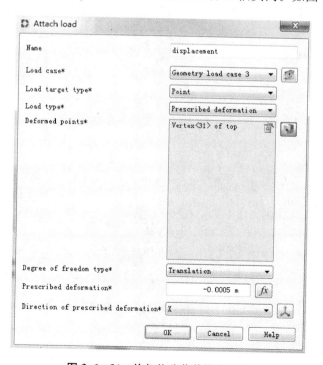

图 3.6-31 施加位移荷载操作图示

创建荷载组合。将重力荷载和均布荷载归为荷载组合 1（Geometry load combination1），位移加载归为荷载组合 2（Geometry load combination2），如图 3.6-32 所示。

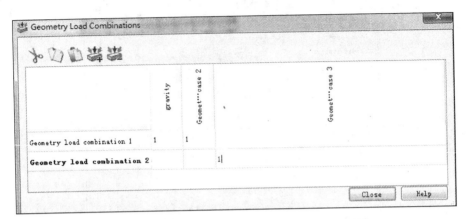

图 3.6-32　荷载组合

待荷载组合添加完毕后，选中整个模型，点击快捷键图标 Set mesh properties of a shape ，弹出如图 3.6-33 所示的网格划分操作界面对话框。操作类型（Operation）选择形状类（Shape），采用单元尺寸（Element size）的网格划分方式，设定所需的尺寸（Desired size）为 0.1 m，网格形状（Mesher type）为六面体/四边形（Hexa/Quad）单元，单元中间节点位置的确定方式（Mid-side node location）为线性插值（Linear interpolation）。

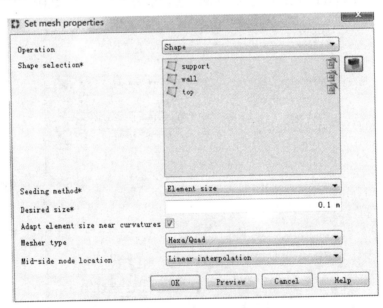

图 3.6-33　网格划分操作界面

点击快捷图标 Generate mesh of a shape 按钮，形成如图 3.6-34 所示的网格划分结果。

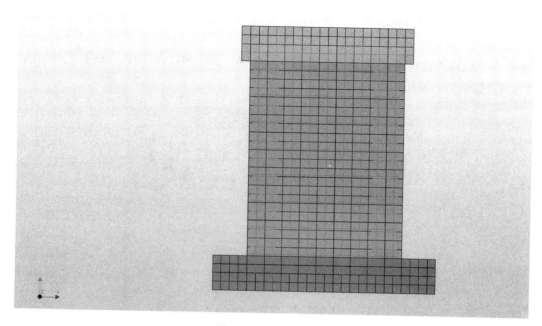

图 3.6-34 网格划分

待网格划分完毕后,点击 Mesh 栏下方的 Element types,确认是我们需要的 2D 平面应力单元 CQ16M。如图 3.6-35 所示。

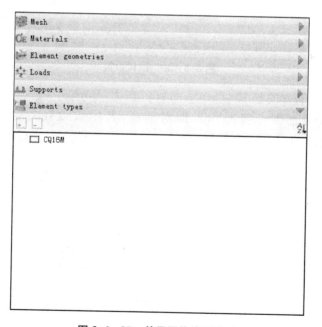

图 3.6-35 单元网格类型查看

在 Analysis 中设置非线性计算模块,点击图标 Add an analysis,创立新的分析工况,右击选择 Add command 栏目下的结构非线性分析(Structural nonlinear)。右击 Structural,点击 Add→Execute steps→Load steps,生成新的执行模块(new execute block),在 new execute block 下的 Load steps 中选择荷载组合 1,荷载步设定为 1,总步长因子为 1,在方程迭代(Equilibrium iteration)中设置最大迭代步数为 50 步,迭代方法采用牛顿—拉夫森(Newton-Raphon),迭代方式为普通迭代(Regular),收敛准则(Convergence norm)勾选位移或力收敛,即上述两个准则中只要有一个达到收敛,即认为该荷载步下的迭代计算达到收敛。如图 3.6-36 所示。

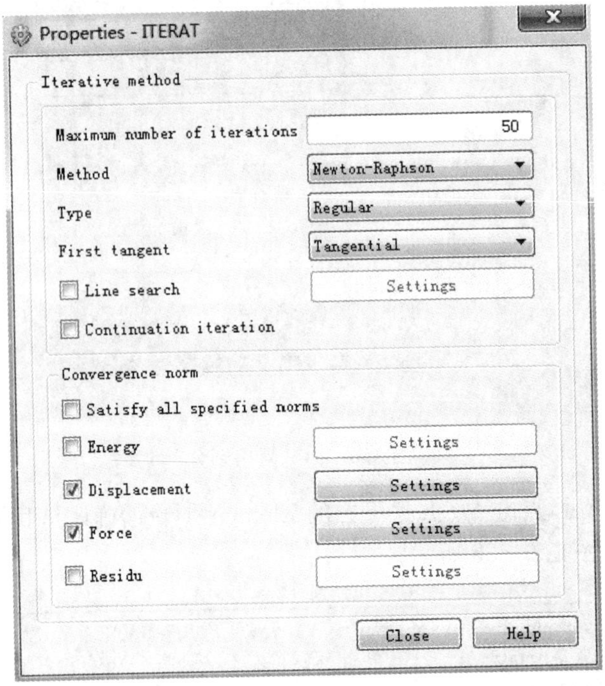

图 3.6-36 迭代设置

点击收敛准则下方的位移(Displacement)和力(Force)右边的 Settings,弹出如图 3.6-37 所示的收敛细则设置操作界面对话框,其中收敛残差容许值(Convergence tolerance)为 0.01,终止迭代标准为 10 000 步,计算未达到收敛时选择继续计算。

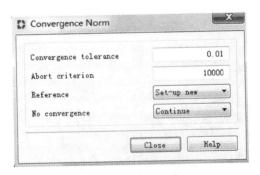

图 3.6-37 荷载组合 1 收敛细则设置操作界面对话框

采用相同的方式创立新的运算模块 2(new execute block2),选择荷载组合 2,最大迭代步数为 20 步,在收敛细则中设置收敛残差容许值(Convergence tolerance)为 0.05,荷载步设置方式为双向逐级加载,且在加载后期荷载步子步长设置较小,位移加载荷载步依次为 1.00000 −1.00000 1.00000(5) −1.00000(5) 1.00000(10) −1.00000(10) 1.00000(20)

−1.00000(20) 0.200000(20) −0.200000(20)，如图 3.6‑38 所示。其余设置与上述相同。迭代方法、迭代步数和收敛准则设置如图 3.6‑39 所示。

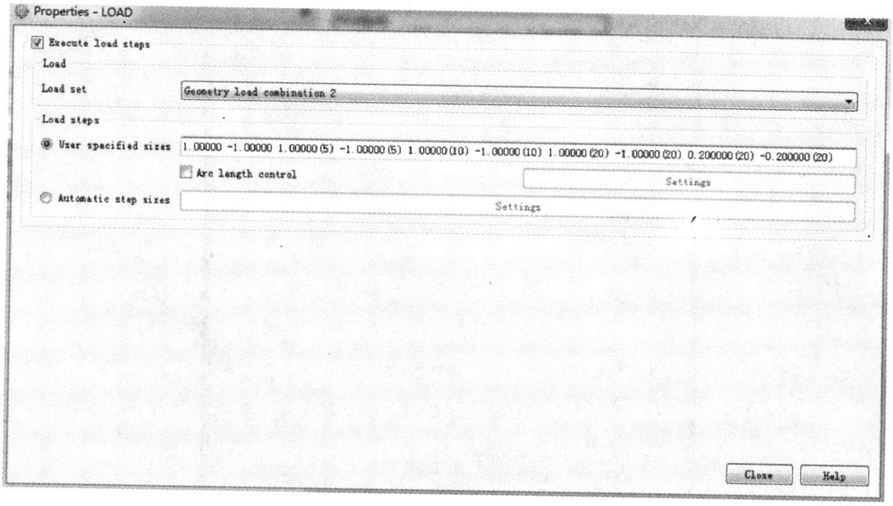

图 3.6‑38　位移加载步长设置

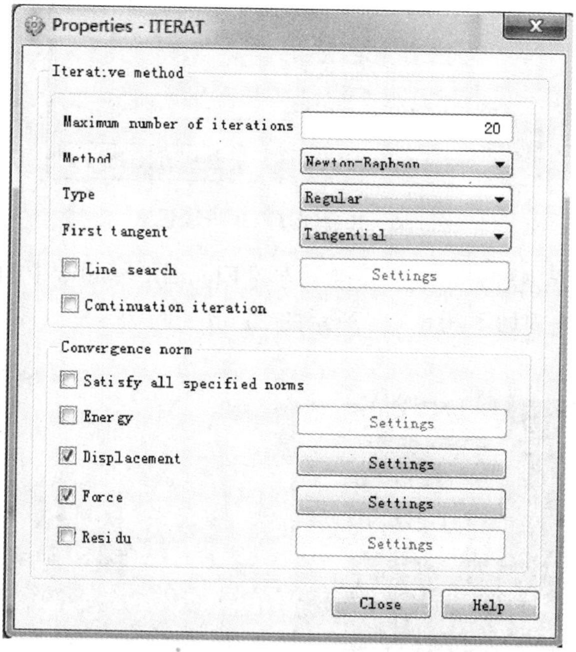

图 3.6‑39　荷载组合 2 收敛方法设置操作界面对话框

对输出结果进行设置，如图 3.6‑40 所示。选择整体坐标系下的位移（DISPLA TOTAL TRANSL GLOBAL）、开裂应变（STRAIN CRACK GREEN）、整体坐标系下的累积开裂应变（STRAIN CRKSUM GREEN GLOBAL）、主应力方向的累积开裂应变（STRAIN CRKSUM GREEN PRINCI）、局部坐标系下的累积开裂应变（STRAIN CRKSUM GREEN LOCAL）、整体坐标系下的累积裂缝宽度（STRAIN CRKWDT GREEN GLOBAL）、局部坐标系下的累积裂缝宽度（STRAIN CRKWDT GREEN LOCAL）。

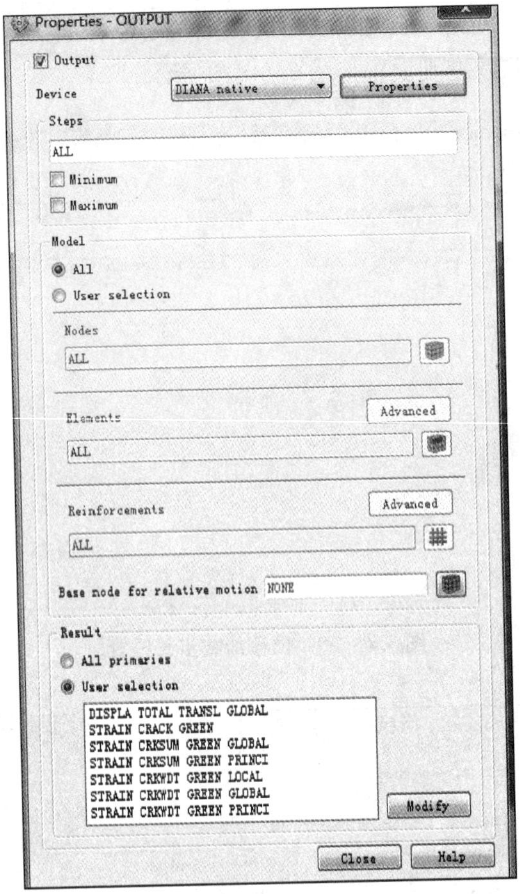

图 3.6-40　OUTPUT 选项设置

选择最终荷载步，点击 Results 结果栏下方的 Element results 中的 Crack Strains 选项，点击其下方的 Eknn，出现如图 3.6-41 所示的 Eknn 云图。

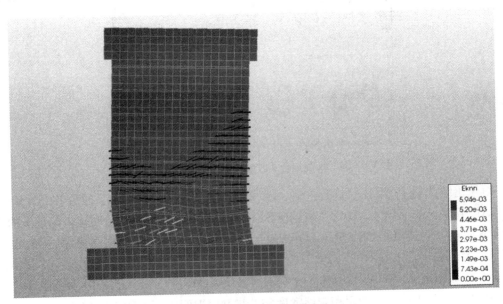

图 3.6-41　Eknn 云图

点击 Summed Crack Strains 下方的 EkXX,查看位移加载完毕后整体坐标系下的 X 方向裂缝累积应变分布云图,如图 3.6-42 所示。再点击下方的 EkYY 选项,查看 EkYY 裂缝应变分布云图。

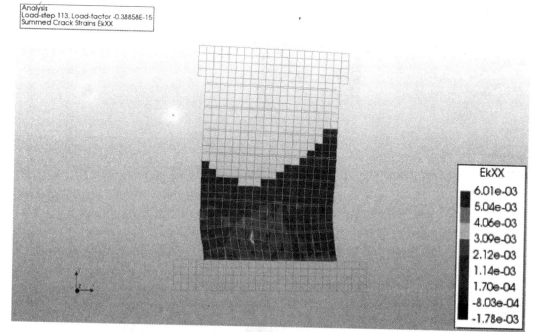

图 3.6-42　最终荷载步时 EkXX 裂缝累积应变分布云图

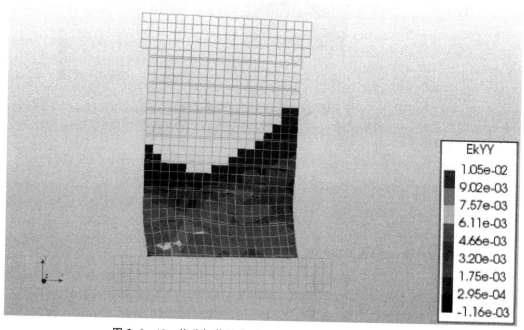

图 3.6-43　位移加载结束时 EkYY 裂缝累积应变分布云图

点击 Element resukts→Crack-widths→Ecw1,查看主应力方向的裂缝宽度分布云图,如图 3.6-44 所示。再查看整体坐标系下的 X 方向裂缝宽度 EcwXX 分布云图,如图 3.6-45 所示。

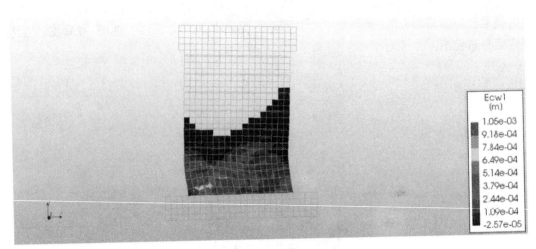

图 3.6-44　Ecw1 主应力方向裂缝宽度云图

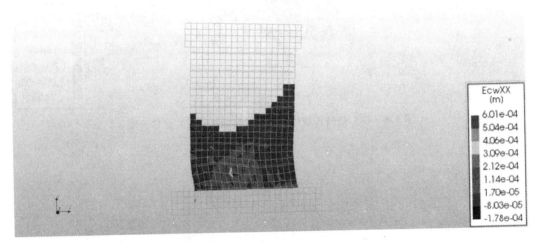

图 3.6-45　EcwXX 裂缝宽度云图

Python 语言如下：

\#

\# DianaIE 10.1 update 2017-04-25 13:38:53

\# Python　3.3.4

\# Session recorded at 2018-04-24 18:30:38

\#

```
newProject( "Shear wall", 10 )
setModelAnalysisAspects( [ "STRUCT" ] )
setModelDimension( "2D" )
setDefaultMeshOrder( "QUADRATIC" )
setDefaultMesherType( "HEXQUAD" )
```

第三章 Diana 土木工程非线性建模案例分析

```
setDefaultMidSideNodeLocation( "LINEAR" )
createSheet( "support", [[ 0, 0, 0 ],[ 2.6, 0, 0 ],[ 2.6, 0.4, 0 ],[ 0, 0.4, 0 ]] )
createSheet( "wall", [[ 0.4, 0.4, 0 ],[ 2.2, 0.4, 0 ],[ 2.2, 2.6, 0 ],[ 0.4, 2.6, 0 ]] )
createSheet( "top", [[ 0.3, 2.6, 0 ],[ 2.3, 2.6, 0 ],[ 2.3, 3, 0 ],[ 0.3, 3, 0 ]] )
createLine( "FRP", [ 0.3, 2.7, 0 ], [ 2.3, 2.7, 0 ] )
createLine( "bar1", [ 1.3, 0.2, 0 ], [ 1.3, 2.8, 0 ] )
arrayCopy( [ "bar1" ], [ -0.2, 0, 0 ], [ 0, 0, 0 ], [ 0, 0, 0 ], 4 )
arrayCopy( [ "bar1" ], [ 0.2, 0, 0 ], [ 0, 0, 0 ], [ 0, 0, 0 ], 4 )
createLine( "stirrup", [ 0.44, 0.5, 0 ], [ 0.74, 0.5, 0 ] )
arrayCopy( [ "stirrup" ], [ 0, 0.2, 0 ], [ 0, 0, 0 ], [ 0, 0, 0 ], 10 )
arrayCopy( [ "stirrup", "stirrup 1", "stirrup 2", "stirrup 3", "stirrup 4", "stirrup 5", "stirrup 6", "stirrup 7", "stirrup 8", "stirrup 9", "stirrup 10" ], [ 1.4, 0, 0 ], [ 0, 0, 0 ], [ 0, 0, 0 ], 1 )
createLine( "barx1", [ 0.42, 0.42, 0 ], [ 2.16, 0.42, 0 ] )
arrayCopy( [ "barx1" ], [ 0, 0.3, 0 ], [ 0, 0, 0 ], [ 0, 0, 0 ], 7 )
addMaterial( "wall", "CONCR", "TSCR", [] )
setParameter( "MATERIAL", "wall", "LINEAR/ELASTI/YOUNG", 3.7e+10 )
setParameter( "MATERIAL", "wall", "LINEAR/ELASTI/POISON", 0.15 )
setParameter( "MATERIAL", "wall", "LINEAR/MASS/DENSIT", 2500 )
setParameter( "MATERIAL", "wall", "TENSIL/TENCRV", "HORDYK" )
setParameter( "MATERIAL", "wall", "TENSIL/TENSTR", 3 )
setParameter( "MATERIAL", "wall", "TENSIL/TENSTR", 2.8 )
setParameter( "MATERIAL", "wall", "TENSIL/TENSTR", 2800000 )
setParameter( "MATERIAL", "wall", "TENSIL/GF1", 200 )
setParameter( "MATERIAL", "wall", "TENSIL/RESTST", 100 )
setParameter( "MATERIAL", "wall", "COMPRS/COMCRV", "MAEKCC" )
setParameter( "MATERIAL", "wall", "COMPRS/COMSTR", 45000000 )
setParameter( "MATERIAL", "wall", "SHEAR/BETA", 0.1 )
addGeometry( "Element geometry 1", "SHEET", "MEMBRA", [] )
rename( "GEOMET", "Element geometry 1", "wall" )
setParameter( "GEOMET", "wall", "THICK", 0.4 )
clearReinforcementAspects( [ "wall" ] )
setElementClassType( "SHAPE", [ "wall" ], "MEMBRA" )
assignMaterial( "wall", "SHAPE", [ "wall" ] )
assignGeometry( "wall", "SHAPE", [ "wall" ] )
resetElementData( "SHAPE", [ "wall" ] )
addMaterial( "beam", "CONCR", "LEI", [] )
setParameter( "MATERIAL", "beam", "LINEAR/ELASTI/YOUNG", 3.25e+10 )
setParameter( "MATERIAL", "beam", "LINEAR/ELASTI/POISON", 0.15 )
```

```
setParameter( "MATERIAL", "beam", "LINEAR /MASS /DENSIT", 2500 )
addGeometry( "Element geometry 2", "SHEET", "MEMBRA", [] )
rename( "GEOMET", "Element geometry 2", "beam" )
setParameter( "GEOMET", "beam", "THICK", 0.6 )
clearReinforcementAspects( [ "support", "top" ] )
setElementClassType( "SHAPE", [ "support", "top" ], "MEMBRA" )
assignMaterial( "beam", "SHAPE", [ "support", "top" ] )
assignGeometry( "beam", "SHAPE", [ "support", "top" ] )
resetElementData( "SHAPE", [ "support", "top" ] )
setParameter( "GEOMET", "beam", "THICK", 0.6 )
addMaterial( "bar", "REINFO", "VMISES", [] )
setParameter( "MATERIAL", "bar", "LINEAR /ELASTI /YOUNG", 2.1e+11 )
setParameter( "MATERIAL", "bar", "PLASTI /HARDI1 /YLDSTR", 4.4e+08 )
addGeometry( "Element geometry 3", "RELINE", "REBAR", [] )
rename( "GEOMET", "Element geometry 3", "bar" )
setParameter( "GEOMET", "bar", "REIEMB /CROSSE", 0.000157 )
setReinforcementAspects( [ "bar1", "bar2", "bar3", "bar4", "bar5", "bar6",
"bar7", "bar8", "bar9" ] )
assignMaterial( "bar", "SHAPE", [ "bar1", "bar2", "bar3", "bar4", "bar5",
"bar6", "bar7", "bar8", "bar9" ] )
assignGeometry( "bar", "SHAPE", [ "bar1", "bar2", "bar3", "bar4", "bar5",
"bar6", "bar7", "bar8", "bar9" ] )
resetElementData( "SHAPE", [ "bar1", "bar2", "bar3", "bar4", "bar5", "bar6",
"bar7", "bar8", "bar9" ] )
setReinforcementDiscretization( [ "bar1", "bar2", "bar3", "bar4", "bar5",
"bar6", "bar7", "bar8", "bar9" ], "SECTION" )
addMaterial( "barx", "REINFO", "VMISES", [] )
setParameter( "MATERIAL", "barx", "LINEAR /ELASTI /YOUNG", 2.1e+11 )
setParameter( "MATERIAL", "barx", "PLASTI /HARDI1 /YLDSTR", 4.4e+08 )
addGeometry( "Element geometry 4", "RELINE", "REBAR", [] )
rename( "GEOMET", "Element geometry 4", "barx" )
setParameter( "GEOMET", "barx", "REIEMB /CROSSE", 0.000157 )
setReinforcementAspects( [ "barx1", "barx2", "barx3", "barx4", "barx5",
"barx6", "barx7", "barx8" ] )
assignMaterial( "barx", "SHAPE", [ "barx1", "barx2", "barx3", "barx4",
"barx5", "barx6", "barx7", "barx8" ] )
assignGeometry( "barx", "SHAPE", [ "barx1", "barx2", "barx3", "barx4",
"barx5", "barx6", "barx7", "barx8" ] )
resetElementData( "SHAPE", [ "barx1", "barx2", "barx3", "barx4", "barx5",
```

```
"barx6", "barx7", "barx8" ] )
    setReinforcementDiscretization( [ "barx1", "barx2", "barx3", "barx4",
"barx5", "barx6", "barx7", "barx8" ], "SECTION" )
    saveProject( )
    saveProject( )
    addMaterial( "stirrup", "REINFO", "VMISES", [] )
    setParameter( "MATERIAL", "stirrup", "LINEAR/ELASTI/YOUNG", 2.1e+11 )
    setParameter( "MATERIAL", "stirrup", "PLASTI/HARDI1/YLDSTR", 4.4e+08 )
    addGeometry( "Element geometry 5", "RELINE", "REBAR", [] )
    rename( "GEOMET", "Element geometry 5", "stirrup" )
    setParameter( "GEOMET", "stirrup", "REIEMB/CROSSE", 1.0053e-04 )
    setReinforcementAspects( [ "stirrup", "stirrup 1", "stirrup 2", "stirrup 3",
"stirrup 4", "stirrup 5", "stirrup 6", "stirrup 7", "stirrup 8", "stirrup 9",
"stirrup 10", "stirrup 11", "stirrup 12", "stirrup 13", "stirrup 14", "stirrup 15",
"stirrup 16", "stirrup 17", "stirrup 18", "stirrup 19", "stirrup 20", "stirrup 21" ]
)
    assignMaterial( "stirrup", "SHAPE", [ "stirrup", "stirrup 1", "stirrup 2",
"stirrup 3", "stirrup 4", "stirrup 5", "stirrup 6", "stirrup 7", "stirrup 8",
"stirrup 9", "stirrup 10", "stirrup 11", "stirrup 12", "stirrup 13", "stirrup 14",
"stirrup 15", "stirrup 16", "stirrup 17", "stirrup 18", "stirrup 19", "stirrup 20",
"stirrup 21" ] )
    assignGeometry( "stirrup", "SHAPE", [ "stirrup", "stirrup 1", "stirrup 2",
"stirrup 3", "stirrup 4", "stirrup 5", "stirrup 6", "stirrup 7", "stirrup 8",
"stirrup 9", "stirrup 10", "stirrup 11", "stirrup 12", "stirrup 13", "stirrup 14",
"stirrup 15", "stirrup 16", "stirrup 17", "stirrup 18", "stirrup 19", "stirrup 20",
"stirrup 21" ] )
    resetElementData( "SHAPE", [ "stirrup", "stirrup 1", "stirrup 2", "stirrup 3",
"stirrup 4", "stirrup 5", "stirrup 6", "stirrup 7", "stirrup 8", "stirrup 9", "stirrup
10", "stirrup 11", "stirrup 12", "stirrup 13", "stirrup 14", "stirrup 15", "stirrup 16",
"stirrup 17", "stirrup 18", "stirrup 19", "stirrup 20", "stirrup 21" ] )
    setReinforcementDiscretization( [ "stirrup", "stirrup 1", "stirrup 2",
"stirrup 3", "stirrup 4", "stirrup 5", "stirrup 6", "stirrup 7", "stirrup 8",
"stirrup 9", "stirrup 10", "stirrup 11", "stirrup 12", "stirrup 13", "stirrup 14",
"stirrup 15", "stirrup 16", "stirrup 17", "stirrup 18", "stirrup 19", "stirrup 20",
"stirrup 21" ], "SECTION" )
    saveProject( )
    addMaterial( "FRP", "REINFO", "VMISES", [] )
    setParameter( "MATERIAL", "FRP", "LINEAR/ELASTI/YOUNG", 1.36e+11 )
    setParameter( "MATERIAL", "FRP", "PLASTI/HARDI1/YLDSTR", 1.2e+09 )
```

```
addGeometry( "Element geometry 6", "RELINE", "REBAR", [ ] )
rename( "GEOMET", "Element geometry 6", "FRP" )
setParameter( "GEOMET", "FRP", "REIEMB /CROSSE", 0.000139 )
setReinforcementAspects( [ "FRP" ] )
assignMaterial( "FRP", "SHAPE", [ "FRP" ] )
assignGeometry( "FRP", "SHAPE", [ "FRP" ] )
resetElementData( "SHAPE", [ "FRP" ] )
setReinforcementDiscretization( [ "FRP" ], "ELEMENT" )
saveProject( )
addSet( "GEOMETRYSUPPORTSET", "Geometry support set 1" )
rename( "GEOMETRYSUPPORTSET", "Geometry support set 1", "co1" )
createLineSupport( "co1", "co1" )
setParameter( "GEOMETRYSUPPORT", "co1", "AXES", [ 1, 2 ] )
setParameter( "GEOMETRYSUPPORT", "co1", "TRANSL", [ 1, 1, 0 ] )
setParameter( "GEOMETRYSUPPORT", "co1", "ROTATI", [ 0, 0, 0 ] )
attach( " GEOMETRYSUPPORT", " co1", " support", [[ 1. 3, 7. 4981504e - 34,
1.2245938e - 17 ]] )
saveProject( )
createVertex( "Vertex 1", [ 2.8, 2.8, 0 ] )
projection( "SHAPEEDGE", "top", [[ 2.3, 2.8, - 5.6944687e - 19 ]], [ "Vertex 1" ],
[ -1, 0, 0 ], True )
removeShape( [ "Vertex 1" ] )
addSet( "GEOMETRYSUPPORTSET", "Geometry support set 2" )
rename( "GEOMETRYSUPPORTSET", "Geometry support set 2", "co2" )
createPointSupport( "co2", "co2" )
setParameter( "GEOMETRYSUPPORT", "co2", "AXES", [ 1, 2 ] )
setParameter( "GEOMETRYSUPPORT", "co2", "TRANSL", [ 1, 0, 0 ] )
setParameter( "GEOMETRYSUPPORT", "co2", "ROTATI", [ 0, 0, 0 ] )
attach( "GEOMETRYSUPPORT", "co2", "top", [[ 2.3, 2.8, -5.6944687e-19 ]] )
saveProject( )
addSet( "GEOMETRYLOADSET", "gravity" )
createModelLoad( "gravity", "gravity" )
saveProject( )
addSet( "GEOMETRYLOADSET", "Geometry load case 2" )
createLineLoad( "load", "Geometry load case 2" )
setParameter( "GEOMETRYLOAD", "load", "FORCE /VALUE", - 30000 )
setParameter( "GEOMETRYLOAD", "load", "FORCE /DIRECT", 2 )
attach( "GEOMETRYLOAD", "load", "top", [[ 1.3, 3, - 1.2245938e - 17 ]] )
saveProject( )
```

```
addSet( "GEOMETRYLOADSET", "Geometry load case 3" )
createPointLoad( "displacement", "Geometry load case 3" )
setParameter( "GEOMETRYLOAD", "displacement", "LODTYP", "DEFORM" )
setParameter( "GEOMETRYLOAD", "displacement", "DEFORM/TR/VALUE", -0.0005 )
setParameter( "GEOMETRYLOAD", "displacement", "DEFORM/TR/DIRECT",1 )
attach( "GEOMETRYLOAD", "displacement", "top", [[ 2.3, 2.8, -5.6944687e-19 ]] )
saveProject( )
setDefaultGeometryLoadCombinations( )
setGeometryLoadCombinationFactor( "Geometry load combination 1", "gravity", 1 )
remove( "GEOMETRYLOADCOMBINATION", "Geometry load combination 1" )
remove( "GEOMETRYLOADCOMBINATION", "Geometry load combination 2" )
remove( "GEOMETRYLOADCOMBINATION", "Geometry load combination 3" )
addGeometryLoadCombination( "" )
setGeometryLoadCombinationFactor( "Geometry load combination 1", "gravity", 1 )
setGeometryLoadCombinationFactor( "Geometry load combination 1", "Geometry load case 2", 1 )
addGeometryLoadCombination( "" )
setGeometryLoadCombinationFactor( "Geometry load combination 2", "Geometry load case 3", 1 )
setGeometryLoadCombinationFactor( "Geometry load combination 1", "gravity", 1 )
setElementSize( [ "support", "wall", "top" ], 0.1, -1, True )
setMesherType( [ "support", "wall", "top" ], "HEXQUAD" )
setMidSideNodeLocation( [ "support", "wall", "top" ], "LINEAR" )
generateMesh( [] )
hideView( "GEOM" )
showView( "MESH" )
addAnalysis( "Analysis7" )
renameAnalysis( "Analysis7", "Analysis" )
addAnalysisCommand( "Analysis", "NONLIN", "Structural nonlinear" )
renameAnalysis( "Analysis", "Analysis" )
setAnalysisCommandDetail( "Analysis", "Structural nonlinear", "EXECUT(1)/ITERAT/MAXITE", 50 )
setAnalysisCommandDetail( "Analysis", "Structural nonlinear", "EXECUT(1)/ITERAT/CONVER/DISPLA/NOCONV", "CONTIN" )
setAnalysisCommandDetail( "Analysis", "Structural nonlinear", "EXECUT(1)/ITERAT/CONVER/FORCE/NOCONV", "CONTIN" )
setAnalysisCommandDetail( "Analysis", "Structural nonlinear", "EXECUT/EXETYP", "LOAD" )
```

```
    setAnalysisCommandDetail( "Analysis", "Structural nonlinear", "EXECUT(2) /
LOAD /LOADNR", 2 )
    setAnalysisCommandDetail( "Analysis", "Structural nonlinear", "EXECUT(2) /
LOAD /STEPS /EXPLIC /SIZES", "1 -1 1(5) -1(5) 1(10) -1(10) 1(20) -1(20) 0.2(20)
 -0.2(20)" )
    setAnalysisCommandDetail( "Analysis", "Structural nonlinear", "EXECUT(2) /
ITERAT /MAXITE", 20 )
    setAnalysisCommandDetail( "Analysis", "Structural nonlinear", "EXECUT(2) /
ITERAT /CONVER /DISPLA /NOCONV", "CONTIN" )
    setAnalysisCommandDetail( "Analysis", "Structural nonlinear", "EXECUT(2) /
ITERAT /CONVER /FORCE /NOCONV", "CONTIN" )
    setAnalysisCommandDetail( "Analysis", "Structural nonlinear", "EXECUT(2) /
ITERAT /CONVER /DISPLA /TOLCON", 0.05 )
    setAnalysisCommandDetail( "Analysis", "Structural nonlinear", "EXECUT(2) /
ITERAT /CONVER /FORCE /TOLCON", 0.05 )
    setAnalysisCommandDetail( "Analysis", "Structural nonlinear", "OUTPUT(1) /
SELTYP", "USER" )
    addAnalysisCommandDetail( "Analysis", "Structural nonlinear", "OUTPUT(1) /
USER" )
    addAnalysisCommandDetail( "Analysis", "Structural nonlinear", "OUTPUT(1) /
USER /DISPLA(1) /TOTAL /TRANSL /GLOBAL" )
    addAnalysisCommandDetail( "Analysis", "Structural nonlinear", "OUTPUT(1) /
USER /STRAIN(1) /CRACK /GREEN" )
    addAnalysisCommandDetail( "Analysis", "Structural nonlinear", "OUTPUT(1) /
USER /STRAIN(2) /CRKSUM /GREEN /GLOBAL" )
    addAnalysisCommandDetail( "Analysis", "Structural nonlinear", "OUTPUT(1) /
USER /STRAIN(3) /CRKSUM /GREEN /PRINCI" )
    addAnalysisCommandDetail( "Analysis", "Structural nonlinear", "OUTPUT(1) /
USER /STRAIN(4) /CRKWDT /GREEN /LOCAL" )
    addAnalysisCommandDetail( "Analysis", "Structural nonlinear", "OUTPUT(1) /
USER /STRAIN(5) /CRKWDT /GREEN /GLOBAL" )
    addAnalysisCommandDetail( "Analysis", "Structural nonlinear", "OUTPUT(1) /
USER /STRAIN(6) /CRKWDT /GREEN /PRINCI" )
    runSolver( "Analysis" )
    showView( "RESULT" )
    setResultCase( [ "Analysis", "Output", "Load - step 113, Load - factor -
0.38858E - 15" ] )
    setResultPlot( "contours", "Crack - widths /node", "Ecwxx" )
    setResultPlot( "cracks", "Crack Strains /mappedcrack", "Eknn" )
```

```
setResultPlot( "contours", "Crack-widths /node", "Ecw3" )
setResultPlot( "contours", "Crack-widths /node", "Ecw2" )
setResultPlot( "contours", "Crack-widths /node", "Ecw1" )
setResultPlot( "contours", "Crack-widths /node", "EcwYY" )
setResultPlot( "contours", "Crack-widths /node", "EcwXX" )
setResultPlot( "contours", "Crack-widths /node", "Ecw1" )
setResultPlot( "contours", "Crack-widths /node", "Ecw2" )
setResultPlot( "contours", "Crack-widths /node", "Ecwxx" )
setResultPlot( "contours", "Summed Crack Strains /node", "EkXX" )
setResultPlot( "contours", "Summed Crack Strains /node", "EkYY" )
setResultPlot( "contours", "Crack-widths /node", "Ecwxx" )
setResultPlot( "contours", "Crack-widths /node", "EcwXX" )
setResultPlot( "contours", "Crack-widths /node", "EcwYY" )
setResultPlot( "contours", "Total Displacements /node", "TDtX" )
setResultPlot( "contours", "Total Displacements /node", "TDtY" )
setResultPlot( "contours", "Summed Crack Strains /node", "Ek1" )
setResultPlot( "contours", "Summed Crack Strains /node", "EkXX" )
setResultPlot( "contours", "Summed Crack Strains /node", "EkYY" )
setResultPlot( "contours", "Crack-widths /node", "EcwYY" )
setResultPlot( "contours", "Crack-widths /node", "Ecw1" )
setResultPlot( "contours", "Crack-widths /node", "Ecw3" )
setResultPlot( "contours", "Crack-widths /node", "EcwXX" )
```

3.7 案例七：预应力筋粘结滑移模型

本案例为一根简单的混凝土简支梁。其中简支梁全长为 20 m，高度为 4 m，直径为 15.24 mm 的抛物线形预应力筋贯穿整个梁体。混凝土单元采用 2D 二次平面应力单元，预应力曲线筋为两端张拉的方式，采用粘结滑移本构模型，弹性模量 $1.95×10^{11}$ N/m^2，屈服强度 1860 MPa。其中抛物线预应力筋最大高度距梁底部为 1.5 m，最低高度距梁底部为 0.7 m。整个简支梁承受逐级等步长加载荷载步数为 5 步，荷载集度共计为 100 kN/m 的均布荷载。混凝土的弹性模量为 $3.25×10^{10}$ N/m^2，单元拉伸断裂能 500 N/m，极限抗拉强度 2.4 MPa，抗压强度为 26.8 MPa，选用 Hordijk 拉伸软化模型。预应力简支梁几何及受力特性如图 3.7-1 所示。

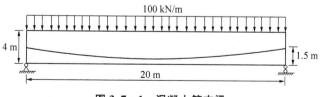

图 3.7-1 混凝土简支梁

学习要义：

(1) 了解均布荷载作用下抛物线预应力筋的布筋方式，学会建立简单形状的抛物线筋几何模型。

(2) 学会定义预应力筋 Bar 单元粘结滑移本构模型界面法向和切向刚度以及截面几何特性。

(3) 学会查看粘结滑移计算下预应力筋塑性变化趋势。

打开 DianaIE，在 G 盘名称为例题的文件夹中创立新的 .dpf 文件，命名为 Bondslip-10.1，建立最大尺寸范围为 100 m 的 2D 模型，网格单元形状为六面体/四边形单元，阶数为二阶。如图 3.7-2 所示。

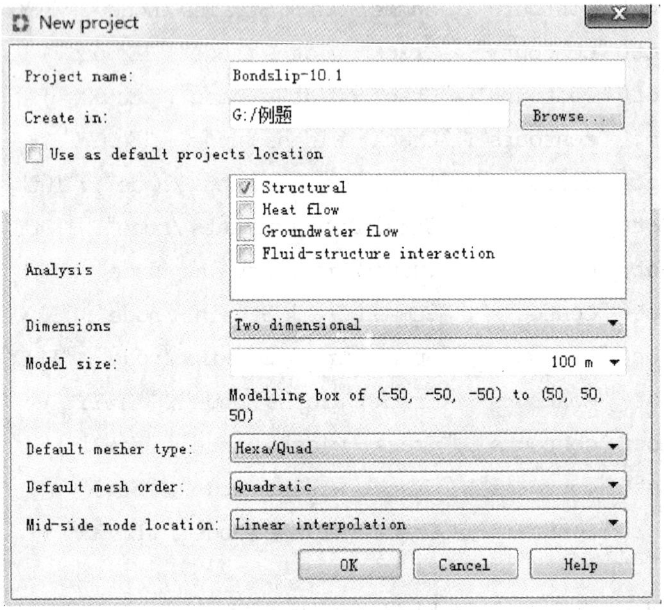

图 3.7-2

点击 Add a sheet 图标，依次输入各点坐标，创立 Sheet1。建立混凝土梁的几何模型，如图 3.7-3 所示。

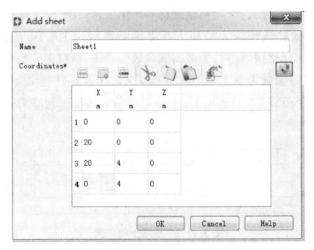

图 3.7-3　各点坐标

建立抛物线形预应力筋的几何模型，点击快捷图标 Add a curve，创立曲线，依次输入坐标(0,1.7),(10,0.6)和(20,1.7),如图 3.7-4 所示。点击 OK，生成曲线预应力筋的几何模型。

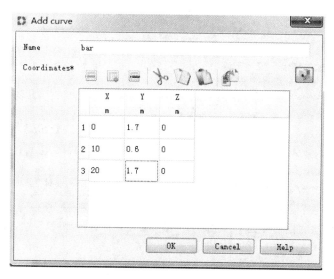

图 3.7-4 曲线各点坐标

点击 Geometry 栏下方的 Sheet1，选中后点击鼠标右键，选择 Property assignment，弹出形状属性编辑界面(Shape property assignments)，下方的单元类型选择普通平面应力单元(Regular Plane Stress)，点击 Material 右侧的小图标，编辑材料属性，将名称命名为 concrete，材料类型为混凝土和砌体结构(Concrete and masonry)类型，材料模型选择为总应变裂缝模型(Total strain based crack model)，裂缝来源(Crack orientation)选择为旋转开裂(Rotating)，拉伸软化曲线选择为 Hordijk 曲线，输入混凝土弹性模量 3.25×10^{10} N/m²，抗压强度 26.8 MPa，极限抗拉强度为 2.4×10^6 N/m²，单元的拉伸断裂能(Mode-I tensile fracture energy)设定为 500 N/m，抗压强度曲线为抛物线型(Parabolic)，抗压断裂能为 4500 N/m，如图 3.7-5—图 3.7-8 所示。

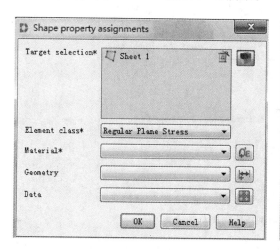

图 3.7-5 平面应力单元

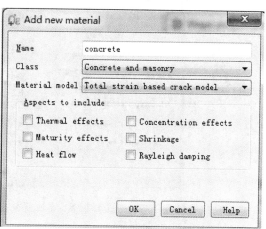

图 3.7-6 总应变裂缝模型界面

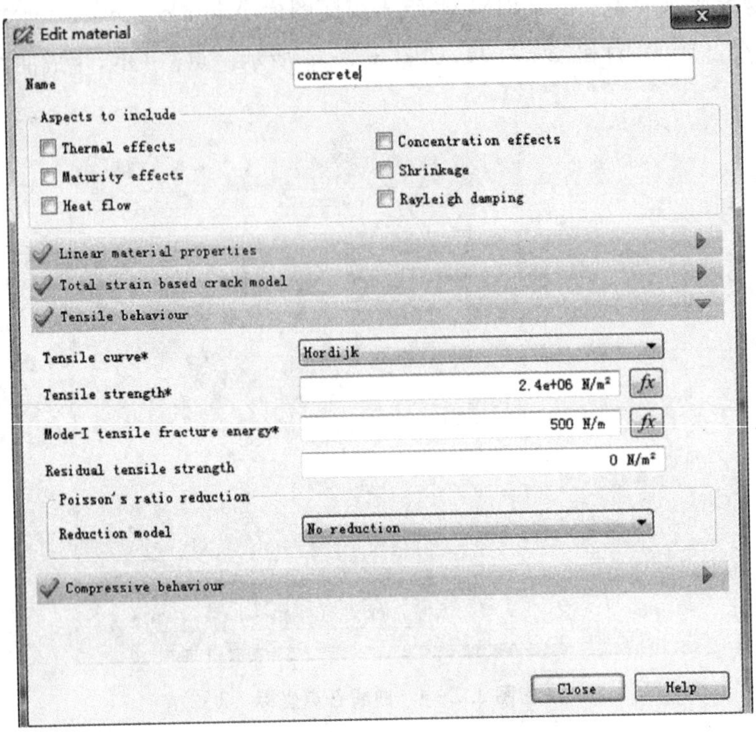

图 3.7-7　总应变裂缝模型拉伸强度参数设置

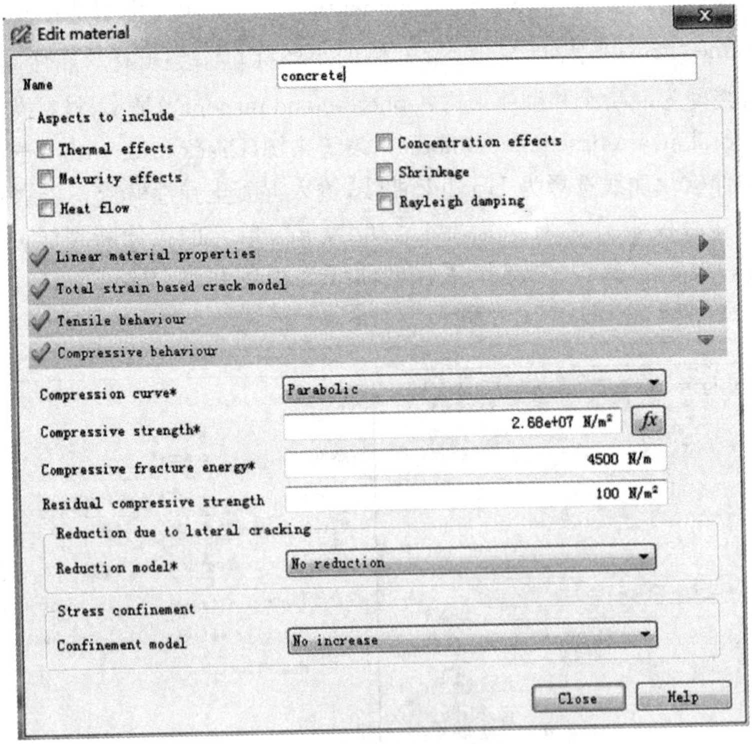

图 3.7-8　总应变裂缝模型抗压强度参数设置

赋予混凝梁截面几何特性,点击图标 ,厚度为 1 m,默认局部坐标系 x 轴对应整体坐标系下的 X 方向。

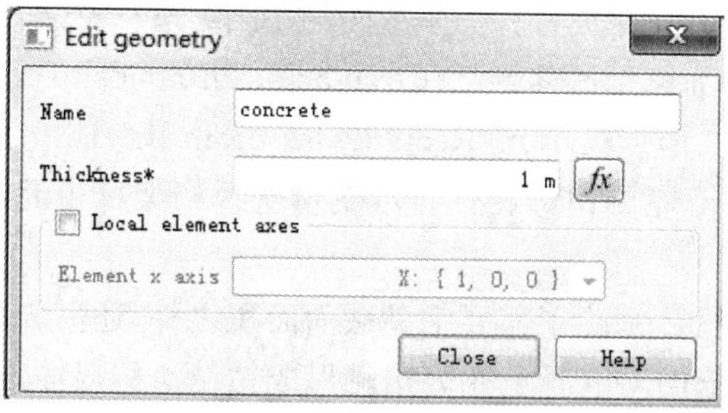

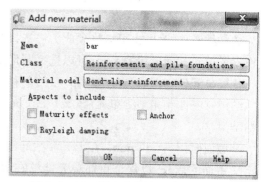

图 3.7-9 截面几何特性

图 3.7-10 预应力粘结—滑移模型材料本构编辑对话框

创立预应力抛物线筋的粘结滑移材料本构特性。选中预应力抛物线筋,右击选择 Reinforcement property assignment 创建预应力筋的材料特性,材料类型选择为粘结滑移钢筋(Bond-slip reinforcement),如图 3.7-10 所示。

设定粘结滑移基本参数,包括钢筋的弹性模量 $1.95\times10^{11}\,\mathrm{N/m^2}$,密度 $7800\,\mathrm{kg/m^3}$。塑性模型选择非塑性。这表明此时粘结滑移状态下的预应力筋为弹性状态,并可以在后续非线性计算中查看预应力筋从弹性进入到塑性的整个完整过程。

点击 OK,弹出如图 3.7-11 所示的粘结滑移界面本构。粘结滑移界面单元(Bond-slip interface)的单元刚度包括法向刚度(Normal stiffness modulus)和剪切刚度(Shear stiffness modulus),通常情况下法向刚度数值较大,在本算例中设定为 $2\times10^{12}\,\mathrm{N/m^3}$,切向刚度相对法向刚度数值较小,设定为 $2\times10^6\,\mathrm{N/m^3}$。在接下来的钢筋粘结滑移界面单元破坏模型(Bond-slip inretface failure model)中选择 Doerr 立方粘结滑移函数(Cubic bond-slip function by Doerr),滑移参数为 $20\,\mathrm{N/m^2}$,剪切滑移开始稳定时的长度(Shear slip at start plateau)为 0.1 m。

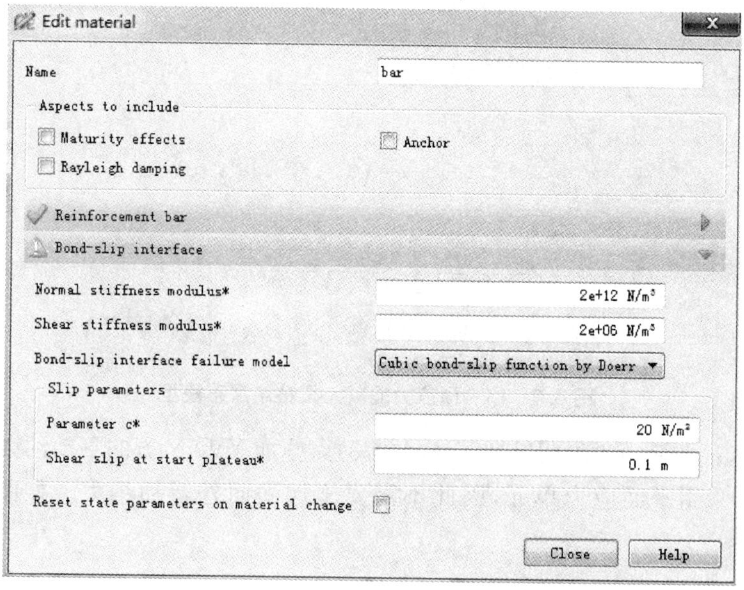

图 3.7-11 Diana10.1 钢筋粘结滑移参数设置

打开 Edit geometry 框,定义粘结滑移几何特性参数,命名为 bar,其中钢筋类型一栏中对于粘结滑移有多种选择,用户既可以将预应力筋赋予桁架杆单元(Truss bondslip)力学行为的粘结滑移特性,也可以将预应力筋选择为具有梁单元杆件力学行为的粘结滑移特性。根据梁单元截面形状的不同,可以选择圆形梁单元粘结滑移特性(Circular beam bondslip)、管状梁单元(Pipe beam bondslip)粘结滑移特性、矩形梁单元(Rectangular beam bondslip)粘结滑移特性和箱型梁单元(Box beam bondslip)粘结滑移特性。本案例选择桁架杆单元粘结滑移。

截面面积为 139 mm²,在 Diana 粘结滑移特性中,除了需要定义截面面积之外,还需要定义钢筋的周长,以保证钢筋单元能够在混凝土内与混凝土充分接触形成粘结滑移界面。在本案例中预应力筋周长为 $2\pi \times 15.24 \approx 95.76$ mm。曲线筋粘结滑移界面几何特性定义界面操作如图 3.7-12 所示。

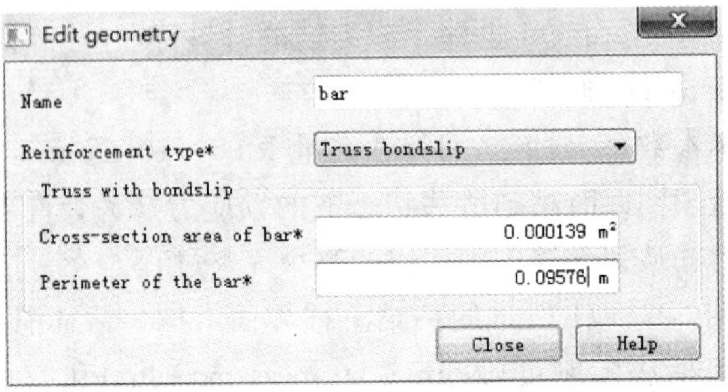

图 3.7-12 曲线筋粘结滑移界面几何特性定义

点击 OK,形成粘结滑移界面单元,同时预应力筋也具有了粘结滑移特性。其中在 Diana 10.1 软件中,粘结滑移界面在各项材料本构参数和几何特性参数定义完毕后会自动与混凝土之间形成粘结滑移界面接触单元,点击 OK,形成如图 3.7-13 所示的模型。

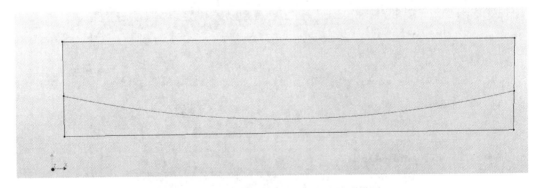

图 3.7-13 预应力混凝土梁粘结滑移模型

添加约束,对坐标(0,0,0)点和坐标(20,0,0)点施加 X 和 Y 方向平动约束,分别命名为 co1,co2。由于采用平面应力单元,因此不需要施加平面外转动约束。如图 3.7-14 和图 3.7-15 所示。

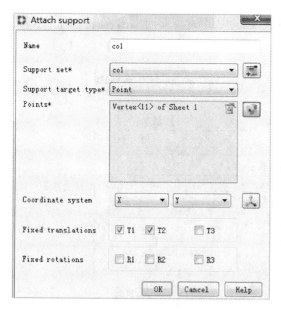
图 3.7-14 施加 co1 约束

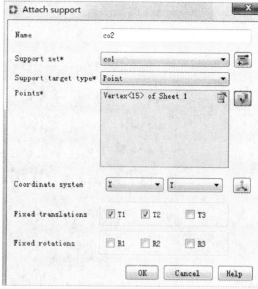
图 3.7-15 施加 co2 约束

点击 Load 栏下方的 Define a global load 图标 ![icon],添加重力荷载,命名为 gravity。添加预应力荷载,采用两端张拉的方式,预应力两端张拉力大小均为 150 kN。两端预应力筋回缩长度 1×10^{-4} m,库仑摩擦系数采用软件默认值 0.22,握裹系数 0.01/m。如图 3.7-16 所示。

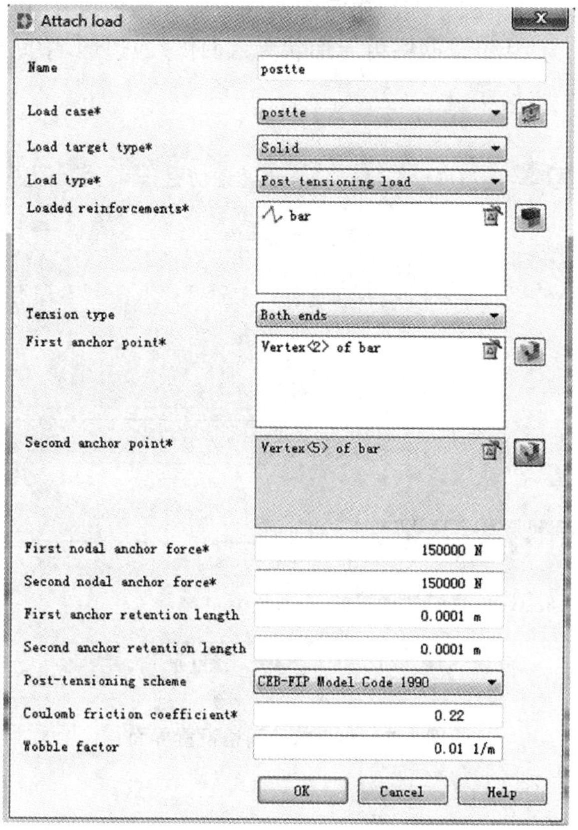

图 3.7-16 施加预应力荷载的操作界面

添加 load 荷载,荷载形式为均布荷载,大小为 20 kN/m,方向沿着 Z 轴向下。如图 3.7-17 所示。

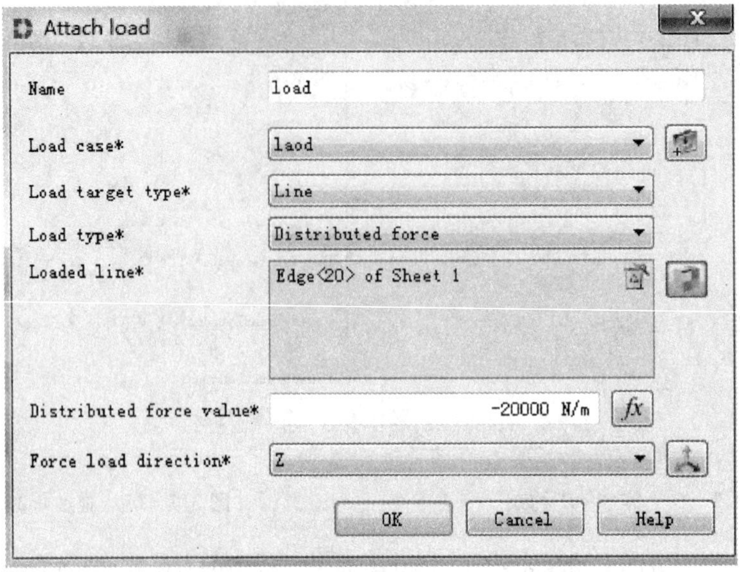

图 3.7-17 施加均布线荷载的操作界面

添加荷载工况组合,将重力 gravity 和预应力荷载 postte 添加为第一组 Load combination1,将 load 荷载单独添加为第二组。设定单元网格尺寸,网格划分方式采用份数(Divisions)划分,份数为 20 份。网格单元形状为六面体/四边形,中间节点位置确定方式为线性插值。如图 3.7-18 所示。

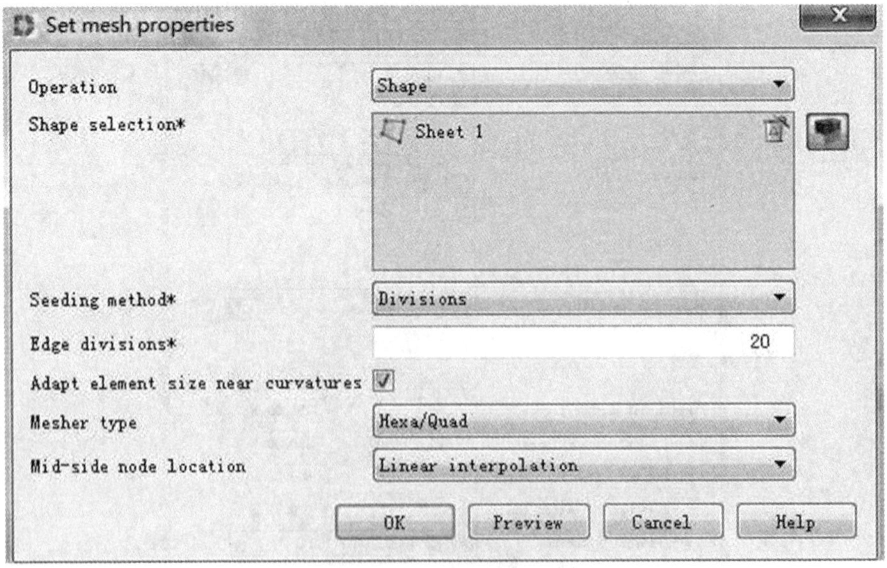

图 3.7-18 单元网格设置界面

点击快捷键图标 Generate mesh of a shape 按钮![img],生成如图 3.7－19 所示的网格图形,在 Mesh 栏下方的 Element types 中查看单元类型名称,确认是需要的二维平面应力单元 CQ16M。

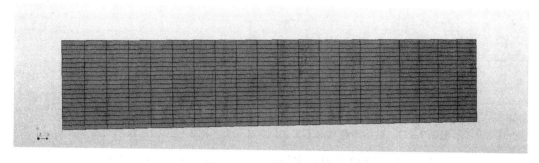

图 3.7－19　单元网格生成

添加分析工况,建立结构非线性分析,将 Load combination1 中的重力荷载和预应力荷载因子设定为初始工况荷载,命名为 postte;将均布荷载设立为 load 荷载,荷载步数为 5 步,子步长因子为 1,即荷载步设置为1.000000(5)。迭代方式均选择普通的牛顿-拉夫森方法,每一个荷载工况下最大迭代步数分别设置为 20 步和 50 步,收敛准则选择力或位移收敛方式,其中收敛容差均设定为 0.01。点击 Run analysis 开始计算

点击 Result 下方的 Reinforcement result 中的 Reinforcement Cross-section Force 查看粘结滑移状态下预应力筋的拉力云图,从图 3.7－20 中可以看出,这时明显看到结构变形图中有反拱现象。

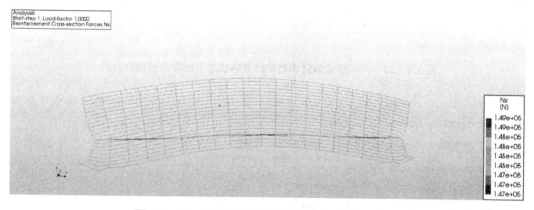

图 3.7－20　初始粘结滑移状态下预应力筋拉力云图

继续查看施加荷载工况后的预应力筋特性,至第 4 荷载步时,查看粘结滑移状态下的预应力筋拉力,可以看出,在粘结滑移状态下预应力筋两端拉力较大,跨中部分拉力数值较小。如图 3.7－21 所示。

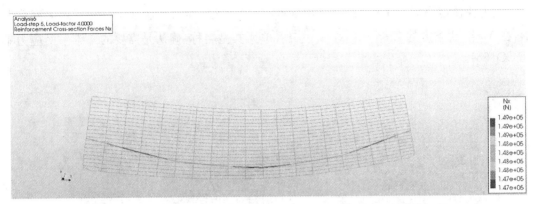

图 3.7-21　粘结滑移状态下施加到第 4 荷载步时的预应力筋拉力云图

继续施加荷载，当最大荷载步施加完毕时，查看粘结滑移状态下的预应力筋的拉力云图，此时可以看到预应力筋各处拉力相等，数值均为 149 kN，由云图可知，此时预应力筋已经进入塑性状态。如图 3.7-22 所示。

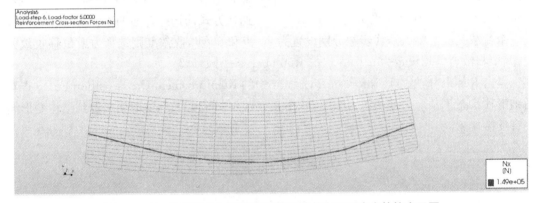

图 3.7-22　荷载步施加完毕时粘结滑移状态下预应力筋拉力云图

Python 语句如下：

```
newProject( "Bondslip", 100 )
setModelAnalysisAspects( [ "STRUCT" ] )
setModelDimension( "2D" )
setDefaultMeshOrder( "QUADRATIC" )
setDefaultMesherType( "HEXQUAD" )
setDefaultMidSideNodeLocation( "LINEAR" )
createSheet( "Sheet 1", [[ 0, 0, 0 ],[ 20, 0, 0 ],[ 20, 4, 0 ],[ 0, 4, 0 ]] )
saveProject(   )
addMaterial( "concrete", "CONCR", "TSCR", [] )
setParameter( "MATERIAL", "concrete", "LINEAR /ELASTI /YOUNG", 3.25e + 10 )
setParameter( "MATERIAL", "concrete", "LINEAR /ELASTI /POISON", 0.15 )
setParameter( "MATERIAL", "concrete", "LINEAR /MASS /DENSIT", 2500 )
setParameter( "MATERIAL", "concrete", "MODTYP /TOTCRK", "ROTATE" )
```

```
setParameter( "MATERIAL", "concrete", "TENSIL /TENCRV", "HORDYK" )
setParameter( "MATERIAL", "concrete", "TENSIL /TENSTR", 2400000 )
setParameter( "MATERIAL", "concrete", "TENSIL /GF1", 500 )
setParameter( "MATERIAL", "concrete", "TENSIL /RESTST", 0 )
setParameter( "MATERIAL", "concrete", "COMPRS /COMCRV", "PARABO" )
setParameter( "MATERIAL", "concrete", "COMPRS /COMSTR", 26800000 )
setParameter( "MATERIAL", "concrete", "COMPRS /GC", 4500 )
setParameter( "MATERIAL", "concrete", "COMPRS /RESCST", 100 )
setParameter( "MATERIAL", "concrete", "COMPRS /RESCST", 100 )
setParameter( "MATERIAL", "concrete", "COMPRS /RESCST", 100 )
addGeometry( "Element geometry 1", "SHEET", "MEMBRA", [] )
setParameter( "GEOMET", "Element geometry 1", "THICK", 1 )
rename( "GEOMET", "Element geometry 1", "concrete" )
clearReinforcementAspects( [ "Sheet 1" ] )
setElementClassType( "SHAPE", [ "Sheet 1" ], "MEMBRA" )
assignMaterial( "concrete", "SHAPE", [ "Sheet 1" ] )
assignGeometry( "concrete", "SHAPE", [ "Sheet 1" ] )
resetElementData( "SHAPE", [ "Sheet 1" ] )
saveProject( )
createCurve( "bar", [[ 0, 1.7, 0 ],[ 10, 0.6, 0 ],[ 20, 1.7, 0 ]] )
addMaterial( "bar", "REINFO", "REBOND", [] )
setParameter( MATERIAL, "bar", "REBARS /ELASTI /YOUNG", 1.95e+11 )
setParameter( MATERIAL, "bar", "REBARS /MASS /DENSIT", 7800 )
setParameter( MATERIAL, "bar", "RESLIP /DSNY", 2e+12 )
setParameter( MATERIAL, "bar", "RESLIP /DSSX", 2000000 )
setParameter( MATERIAL, "bar", "RESLIP /SHFTYP", "BONDS1" )
setParameter( MATERIAL, "bar", "RESLIP /BONDS1 /SLPVAL", [ 20, 0.1 ] )
addGeometry( "Element geometry 1", "RELINE", "REBAR", [] )
rename( GEOMET, "Element geometry 1", "bar" )
setParameter( GEOMET, "bar", "REITYP", "REITRU" )
setParameter( GEOMET, "bar", "REITRU /CROSSE", 0.000139 )
setParameter( GEOMET, "bar", "REITRU /CROSSE", 0.000139 )
setParameter( GEOMET, "bar", "REITRU /PERIME", 0.09576 )
setReinforcementAspects( [ "bar" ] )
assignMaterial( "bar", SHAPE, [ "bar" ] )
assignGeometry( "bar", SHAPE, [ "bar" ] )
resetElementData( SHAPE, [ "bar" ] )
setReinforcementDiscretization( [ "bar" ], "SECTION" )
```

```
saveProject( )
addSet( "GEOMETRYSUPPORTSET", "co1" )
createPointSupport( "co1", "co1" )
setParameter( "GEOMETRYSUPPORT", "co1", "AXES", [ 1, 2 ] )
setParameter( "GEOMETRYSUPPORT", "co1", "TRANSL", [ 1, 1, 0 ] )
setParameter( "GEOMETRYSUPPORT", "co1", "ROTATI", [ 0, 0, 0 ] )
attach( "GEOMETRYSUPPORT", "co1", "Sheet 1", [[ 7.2976811e-35, 7.8468209e-33, 1.2815385e-16 ]] )
saveProject( )
createPointSupport( "co2", "co1" )
setParameter( "GEOMETRYSUPPORT", "co2", "AXES", [ 1, 2 ] )
setParameter( "GEOMETRYSUPPORT", "co2", "TRANSL", [ 1, 1, 0 ] )
setParameter( "GEOMETRYSUPPORT", "co2", "ROTATI", [ 0, 0, 0 ] )
attach( "GEOMETRYSUPPORT", "co2", "Sheet 1", [[ 20, 7.1494798e-33, 1.1676492e-16 ]] )
saveProject( )
addSet( "GEOMETRYLOADSET", "gravity" )
createModelLoad( "gravity", "gravity" )
saveProject( )
addSet( "GEOMETRYLOADSET", "Geometry load case 2" )
rename( "GEOMETRYLOADSET", "Geometry load case 2", "postte" )
createBodyLoad( "postte", "postte" )
setParameter( "GEOMETRYLOAD", "postte", "LODTYP", "POSTEN" )
setParameter( "GEOMETRYLOAD", "postte", "POSTEN /BOTHEN /FORCE1", 150000 )
setParameter( "GEOMETRYLOAD", "postte", "POSTEN /BOTHEN /FORCE2", 150000 )
setParameter( "GEOMETRYLOAD", "postte", "POSTEN /BOTHEN /RETLE1", 0.0001 )
setParameter( "GEOMETRYLOAD", "postte", "POSTEN /BOTHEN /RETLE2", 0.0001 )
setParameter( "GEOMETRYLOAD", "postte", "POSTEN /SHEAR", 0.22 )
setParameter( "GEOMETRYLOAD", "postte", "POSTEN /WOBBLE", 0.01 )
attach( "GEOMETRYLOAD", "postte", [ "bar" ] )
attachTo( "GEOMETRYLOAD", "postte", "POSTEN /BOTHEN /PNTS1", "bar", [[ 0, 0.7, 0 ]] )
attachTo( "GEOMETRYLOAD", "postte", "POSTEN /BOTHEN /PNTS2", "bar", [[ 20, 0.7, 0 ]] )
saveProject( )
addSet( "GEOMETRYLOADSET", "Geometry load case 3" )
rename( "GEOMETRYLOADSET", "Geometry load case 3", "laod" )
createLineLoad( "load", "laod" )
```

```
setParameter( "GEOMETRYLOAD", "load", "FORCE/VALUE", -20000 )
setParameter( "GEOMETRYLOAD", "load", "FORCE/DIRECT", 2 )
attach( "GEOMETRYLOAD", "load", "Sheet 1", [[ 10, 4, -1.2245938e-16 ]] )
saveProject( )
setEdgeMeshSeed( [ "Sheet 1" ], 20 )
setMesherType( [ "Sheet 1" ], "HEXQUAD" )
setMidSideNodeLocation( [ "Sheet 1" ], "LINEAR" )
saveProject( )
setDefaultGeometryLoadCombinations( )
setGeometryLoadCombinationFactor( "Geometry load combination 1", "gravity", 1 )
remove( "GEOMETRYLOADCOMBINATION", "Geometry load combination 1" )
remove( "GEOMETRYLOADCOMBINATION", "Geometry load combination 2" )
setGeometryLoadCombinationFactor( "Geometry load combination 3", "gravity", 1 )
addGeometryLoadCombination( "" )
setGeometryLoadCombinationFactor( "Geometry load combination 3", "postte", 1 )
setGeometryLoadCombinationFactor( "Geometry load combination 2", "laod", 1 )
setGeometryLoadCombinationFactor( "Geometry load combination 3", "laod", 1 )
remove( "GEOMETRYLOADCOMBINATION", "Geometry load combination 2" )
setGeometryLoadCombinationFactor( "Geometry load combination 3", "gravity", 1 )
remove( "GEOMETRYLOADCOMBINATION", "Geometry load combination 3" )
addGeometryLoadCombination( "" )
setGeometryLoadCombinationFactor( "Geometry load combination 1", "gravity", 1 )
setGeometryLoadCombinationFactor( "Geometry load combination 1", "postte", 1 )
addGeometryLoadCombination( "" )
setGeometryLoadCombinationFactor( "Geometry load combination 2", "laod", 1 )
setGeometryLoadCombinationFactor( "Geometry load combination 1", "gravity", 1 )
saveProject( )
generateMesh( [] )
hideView( "GEOM" )
showView( "MESH" )
addAnalysis( "Analysis6" )
addAnalysisCommand( "Analysis6", "NONLIN", "Structural nonlinear" )
renameAnalysis( "Analysis6", "Analysis6" )
removeAnalysisCommandDetail( "Analysis6", "Structural nonlinear", "EXECUT(1)" )
setAnalysisCommandDetail ( "Analysis6", "Structural nonlinear", "EXECUT/EXETYP", "START" )
renameAnalysisCommandDetail( "Analysis6", "Structural nonlinear", "EXECUT(1)", "tenin" )
```

```
    addAnalysisCommandDetail( "Analysis6", "Structural nonlinear", "EXECUT(1) /
START /INITIA /STRESS" )
    setAnalysisCommandDetail( "Analysis6", "Structural nonlinear", "EXECUT(1) /
START /INITIA /STRESS", True )
    setAnalysisCommandDetail( "Analysis6", "Structural nonlinear", "EXECUT(1) /
START /LOAD /PREVIO", False )
    saveProject( )
    setAnalysisCommandDetail( "Analysis6", "Structural nonlinear", "EXECUT(1) /
ITERAT /MAXITE", 20 )
    setAnalysisCommandDetail( "Analysis6", "Structural nonlinear", "EXECUT(1) /
ITERAT /CONVER /DISPLA /NOCONV", "CONTIN" )
    setAnalysisCommandDetail( "Analysis6", "Structural nonlinear", "EXECUT(1) /
ITERAT /CONVER /FORCE /NOCONV", "TERMIN" )
    setAnalysisCommandDetail( "Analysis6", "Structural nonlinear", "EXECUT(1) /
ITERAT /CONVER /FORCE /NOCONV", "CONTIN" )
    saveProject( )
    setAnalysisCommandDetail( "Analysis6", "Structural nonlinear", "EXECUT /
EXETYP", "LOAD" )
    renameAnalysisCommandDetail( "Analysis6", "Structural nonlinear", "EXECUT
(2)", "load" )
    setAnalysisCommandDetail( "Analysis6", "Structural nonlinear", "EXECUT(2) /
LOAD /STEPS /EXPLIC /SIZES", "1.00000" )
    setAnalysisCommandDetail( "Analysis6", "Structural nonlinear", "EXECUT(2) /
LOAD /LOADNR", 2 )
    setAnalysisCommandDetail( "Analysis6", "Structural nonlinear", "EXECUT(2) /
LOAD /STEPS /EXPLIC /SIZES", "1.00000(5)" )
    saveProject( )
    setAnalysisCommandDetail( "Analysis6", "Structural nonlinear", "OUTPUT(1) /
SELTYP", "PRIMAR" )
    saveProject( )
    runSolver( "Analysis6" )
```

3.8 案例八:钢筋混凝土结构非线性动力分析

特别说明:本节案例来自于作者本人在 2018 年 5 月东南大学高等结构动力学课程期末论文有限元建模部分摸索出的非线性动力分析数值模拟经验。鉴于国内尚无此类著作系统介绍 Diana10.1 软件在动力学领域的数值模拟操作流程,在此作者将期末论文中独立摸索的 Diana 非线性动力分析经验和操作流程分享出来,进行系统的提炼和介绍,并提供两种建

第三章　Diana 土木工程非线性建模案例分析

模文件命令流以供读者参考。相信本节内容对使用 Diana 软件进行动力非线性分析的初学者入门有很大的帮助。其中书中内容和命令流语句.py 文件仅供使用 Diana 软件的用户在学习中使用,版权为作者独有。用户不得将本节.py 语言命令流和相关内容私自上传、扩散、抄袭或剽窃,一经发现,作者本人将保留追究责任的权利。

本节案例以本书 3.5 节的门式框架模型的左半部分单独取出作为非线性动力时程分析的整个几何尺寸模型。鉴于 3.5 节中的模型尺寸较小,因此将长度、高度和厚度尺寸均放大十倍,如图 3.8-1 所示。分别采用 Newmark-β 法和 Wilson-θ 法进行非线性动力分析的时间隐式积分算法,混凝土开裂模型为总应变固定裂缝模型,受压模型采用 Diana 软件中的前川混凝土开裂曲线(Maekawa Cracked Concrete curves)模型,拉伸软化曲线采用欧洲 2010 混凝土结构规范(fib Model Code for Concrete Structures 2010),混凝土弹性模量设定为 $3.45 \times 10^{10} \text{N/m}^2$,抗压强度为 32.5 MPa。地震波选取 72448 地震波的前 5 秒数据,并以 0.1 s 作为输入时间—荷载曲线的一个地震波子步长。钢筋同样采用塑性模型,瑞雷阻尼中的质量阵和刚度矩阵的阻尼系数值采用 R.克拉夫和 J.彭津《结构动力学》第二版一书例题中的 1.10412 和 0.00165,探究结构在地震作用下的开裂问题,并对两种方法下的地震作用计算结果进行比较。这两种方法在抗风和抗震等动力分析问题中较为常用。

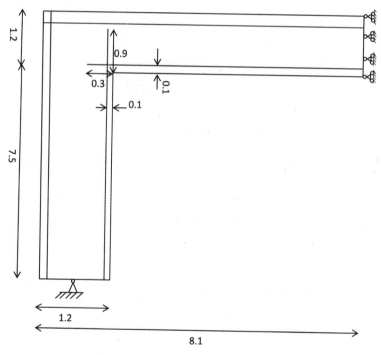

图 3.8-1　模型几何尺寸图(单位:m)

学习要义:

(1) 学习 fib Model Code for Concrete Structures 2010 拉伸软化模型和混凝土前川受压模型的参数设置。

(2) 学习定义地震波时间—荷载曲线在时变模型下的设置方式。

(3) 学习 Newmark-β 法和 Wilson-θ 法的非线性计算荷载步及参数设置。

启动 DianaIE,点击上方的快捷菜单 File→New,弹出 New project 对话框,命名为 ChaiShun-Newmark,选择分析类型为结构类(Structural),维数为 2 维平面(Two dimensional),模型最大尺寸为 100 m,网格形状类型为六面体/四边形,网格默认阶数为二次(Quadratic),中间节点位置确定方式与之前各节相同,为线性插值(Linear interpolation)。

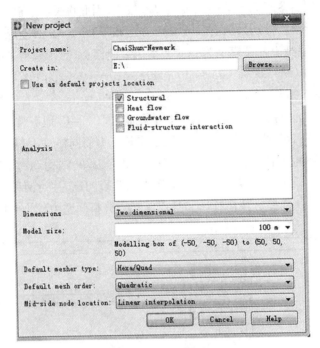

图 3.8-2　New project 对话框

点击快捷图标 Adds a sheet ,分别输入坐标(0,0,0),(0.6,0,0),(1.2,0,0),(1.2,7.5,0),(8.1,7.5,0),)(1,8.7,0),(0,8.7,0),创立 Sheet1。点击 OK,生成 Sheet1,再点击快捷图标 Adds a line,创立钢筋几何模型,分别输入表 3-7 中坐标,生成各个钢筋的几何模型。整个结构的几何模型如图 3.8-3 所示。

表 3-7　各个钢筋的几何坐标

bar1	[0.1, 0, 0], [0.1, 8.7, 0]
bar2	[1.1, 0, 0], [1.1, 8.4, 0]
bar3	[0.9, 7.6, 0], [8.1, 7.6, 0]
bar4	[0, 8.6, 0], [8.1, 8.6, 0]

在本节中,对混凝土模型各参数重新进行定义,其中混凝土弹性模量输入值为 $3.45 \times 10^{10} N/m^2$,泊松比为 0.15。由于在模拟中需要考虑瑞雷阻尼系数,因此混凝土开裂模型为总应变固定裂缝模型而不是多向固定裂缝模型,拉伸软化曲线选择欧洲 2010 混凝土结构规范(fib Model Code for Concrete Structures 2010),抗拉强度为 2.6 MPa,拉伸 I 型断裂能设定为 500 N/m,如图 3.8-4 所示。混凝土抗压模型为前川受压模型,如图 3.8-5 所示。这

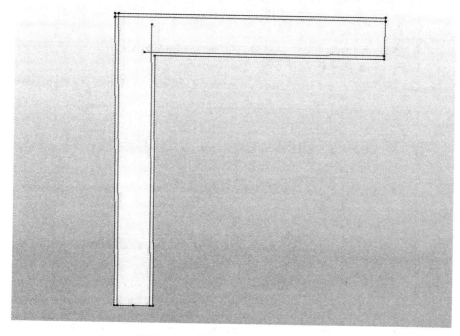

图 3.8-3　混凝土和钢筋几何模型图

种受压模型适用于低周循环加载和动力学非线性分析领域,并且可以与瑞雷阻尼模块完美结合。瑞雷阻尼质量矩阵前系数选择 1.10412,刚度矩阵前系数选择 0.00165。混凝土的截面几何属性定义模块,输入厚度值 0.7 m。

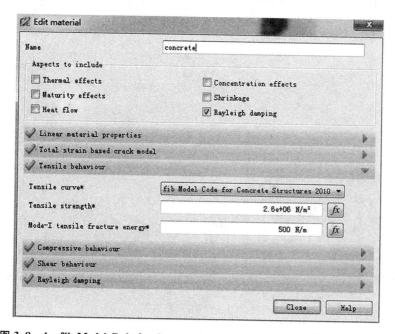

图 3.8-4　fib Model Code for Concrete Structures 2010 曲线模型参数设置界面

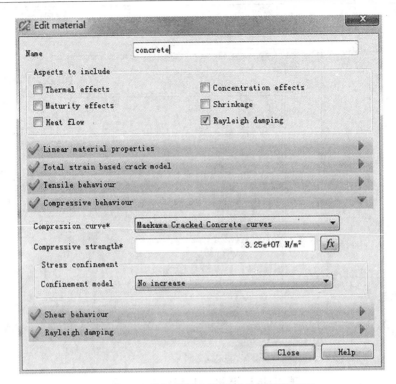

图 3.8-5 前川受压曲线模型参数设置界面

定义钢筋材料属性,选择瑞雷阻尼模块,钢筋本构模型采用 Von Mises 塑性屈服模型,屈服强度为 4.5×10^8 N/m²,这里不再赘述。

选择底边的中点作为荷载施加点,施加 X 向和 Y 向的平动约束,选择右上方所在的边施加水平 X 向的平动约束,施加约束后如图 3.8-6 所示。

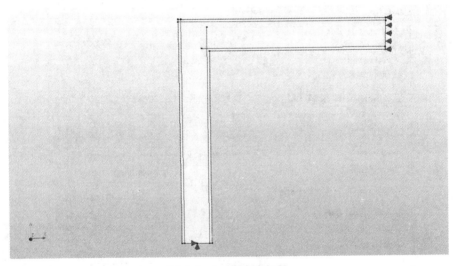

图 3.8-6 约束施加图

定义重力荷载,这步流程与之前操作相同,这里不再赘述。定义地震荷载,命名为 lo2,目标加载类型选择为实体(Solid),荷载形式为等效加速度(Equivalent acceleration)。加载对象为 Sheet1 整个结构,由于在后续的时间—荷载因子曲线中需要输入地震波数据中的加速度因子以 g 为单位,因此等效加速度数值为 g,方向沿着 X 轴负向,如图 3.8-7 所示。

注：根据牛顿力学第一定律,当地面出现水平运动时,建筑物由于惯性作用并未与地面共同运动,这种运动不协调所产生的作用力就是地震作用。因而在本书实际计算中采用等效加速度原则,将地面运动所产生的惯性力等效为地面不动而施加到结构上的力,大小按 **F = ma** 计算,a 为地面往复运动等效加速度。

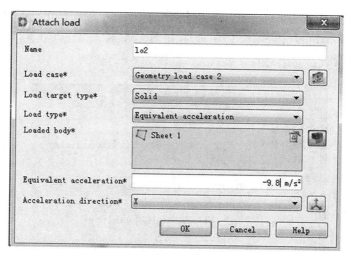

图 3.8-7　等效加速度荷载定义界面

设置荷载组合,定义重力荷载为荷载组合 1,等效加速度荷载为荷载组合 2,对荷载组合 1 定义荷载组合曲线为 1 天之内荷载不随时间变化,荷载因子大小始终为 1。荷载组合 2 定义采用 72448 地震波的前 5 秒数据,期中荷载因子系数表征地震波的等效加速度与重力加速度的关系,选取每隔 0.1 s 的数据,如图 3.8-8 所示。

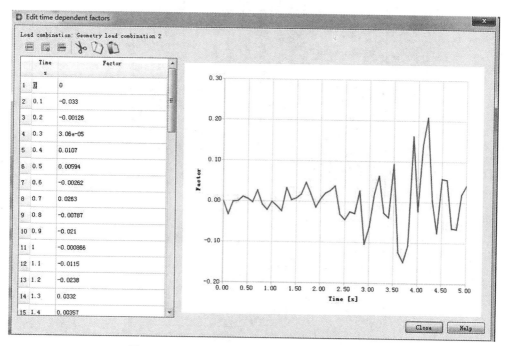

图 3.8-8　72448 地震波的时间选取(间隔 **0.1 s**)

待时间荷载曲线定义完毕后,划分网格单元并生成网格,网格形状为六面体/四边形,网格尺寸大小为 0.1 m,具体操作流程与书中其他各章节操作流程类似,这里不再赘述。

添加分析工况 Analysis1,选择非线性分析类型,在非线性分析类型下的非线性效果(Nonlinear effects)一栏中勾选物理非线性(Physically nonliear)和暂态效应(Transient effects)分析,如图 3.8-9 所示。点击右边的设置(settings),弹出暂态效应设置对话框,选择 Newmark-β 法,该方法下的 β 和 γ 系数分别采用软件中的默认值 0.25 和 0.5。勾选下方的动力分析(Dynamic effects),其中质量矩阵(Mass matrix)和阻尼矩阵(Damping matrix)均选择为常数。勾选时间导数效应。如图 3.8-10 所示。

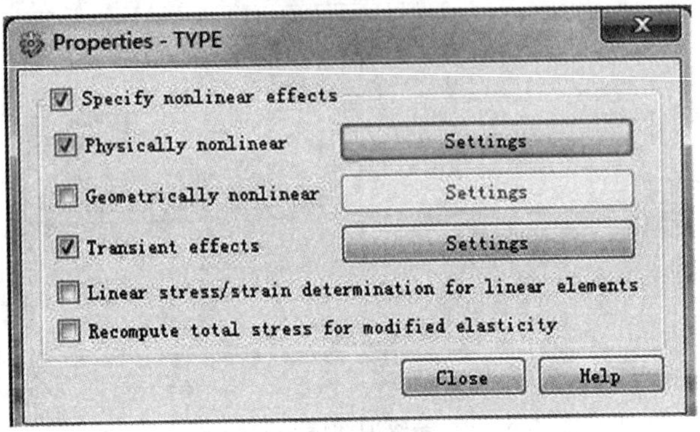

图 3.8-9 选择暂态效应

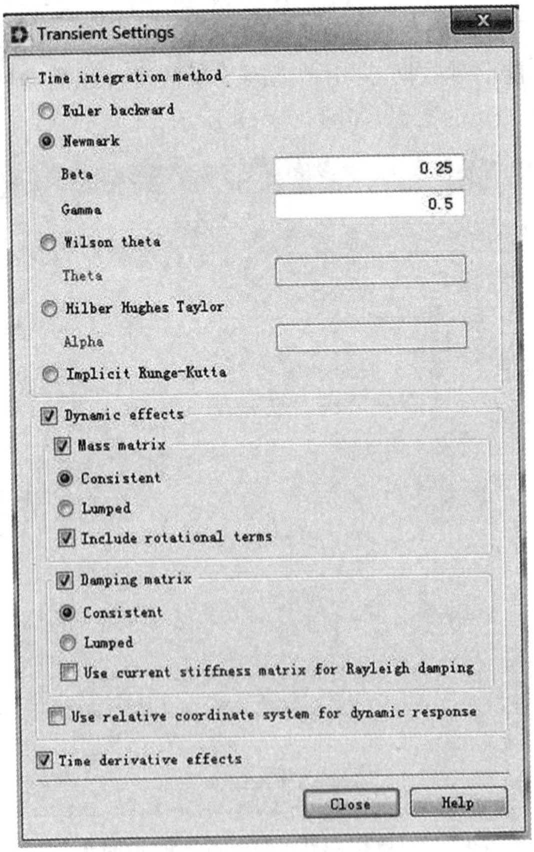

图 3.8-10 Newmark-β 法定义界面

将重力荷载所在的荷载组合 1 设置为荷载工况 1(new execute block),迭代方法为牛顿-拉夫森方法(Newton-Raphson),选择力和位移收敛准则,收敛容差为 0.01。用同样的方式定义荷载工况 2 和时间荷载,其中时间荷载步的步数和子步长设置大小为 0.100000 (50)。在结果输出选项中选择整体坐标系下的各个方向平动位移(DISPLA TOTAL TRANSL GLOBAL)、整体坐标系下的累积开裂应变(STRAIN CRKSUM GREEN GLOBAL)、主应力方向上的的累积开裂应变(STRAIN CRKSUM GREEN PRINCI)、开裂应变(STRAIN CRACK GREEN)、整体坐标系/局部坐标系/主应力方向裂缝宽度(STRAIN CRKWDT GREEN GLOBAL/LOCAL/PRINCI)等选项作为输出结果。

点击 Run analysis,待计算完毕后查看 5 s 地震反应结束后最后一荷载步下各项结果如图 3.8-11—图 3.8-15 中。

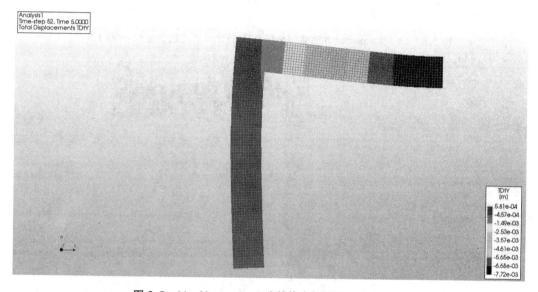

图 3.8-11　Newmark-β 法整体坐标系下 Y 方向位移

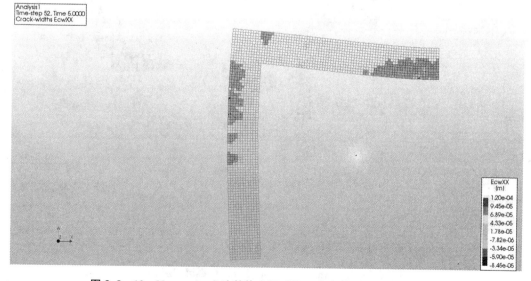

图 3.8-12　Newmark-β 法整体坐标系下 X 方向的裂缝宽度(EcwXX)

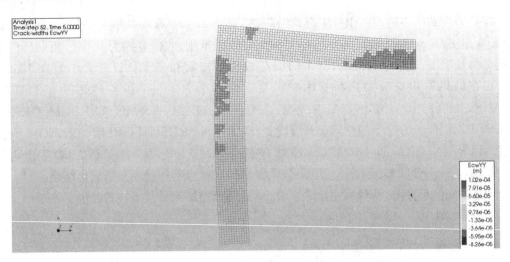

图 3.8-13　Newmark-β 法整体坐标系下 Y 方向的裂缝宽度（EcwYY）

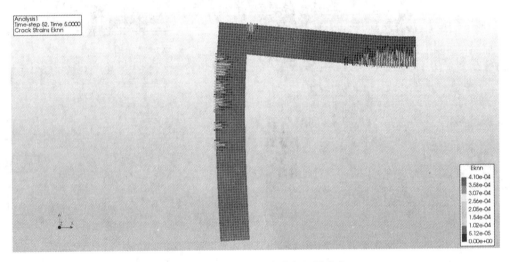

图 3.8-14　Newmark-β 法法向开裂应变 Eknn

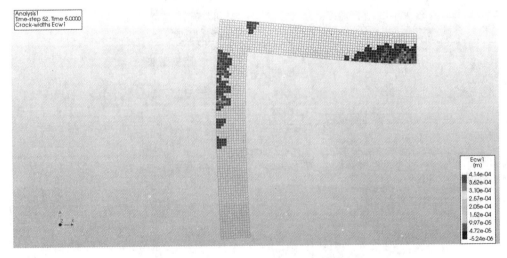

图 3.8-15　Newmark-β 法主应力 X 向裂缝宽度 Ecw1

其他设置不变,选择 Wilson-θ法,θ值选择软件中默认值1.4,如图3.8-16所示。重新开始计算,得到如图3.8-17—图3.8-19各项计算结果。

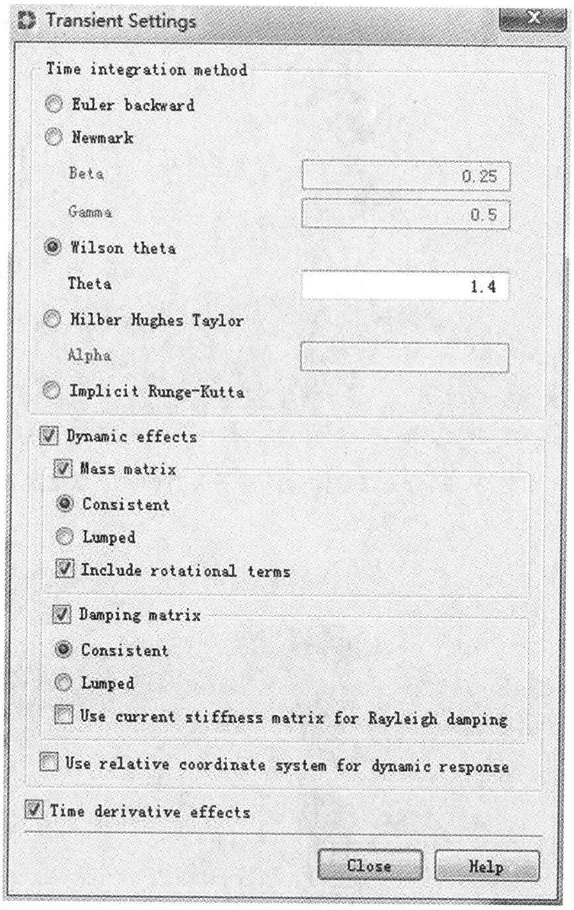

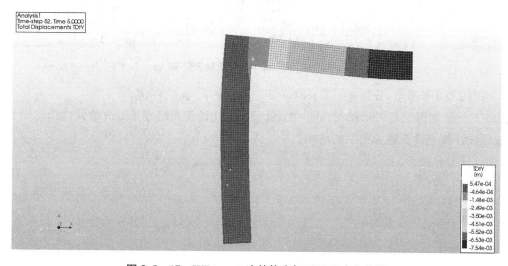

图 3.8-16 Wilson-θ法定义界面

图 3.8-17 Wilson-θ法整体坐标系下 Y 方向位移

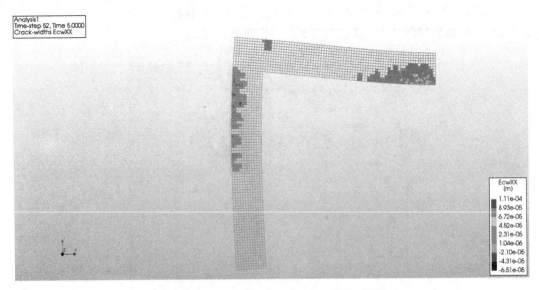

图 3.8-18 Wilson-θ 法整体坐标系下 X 方向的裂缝宽度 EcwXX

图 3.8-19 Wilson-θ 法开裂应变 Eknn

选取 2289 号节点，分别查看 5 s 地震反应之后的 Y 向位移值，分别为 7.65648 mm 和 7.49014 mm，如图 3.8-20 和图 3.8-21 所示。对比验证可证明 Diana 软件在两种动力学非线性法计算上具有较为相似的结果。

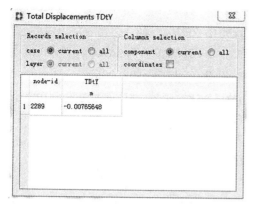
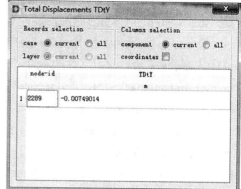

图 3.8-20　Newmark-β 法 2289 号节点位移　　图 3.8-21　Wilson-θ 法 2289 号节点位移

分别附上 Newmark-β 法和 Wilson-θ 法下的 Python 语言命令流：
Newmark-β 法命令流：
newProject("ChaiShun-Newmark", 100)
setModelAnalysisAspects(["STRUCT"])
setModelDimension("2D")
setDefaultMeshOrder("QUADRATIC")
setDefaultMesherType("HEXQUAD")
setDefaultMidSideNodeLocation("ONSHAP")
setUnit("TEMPER", "CELSIU")
setUnit("ANGLE", "DEGREE")
saveProject( )
createSheet("Sheet 1", [[0, 0, 0],[0.6, 0, 0],[1.2, 0, 0],[1.2, 7.5, 0],[8.1, 7.5, 0],[8.1, 8.7, 0],[0, 8.7, 0]])
createLine("bar1", [0.1, 0, 0], [0.1, 8.7, 0])
saveProject( )
createLine("bar2", [1.1, 0, 0], [1.1, 8.4, 0])
saveProject( )
createLine("bar3", [0.9, 7.6, 0], [0.9, 8.1, 0])
removeShape(["bar3"])
createLine("bar3", [0.9, 7.6, 0], [8.1, 7.6, 0])
saveProject( )
createLine("bar4", [0, 8.6, 0], [8.1, 8.6, 0])
saveProject( )
addMaterial("concrete", "CONCR", "TSCR", ["RAYDAM"])
setParameter(MATERIAL, "concrete", "LINEAR/ELASTI/YOUNG", 3.45e+10)
setParameter(MATERIAL, "concrete", "LINEAR/ELASTI/POISON", 0.15)
setParameter(MATERIAL, "concrete", "LINEAR/MASS/DENSIT", 2500)
setParameter(MATERIAL, "concrete", "TENSIL/TENCRV", "MC2010")

```
setParameter( MATERIAL, "concrete", "TENSIL /TENSTR", 2600000 )
setParameter( MATERIAL, "concrete", "TENSIL /GF1", 500 )
setParameter( MATERIAL, "concrete", "COMPRS /COMCRV", "MAEKCC" )
setParameter( MATERIAL, "concrete", "COMPRS /COMSTR", 32500000 )
setParameter( MATERIAL, "concrete", "RAYDAM /RAYLEI", [ 1.1042, 0.00165 ] )
addGeometry( "Element geometry 1", "SHEET", "MEMBRA", [] )
rename( GEOMET, "Element geometry 1", "concrete" )
setParameter( GEOMET, "concrete", "THICK", 0.7 )
clearReinforcementAspects( [ "Sheet 1" ] )
setElementClassType( SHAPE, [ "Sheet 1" ], "MEMBRA" )
assignMaterial( "concrete", SHAPE, [ "Sheet 1" ] )
assignGeometry( "concrete", SHAPE, [ "Sheet 1" ] )
resetElementData( SHAPE, [ "Sheet 1" ] )
addMaterial( "bar", "REINFO", "VMISES", [ "RAYDAM" ] )
setParameter( MATERIAL, "bar", "LINEAR /ELASTI /YOUNG", 2.1e+11 )
setParameter( MATERIAL, "bar", "PLASTI /HARDI1 /YLDSTR", 4.5e+08 )
setParameter( MATERIAL, "bar", "RAYDAM /RAYLEI", [ 1.1042, 0.00165 ] )
addGeometry( "Element geometry 2", "RELINE", "REBAR", [] )
rename( GEOMET, "Element geometry 2", "bar" )
setParameter( GEOMET, "bar", "REITYP", "REITRU" )
setParameter( GEOMET, "bar", "REITRU /CROSSE", 0.000157 )
setParameter( GEOMET, "bar", "REITRU /PERIME", 0.012 )
setReinforcementAspects( [ "bar1", "bar2", "bar3", "bar4" ] )
assignMaterial( "bar", SHAPE, [ "bar1", "bar2", "bar3", "bar4" ] )
assignGeometry( "bar", SHAPE, [ "bar1", "bar2", "bar3", "bar4" ] )
resetElementData( SHAPE, [ "bar1", "bar2", "bar3", "bar4" ] )
setReinforcementDiscretization( [ " bar1", " bar2", " bar3", " bar4" ],
"ELEMENT" )
saveProject(  )
addSet( "GEOMETRYSUPPORTSET", "co1" )
createPointSupport( "co1", "co1" )
setParameter( "GEOMETRYSUPPORT", "co1", "AXES", [ 1, 2 ] )
setParameter( "GEOMETRYSUPPORT", "co1", "TRANSL", [ 1, 1, 0 ] )
setParameter( "GEOMETRYSUPPORT", "co1", "ROTATI", [ 0, 0, 0 ] )
attach( " GEOMETRYSUPPORT", " co1", " Sheet 1", [[ 0.6, 1.8378494e-34,
-5.2871687e-18 ]] )
addSet( "GEOMETRYSUPPORTSET", "Geometry support set 2" )
rename( "GEOMETRYSUPPORTSET", "Geometry support set 2", "co2" )
createLineSupport( "co2", "co2" )
```

```
    setParameter( "GEOMETRYSUPPORT", "co2", "AXES", [ 1, 2 ] )
    setParameter( "GEOMETRYSUPPORT", "co2", "TRANSL", [ 1, 0, 0 ] )
    setParameter( "GEOMETRYSUPPORT", "co2", "ROTATI", [ 0, 0, 0 ] )
    attach( "GEOMETRYSUPPORT", "co2", "Sheet 1", [[ 8.1, 8.1, -1.4938131e-17
]] )
    saveProject( )
    addSet( "GEOMETRYLOADSET", "gravity" )
    createModelLoad( "gravity", "gravity" )
    createBodyLoad( "lo2", "Geometry load case 2" )
    setParameter( GEOMETRYLOAD, "lo2", "LODTYP", "EQUIAC" )
    setParameter( GEOMETRYLOAD, "lo2", "EQUIAC/ACCELE", -9.8 )
    setParameter( GEOMETRYLOAD, "lo2", "EQUIAC/DIRECT", 1 )
    attach( GEOMETRYLOAD, "lo2", [ "Sheet 1" ] )
    setDefaultGeometryLoadCombinations( )
    setGeometryLoadCombinationFactor( "Geometry load combination 1", "gravity",
1 )
    remove( GEOMETRYLOADCOMBINATION, "Geometry load combination 2" )
    setGeometryLoadCombinationFactor( "Geometry load combination 1", "gravity",
1 )
    remove( GEOMETRYLOADCOMBINATION, "Geometry load combination 1" )
    addGeometryLoadCombination( "" )
    setGeometryLoadCombinationFactor( "Geometry load combination 1", "gravity",
1 )
    addGeometryLoadCombination( "" )
    setGeometryLoadCombinationFactor( "Geometry load combination 2", "Geometry
load case 2", 1 )
    setTimeDependentLoadFactors ( GEOMETRYLOADCOMBINATION, " Geometry load
combination 1", [ 0, 86400 ], [ 1, 1 ] )
    setTimeDependentLoadFactors ( GEOMETRYLOADCOMBINATION, " Geometry load
combination 2", [ 0, 0.1, 0.2, 0.3, 0.4, 0.5, 0.6, 0.7, 0.8, 0.9, 1, 1.1, 1.2, 1.3,
1.4, 1.5, 1.6, 1.7, 1.8, 1.9, 2, 2.1, 2.2, 2.3, 2.4, 2.5, 2.6, 2.7, 2.8, 2.9, 3, 3.1,
3.2, 3.3, 3.4, 3.5, 3.6, 3.7, 3.8, 3.9, 4, 4.1, 4.2, 4.3, 4.4, 4.5, 4.6, 4.7, 4.8,
4.9, 5 ], [ 0, -0.033, -0.00126, 3.06e-05, 0.0107, 0.00594, -0.00262, 0.0263,
-0.00787, -0.021, -0.000866, -0.0115, -0.0238, 0.0332, 0.00357, 0.00819,
0.0176, 0.0468, 0.018, -0.0139, 0.0062, 0.0206, 0.0271, 0.0385, -0.0307,
-0.0442, -0.0246, -0.0288, 0.0267, -0.104, -0.0613, 0.0187, 0.0637,
-0.0269, -0.0381, 0.0935, -0.124, -0.148, -0.107, 0.162, -0.0218, 0.141,
0.208, 0.0046, -0.0751, 0.0576, 0.0553, -0.0639, -0.0653, 0.0194, 0.041 ] )
    setElementSize( [ "Sheet 1" ], 0.1, -1, True )
```

```
setMesherType( [ "Sheet 1" ], "HEXQUAD" )
setMidSideNodeLocation( [ "Sheet 1" ], "LINEAR" )
generateMesh( [] )
hideView( "GEOM" )
showView( "MESH" )
addAnalysis( "Analysis1" )
addAnalysisCommand( "Analysis1", "NONLIN", "Structural nonlinear" )
renameAnalysis( "Analysis1", "Analysis1" )
setAnalysisCommandDetail( "Analysis1", "Structural nonlinear", "EXECUT(1) /
ITERAT /MAXITE", 50 )
setAnalysisCommandDetail( "Analysis1", "Structural nonlinear", "EXECUT(1) /
ITERAT /CONVER /DISPLA /NOCONV", "CONTIN" )
setAnalysisCommandDetail( "Analysis1", "Structural nonlinear", "EXECUT(1) /
ITERAT /CONVER /FORCE /NOCONV", "CONTIN" )
setAnalysisCommandDetail( "Analysis1", "Structural nonlinear", "EXECUT(1) /
ITERAT /MAXITE", 50 )
addAnalysisCommandDetail( "Analysis1", "Structural nonlinear", "TYPE /TRANSI" )
setAnalysisCommandDetail ( " Analysis1 ", " Structural nonlinear ", " TYPE /
TRANSI", True )
addAnalysisCommandDetail( "Analysis1", "Structural nonlinear", "TYPE /TRANSI /
DYNAMI" )
setAnalysisCommandDetail( "Analysis1", "Structural nonlinear", "TYPE /TRANSI /
DYNAMI", True )
setAnalysisCommandDetail( "Analysis1", "Structural nonlinear", "EXECUT(1) /
ITERAT /MAXITE", 50 )
setAnalysisCommandDetail ( " Analysis1 ", " Structural nonlinear ", " EXECUT /
EXETYP", "LOAD" )
renameAnalysisCommand ( " Analysis1 ", " Structural nonlinear ", " Structural
nonlinear" )
setAnalysisCommandDetail( "Analysis1", "Structural nonlinear", "EXECUT(2) /
ITERAT /MAXITE", 10 )
setAnalysisCommandDetail( "Analysis1", "Structural nonlinear", "EXECUT(2) /
ITERAT /MAXITE", 50 )
setAnalysisCommandDetail( "Analysis1", "Structural nonlinear", "EXECUT(2) /
ITERAT /CONVER /DISPLA /NOCONV", "CONTIN" )
setAnalysisCommandDetail( "Analysis1", "Structural nonlinear", "EXECUT(2) /
ITERAT /CONVER /FORCE /NOCONV", "CONTIN" )
setAnalysisCommandDetail( "Analysis1", "Structural nonlinear", "EXECUT(2) /
ITERAT /MAXITE", 50 )
```

```
    setAnalysisCommandDetail( "Analysis1", "Structural nonlinear", "EXECUT(2) /
ITERAT /MAXITE", 50 )
    setAnalysisCommandDetail( "Analysis1", "Structural nonlinear", "EXECUT(1) /
ITERAT /MAXITE", 50 )
    addAnalysisCommandDetail( "Analysis1", "Structural nonlinear", "TYPE /TRANSI /
DYNAMI /DAMPIN" )
    setAnalysisCommandDetail( "Analysis1", "Structural nonlinear", "TYPE /TRANSI /
DYNAMI /DAMPIN", True )
    setAnalysisCommandDetail( "Analysis1", "Structural nonlinear", "EXECUT(2) /
LOAD /LOADNR", 2 )
    setAnalysisCommandDetail( "Analysis1", "Structural nonlinear", "EXECUT /
EXETYP", "TIME" )
    renameAnalysisCommandDetail( "Analysis1", "Structural nonlinear", "EXECUT
(3)", "new execute block 3" )
    setAnalysisCommandDetail( "Analysis1", "Structural nonlinear", "EXECUT(3) /
TIME /STEPS /EXPLIC /SIZES", "0.1(50)" )
    setAnalysisCommandDetail( "Analysis1", "Structural nonlinear", "EXECUT(3) /
ITERAT /MAXITE", 50 )
    setAnalysisCommandDetail( "Analysis1", "Structural nonlinear", "EXECUT(3) /
ITERAT /CONVER /DISPLA /NOCONV", "CONTIN" )
    setAnalysisCommandDetail( "Analysis1", "Structural nonlinear", "EXECUT(3) /
ITERAT /CONVER /FORCE /NOCONV", "CONTIN" )
    setAnalysisCommandDetail( "Analysis1", "Structural nonlinear", "OUTPUT(1) /
SELTYP", "USER" )
    addAnalysisCommandDetail( "Analysis1", "Structural nonlinear", "OUTPUT(1) /
USER" )
    addAnalysisCommandDetail( "Analysis1", "Structural nonlinear", "OUTPUT(1) /
USER /DISPLA(1) /TOTAL /TRANSL /GLOBAL" )
    addAnalysisCommandDetail( "Analysis1", "Structural nonlinear", "OUTPUT(1) /
USER /STRAIN(1) /CRKSUM /GREEN /GLOBAL" )
    addAnalysisCommandDetail( "Analysis1", "Structural nonlinear", "OUTPUT(1) /
USER /STRAIN(2) /CRKSUM /GREEN /PRINCI" )
    addAnalysisCommandDetail( "Analysis1", "Structural nonlinear", "OUTPUT(1) /
USER /STRAIN(3) /CRACK /GREEN" )
    addAnalysisCommandDetail( "Analysis1", "Structural nonlinear", "OUTPUT(1) /
USER /STRAIN(4) /CRKWDT /GREEN /GLOBAL" )
    addAnalysisCommandDetail( "Analysis1", "Structural nonlinear", "OUTPUT(1) /
USER /STRAIN(5) /CRKWDT /GREEN /LOCAL" )
    addAnalysisCommandDetail( "Analysis1", "Structural nonlinear", "OUTPUT(1) /
```

```
USER /STRAIN(6) /CRKWDT /GREEN /PRINCI" )
    runSolver( "Analysis1" )
    Wilson-θ法命令流：
    newProject( "ChaiShun-Wilson", 100 )
    setModelAnalysisAspects( [ "STRUCT" ] )
    setModelDimension( "2D" )
    setDefaultMeshOrder( "QUADRATIC" )
    setDefaultMesherType( "HEXQUAD" )
    setDefaultMidSideNodeLocation( "ONSHAP" )
    setUnit( "TEMPER", "CELSIU" )
    setUnit( "ANGLE", "DEGREE" )
    saveProject( )
    createSheet( "Sheet 1", [[ 0, 0, 0 ],[ 0.6, 0, 0 ],[ 1.2, 0, 0 ],[ 1.2, 7.5, 0 ],
[ 8.1, 7.5, 0 ],[ 8.1, 8.7, 0 ],[ 0, 8.7, 0 ]] )
    createLine( "bar1", [ 0.1, 0, 0 ], [ 0.1, 8.7, 0 ] )
    saveProject( )
    createLine( "bar2", [ 1.1, 0, 0 ], [ 1.1, 8.4, 0 ] )
    saveProject( )
    createLine( "bar3", [ 0.9, 7.6, 0 ], [ 0.9, 8.1, 0 ] )
    removeShape( [ "bar3" ] )
    createLine( "bar3", [ 0.9, 7.6, 0 ], [ 8.1, 7.6, 0 ] )
    saveProject( )
    createLine( "bar4", [ 0, 8.6, 0 ], [ 8.1, 8.6, 0 ] )
    saveProject( )
    addMaterial( "concrete", "CONCR", "TSCR", [ "RAYDAM" ] )
    setParameter( MATERIAL, "concrete", "LINEAR /ELASTI /YOUNG", 3.45e+10 )
    setParameter( MATERIAL, "concrete", "LINEAR /ELASTI /POISON", 0.15 )
    setParameter( MATERIAL, "concrete", "LINEAR /MASS /DENSIT", 2500 )
    setParameter( MATERIAL, "concrete", "TENSIL /TENCRV", "MC2010" )
    setParameter( MATERIAL, "concrete", "TENSIL /TENSTR", 2600000 )
    setParameter( MATERIAL, "concrete", "TENSIL /GF1", 500 )
    setParameter( MATERIAL, "concrete", "COMPRS /COMCRV", "MAEKCC" )
    setParameter( MATERIAL, "concrete", "COMPRS /COMSTR", 32500000 )
    setParameter( MATERIAL, "concrete", "RAYDAM /RAYLEI", [ 1.1042, 0.00165 ] )
    addGeometry( "Element geometry 1", "SHEET", "MEMBRA", [] )
    rename( GEOMET, "Element geometry 1", "concrete" )
    setParameter( GEOMET, "concrete", "THICK", 0.7 )
    clearReinforcementAspects( [ "Sheet 1" ] )
    setElementClassType( SHAPE, [ "Sheet 1" ], "MEMBRA" )
```

```
assignMaterial( "concrete", SHAPE, [ "Sheet 1" ] )
assignGeometry( "concrete", SHAPE, [ "Sheet 1" ] )
resetElementData( SHAPE, [ "Sheet 1" ] )
addMaterial( "bar", "REINFO", "VMISES", [ "RAYDAM" ] )
setParameter( MATERIAL, "bar", "LINEAR /ELASTI /YOUNG", 2.1e+11 )
setParameter( MATERIAL, "bar", "PLASTI /HARDI1 /YLDSTR", 4.5e+08 )
setParameter( MATERIAL, "bar", "RAYDAM /RAYLEI", [ 1.1042, 0.00165 ] )
addGeometry( "Element geometry 2", "RELINE", "REBAR", [] )
rename( GEOMET, "Element geometry 2", "bar" )
setParameter( GEOMET, "bar", "REITYP", "REITRU" )
setParameter( GEOMET, "bar", "REITRU /CROSSE", 0.000157 )
setParameter( GEOMET, "bar", "REITRU /PERIME", 0.012 )
setReinforcementAspects( [ "bar1", "bar2", "bar3", "bar4" ] )
assignMaterial( "bar", SHAPE, [ "bar1", "bar2", "bar3", "bar4" ] )
assignGeometry( "bar", SHAPE, [ "bar1", "bar2", "bar3", "bar4" ] )
resetElementData( SHAPE, [ "bar1", "bar2", "bar3", "bar4" ] )
setReinforcementDiscretization ( [ " bar1", " bar2", " bar3", " bar4" ],
"ELEMENT" )
saveProject( )
addSet( "GEOMETRYSUPPORTSET", "co1" )
createPointSupport( "co1", "co1" )
setParameter( "GEOMETRYSUPPORT", "co1", "AXES", [ 1, 2 ] )
setParameter( "GEOMETRYSUPPORT", "co1", "TRANSL", [ 1, 1, 0 ] )
setParameter( "GEOMETRYSUPPORT", "co1", "ROTATI", [ 0, 0, 0 ] )
attach( " GEOMETRYSUPPORT", " co1", " Sheet 1", [[ 0.6, 1.8378494e-34,
-5.2871687e-18 ]] )
addSet( "GEOMETRYSUPPORTSET", "Geometry support set 2" )
rename( "GEOMETRYSUPPORTSET", "Geometry support set 2", "co2" )
createLineSupport( "co2", "co2" )
setParameter( "GEOMETRYSUPPORT", "co2", "AXES", [ 1, 2 ] )
setParameter( "GEOMETRYSUPPORT", "co2", "TRANSL", [ 1, 0, 0 ] )
setParameter( "GEOMETRYSUPPORT", "co2", "ROTATI", [ 0, 0, 0 ] )
attach( "GEOMETRYSUPPORT", "co2", "Sheet 1", [[ 8.1, 8.1, -1.4938131e-17 ]] )
saveProject( )
addSet( "GEOMETRYLOADSET", "gravity" )
createModelLoad( "gravity", "gravity" )
createBodyLoad( "lo2", "Geometry load case 2" )
setParameter( GEOMETRYLOAD, "lo2", "LODTYP", "EQUIAC" )
setParameter( GEOMETRYLOAD, "lo2", "EQUIAC /ACCELE", -9.8 )
```

```
setParameter( GEOMETRYLOAD, "lo2", "EQUIAC /DIRECT", 1 )
attach( GEOMETRYLOAD, "lo2", [ "Sheet 1" ] )
setDefaultGeometryLoadCombinations( )
setGeometryLoadCombinationFactor( "Geometry load combination 1", "gravity", 1 )
remove( GEOMETRYLOADCOMBINATION, "Geometry load combination 2" )
setGeometryLoadCombinationFactor( "Geometry load combination 1", "gravity", 1 )
remove( GEOMETRYLOADCOMBINATION, "Geometry load combination 1" )
addGeometryLoadCombination( "" )
setGeometryLoadCombinationFactor( "Geometry load combination 1", "gravity", 1 )
addGeometryLoadCombination( "" )
setGeometryLoadCombinationFactor( "Geometry load combination 2", "Geometry load case 2", 1 )
setTimeDependentLoadFactors ( GEOMETRYLOADCOMBINATION, " Geometry load combination 1", [ 0, 86400 ], [ 1, 1 ] )
setTimeDependentLoadFactors ( GEOMETRYLOADCOMBINATION, " Geometry load combination 2", [ 0, 0.1, 0.2, 0.3, 0.4, 0.5, 0.6, 0.7, 0.8, 0.9, 1, 1.1, 1.2, 1.3, 1.4, 1.5, 1.6, 1.7, 1.8, 1.9, 2, 2.1, 2.2, 2.3, 2.4, 2.5, 2.6, 2.7, 2.8, 2.9, 3, 3.1, 3.2, 3.3, 3.4, 3.5, 3.6, 3.7, 3.8, 3.9, 4, 4.1, 4.2, 4.3, 4.4, 4.5, 4.6, 4.7, 4.8, 4.9, 5 ], [ 0, -0.033, -0.00126, 3.06e-05, 0.0107, 0.00594, -0.00262, 0.0263, -0.00787, -0.021, -0.000866, -0.0115, -0.0238, 0.0332, 0.00357, 0.00819, 0.0176, 0.0468, 0.018, -0.0139, 0.0062, 0.0206, 0.0271, 0.0385, -0.0307, -0.0442, -0.0246, -0.0288, 0.0267, -0.104, -0.0613, 0.0187, 0.0637, -0.0269, -0.0381, 0.0935, -0.124, -0.148, -0.107, 0.162, -0.0218, 0.141, 0.208, 0.0046, -0.0751, 0.0576, 0.0553, -0.0639, -0.0653, 0.0194, 0.041 ] )
setElementSize( [ "Sheet 1" ], 0.1, -1, True )
setMesherType( [ "Sheet 1" ], "HEXQUAD" )
setMidSideNodeLocation( [ "Sheet 1" ], "LINEAR" )
generateMesh( [] )
hideView( "GEOM" )
showView( "MESH" )
addAnalysis( "Analysis1" )
addAnalysisCommand( "Analysis1", "NONLIN", "Structural nonlinear" )
renameAnalysis( "Analysis1", "Analysis1" )
setAnalysisCommandDetail( "Analysis1", "Structural nonlinear", "TYPE /TRANSI / METHOD /INTTYP", "WILSON" )
setAnalysisCommandDetail( "Analysis1", "Structural nonlinear", "EXECUT(1) /
```

```
ITERAT/MAXITE", 50 )
    setAnalysisCommandDetail( "Analysis1", "Structural nonlinear", "EXECUT(1)/
ITERAT/CONVER/DISPLA/NOCONV", "CONTIN" )
    setAnalysisCommandDetail( "Analysis1", "Structural nonlinear", "EXECUT(1)/
ITERAT/CONVER/FORCE/NOCONV", "CONTIN" )
    setAnalysisCommandDetail( "Analysis1", "Structural nonlinear", "EXECUT(1)/
ITERAT/MAXITE", 50 )
    addAnalysisCommandDetail( "Analysis1", "Structural nonlinear", "TYPE/TRANSI" )
    setAnalysisCommandDetail( "Analysis1", "Structural nonlinear", "TYPE/
TRANSI", True )
    addAnalysisCommandDetail( "Analysis1", "Structural nonlinear", "TYPE/TRANSI/
DYNAMI" )
    setAnalysisCommandDetail( "Analysis1", "Structural nonlinear", "TYPE/TRANSI/
DYNAMI", True )
    setAnalysisCommandDetail( "Analysis1", "Structural nonlinear", "EXECUT(1)/
ITERAT/MAXITE", 50 )
    setAnalysisCommandDetail( "Analysis1", "Structural nonlinear", "EXECUT/
EXETYP", "LOAD" )
    renameAnalysisCommand( "Analysis1", "Structural nonlinear", "Structural
nonlinear" )
    setAnalysisCommandDetail( "Analysis1", "Structural nonlinear", "EXECUT(2)/
ITERAT/MAXITE", 10 )
    setAnalysisCommandDetail( "Analysis1", "Structural nonlinear", "EXECUT(2)/
ITERAT/MAXITE", 50 )
    setAnalysisCommandDetail( "Analysis1", "Structural nonlinear", "EXECUT(2)/
ITERAT/CONVER/DISPLA/NOCONV", "CONTIN" )
    setAnalysisCommandDetail( "Analysis1", "Structural nonlinear", "EXECUT(2)/
ITERAT/CONVER/FORCE/NOCONV", "CONTIN" )
    setAnalysisCommandDetail( "Analysis1", "Structural nonlinear", "EXECUT(2)/
ITERAT/MAXITE", 50 )
    setAnalysisCommandDetail( "Analysis1", "Structural nonlinear", "EXECUT(2)/
ITERAT/MAXITE", 50 )
    setAnalysisCommandDetail( "Analysis1", "Structural nonlinear", "EXECUT(1)/
ITERAT/MAXITE", 50 )
    addAnalysisCommandDetail( "Analysis1", "Structural nonlinear", "TYPE/TRANSI/
DYNAMI/DAMPIN" )
    setAnalysisCommandDetail( "Analysis1", "Structural nonlinear", "TYPE/TRANSI/
DYNAMI/DAMPIN", True )
    setAnalysisCommandDetail( "Analysis1", "Structural nonlinear", "EXECUT(2)/
```

```
LOAD /LOADNR", 2 )
    setAnalysisCommandDetail( "Analysis1", "Structural nonlinear", "EXECUT /
EXETYP", "TIME" )
    renameAnalysisCommandDetail( "Analysis1", "Structural nonlinear", "EXECUT
(3)", "new execute block 3" )
    setAnalysisCommandDetail( "Analysis1", "Structural nonlinear", "EXECUT(3) /
TIME /STEPS /EXPLIC /SIZES", "0.1(50)" )
    setAnalysisCommandDetail( "Analysis1", "Structural nonlinear", "EXECUT(3) /
ITERAT /MAXITE", 50 )
    setAnalysisCommandDetail( "Analysis1", "Structural nonlinear", "EXECUT(3) /
ITERAT /CONVER /DISPLA /NOCONV", "CONTIN" )
    setAnalysisCommandDetail( "Analysis1", "Structural nonlinear", "EXECUT(3) /
ITERAT /CONVER /FORCE /NOCONV", "CONTIN" )
    setAnalysisCommandDetail( "Analysis1", "Structural nonlinear", "OUTPUT(1) /
SELTYP", "USER" )
    addAnalysisCommandDetail( "Analysis1", "Structural nonlinear", "OUTPUT(1) /
USER" )
    addAnalysisCommandDetail( "Analysis1", "Structural nonlinear", "OUTPUT(1) /
USER /DISPLA(1) /TOTAL /TRANSL /GLOBAL" )
    addAnalysisCommandDetail( "Analysis1", "Structural nonlinear", "OUTPUT(1) /
USER /STRAIN(1) /CRKSUM /GREEN /GLOBAL" )
    addAnalysisCommandDetail( "Analysis1", "Structural nonlinear", "OUTPUT(1) /
USER /STRAIN(2) /CRKSUM /GREEN /PRINCI" )
    addAnalysisCommandDetail( "Analysis1", "Structural nonlinear", "OUTPUT(1) /
USER /STRAIN(3) /CRACK /GREEN" )
    addAnalysisCommandDetail( "Analysis1", "Structural nonlinear", "OUTPUT(1) /
USER /STRAIN(4) /CRKWDT /GREEN /GLOBAL" )
    addAnalysisCommandDetail( "Analysis1", "Structural nonlinear", "OUTPUT(1) /
USER /STRAIN(5) /CRKWDT /GREEN /LOCAL" )
    addAnalysisCommandDetail( "Analysis1", "Structural nonlinear", "OUTPUT(1) /
USER /STRAIN(6) /CRKWDT /GREEN /PRINCI" )
    runSolver( "Analysis1" )
```

第四章 水化热反应

4.1 案例一：管廊节段瞬态水化热分析案例

本模型为工程中一段未张拉预应力筋的简化空心管廊节段模型，管廊长 5.7 m，宽 3.5 m，高 1.5 m，管廊为空心并配有箍筋，在 Diana 中采用钢筋网片单元 Grid 进行模拟，其中钢筋网片的间距为 2.45 m。采用 Solid 实体单元模拟大体积混凝土施工过程中水化热作用对混凝土的影响。混凝土及钢筋材料特性见表 4-1，立体图形和配筋立面如图 4.1-1 和图 4.1-2 所示。

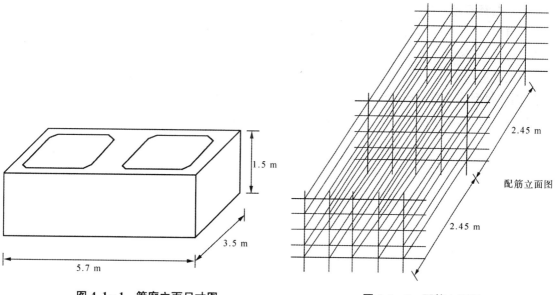

图 4.1-1 管廊立面尺寸图　　　　图 4.1-2 配筋立面图

表 4-1

混凝土	参　数	单　位
弹性模量	3.45×10^{10}	N/m^2
泊松比	0.15	—
热膨胀系数	1×10^{-5}	$1/℃$
热传导系数	1.73×10^{15}	$kg \cdot m/day^3 ℃$
热容量	2.0×10^{16}	$kg/m\ day^2 ℃$

(续表)

钢 筋	参 数	单 位
弹性模量	2.1×10^{11}	$kg/mday^2$
密度	7800	kg/m^3
屈服强度	4.4×10^8	$kg/mday^2$

学习要义：
(1) 使用图形的布尔加减运算生成空心平面图。
(2) 学习 Grid 钢筋网片单元的直接输入建模方式。
(3) 采用 Extrude 拉伸功能将平面图形拉伸成立体形状。
(4) 实体单元的建模方式及热对流材料属性赋予。
(5) 热对流边界条件的添加。
(6) 初始温度的添加。
(7) 温度非线性分析中荷载工况下的水化热反应时间荷载步和收敛准则设置。

注：作为 **Diana10.1** 的分析案例，本例题分析模型及分析条件全部是假定值，分析结果妥当性另当别论，望谅解。另为方便操作演示，对于大型混凝土结构在施工中预埋冷却管的问题本书暂不做考虑。

首先，打开 Diana10.1 界面，点击 File→New，在 F 盘的名称为例题的文件夹中创建新的文件，命名为管廊水化热分析，由于需要考虑水化热对结构的影响，因此在默认项类型设置时同时选择结构(Structural)和热流(Heat flow)两个模块，采用 Solid 实体单元进行模拟，因此模型的维数选择三维，模型的尺寸范围选择 100 m，默认的划分类型选择六面体/矩形单元，模型的划分阶数选择二次。如图 4.1-3 所示。

图 4.1-3 模型参数赋予

首先创建外部的大平面,作为管廊的底面。点击应用菜单栏 Add a sheet,创建一个平面,输入坐标(0,0,0),(0,5.7,0),(0,5.7,3.5),(0,0,3.5)。点击 OK 生成该平面。

建立内边缘的左边外形轮廓,命名为 Sheet 2,其内部形状的坐标尺寸如图 4.1-4 所示。

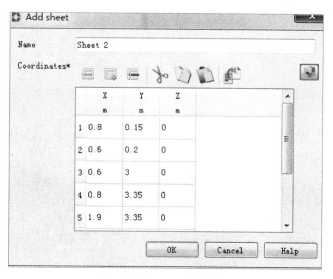

图 4.1-4　Sheet 2 内部形状坐标

选中 Sheet 2,右击采用 Array copy 方式对已经建成的左边内部图形通过复制平移方式生成右边图形轮廓,平移方向为沿 X 轴正方向,平移距离为 3 m,命名为 Sheet 3,复制份数为 1 份。Array copy 的方式如图 4.1-5 所示。

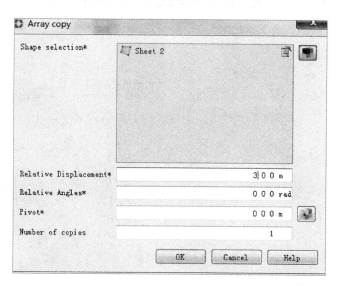

图 4.1-5　Array copy 创建 Sheet 3

采用布尔逻辑运算。如图 4.1-6 所示,选中外边的 Sheet 1 作为布尔逻辑运算选择目标(Target selection),将 Sheet 2 和 Sheet 3 作为选择的工具(Tool selection),布尔逻辑运算操作方式为相减(Subtract),选择相减后的轮廓进行合并(Merge)。在 Diana 软件的布尔逻辑运算中,目标作为布尔运算中被减对象,工具是图形中整个要减去的对象。将整个平面图形扣去 Sheet 2,Sheet 3 左右两个图形后生成的平面轮廓如图 4.1-7 所示。

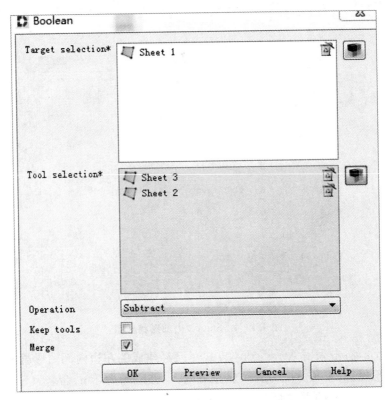

图 4.1-6　布尔逻辑运算操作界面

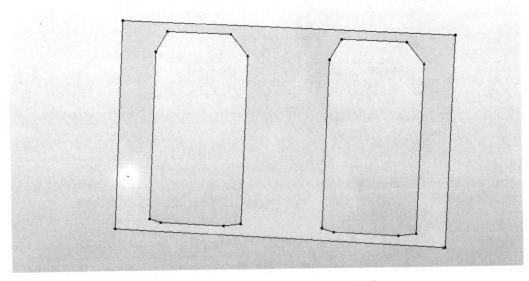

图 4.1-7　布尔逻辑运算后生成的平面轮廓

点击快捷键 Extrude 拉伸功能,弹出如图 4.1-8 所示的界面,选择在 Z 向拉伸形成整个立体单元,拉伸距离为 1.5 m。拉伸后生成的立体图形如图 4.1-9 所示。

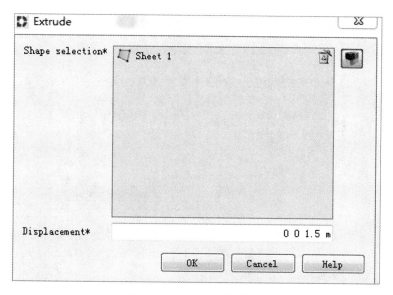

图 4.1-8　Extrude 拉伸操作

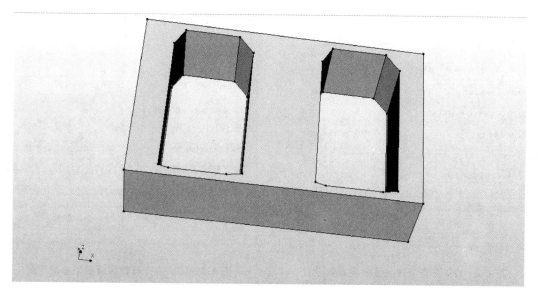

图 4.1-9　拉伸生成的立体图形

选中 Sheet 1,右击选择 Property assignments 赋予材料属性。这里选择混凝土材料,单元类型选择 Structural Solid 实体单元类型,命名为 concrete,材料模型选择弥散开裂下总应变裂缝模型中的正交旋转裂缝模型,并且在模型中考虑热流影响。设定混凝土弹性模量为 $3.45 \times 10^{10}\,\mathrm{N/m^2}$,泊松比为 0.15,混凝土本构的拉伸软化曲线选择 Hordijk 曲线,压缩曲线符合欧洲 1990 规范,拉伸强度为 $3.7 \times 10^6\,\mathrm{N/m^2}$,Hordijk 模型下第一型拉伸软化曲线的断裂能设定为 150 N/m,抗压强度为 $5.8 \times 10^7\,\mathrm{N/m^2}$。热流条件下的正交旋转裂缝模型设置参数如图 4.1-10—图 4.1-13 所示。

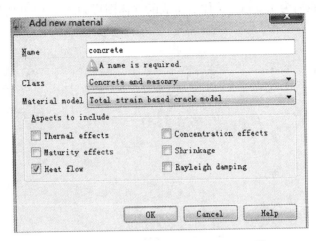

图 4.1-10　正交旋转裂缝模型参数设置

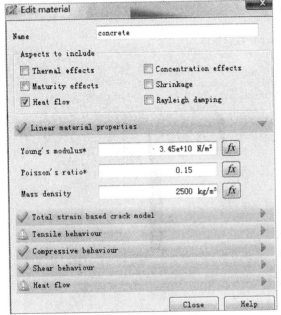

图 4.1-11　正交旋转裂缝模型参数设置

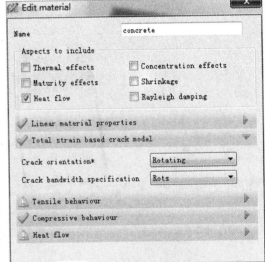

图 4.1-12　正交旋转裂缝模型参数设置

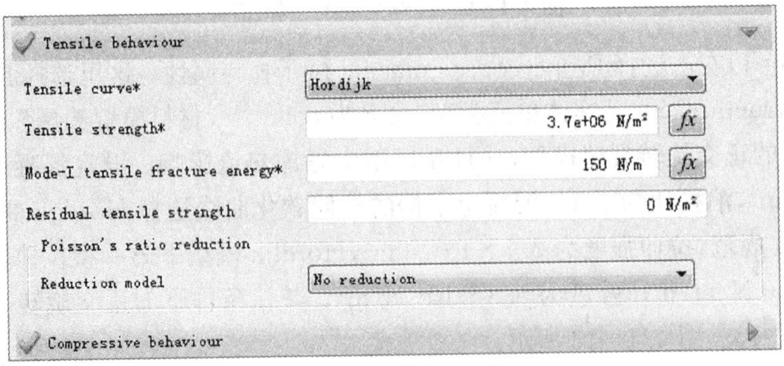

图 4.1-13　正交旋转裂缝模型参数设置

注：以上混凝土参数设置对水化热影响甚微，水化热模拟效果的关键在于热传导模块的参数设置。

设置单位。设置温度单位为摄氏度（celsius），时间为天（day），力的单位自动换算为 kgm/day^2。如图 4.1-14 所示。

设置热传导率和热容量参数。热传导率和热容量均可设置成与单元龄期、时间、温度等有关的函数，但在这里我们认为热传导率以及热容量均为常数，因此选择 No dependency。在水化热方法一栏中，有预处理（Preprocessing）、直接输入（Direct Input）以及用户自定义二次开发（User-supplied）三种方式，其中界面操作中提供了前两种方式，在本案例中，选择 Preprocessing 预处理方法。热传导

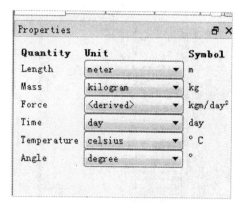

图 4.1-14 单位设置

系数设置为 $1.73×10^{15}\,kg·m/day^3℃$，热容量 $2×10^{16}\,kg/m\,day^2℃$。参考温度和阿列纽斯系数选择默认设定值。在绝热发展（Adiabatic heat development）模块中定义龄期时间（Age）—温度（Temperature）绝热温升曲线，其中混凝土的龄期时间范围设定为 60 天，温度从 0℃逐渐上升到 70℃。考虑到在实际情况下，大体积混凝土在最初几天内水化热反应温度变化较为明显，参考规范设置，定义前 10 天温度上升速度较快，龄期至第 10 天时温度逐渐升高至 70℃，在混凝土龄期 10~60 天内温度稳定在 70℃。热流模块参数设置和龄期—时间曲线定义如图 4.1-15—图 4.1-16 所示。

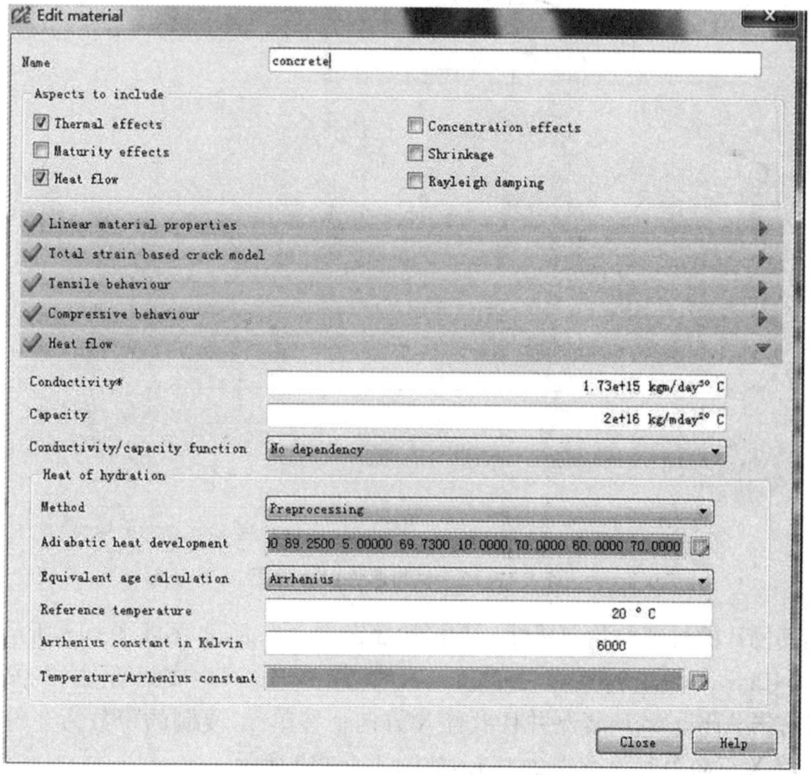

图 4.1-15 热流模块参数设置

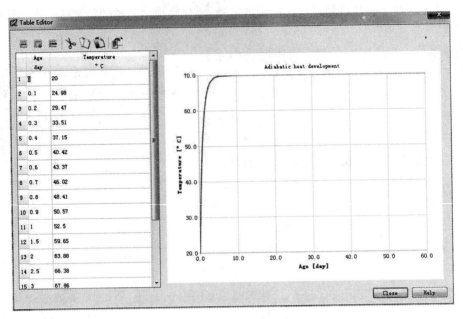

图 4.1-16　绝热温升曲线龄期—温度设置

由于实体单元不需要定义截面几何特性,因此在图 4.1-17 操作界面上定义完材料属性后直接点击 OK,确认生成即可。

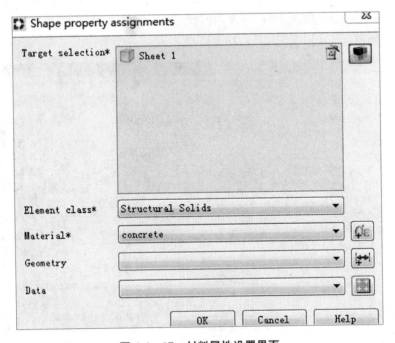

图 4.1-17　材料属性设置界面

定义钢筋网片的材料和几何属性。点击快捷键创立平面,如图 4.1-18 所示,输入坐标 (0.5,0,0),(0.5,0,1.5),(0.5,3.5,1.5),(0.5,3.5,0),生成钢筋网片的平面,命名为 Grid1,并且右击 Array copy 复制并且沿着 X 方向平移单元,复制的份数为 2 份,平移方向和距离为沿着 X 正向等间距平移 2.45 m,如图 4.1-19 所示。

第四章 水化热反应

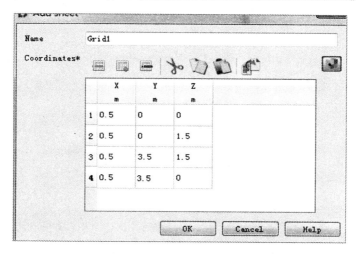

图 4.1-18　Grid1 几何模型各点坐标

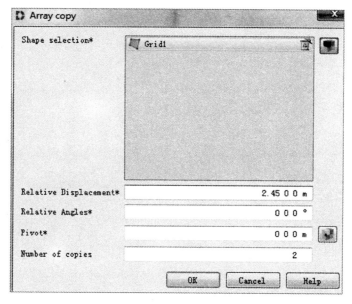

图 4.1-19　Array copy Grid1

输入表 4-2 中坐标,生成侧面的钢筋网片,分别命名为 Grid4,Grid5,Grid6,Grid7。生成各面之后,选中 Grid4,Grid5,右击 Move shape 沿着 Y 向正向移动 0.1 m,选中 Grid6,Grid7 沿着 Y 向负向移动 0.1 m,生成侧面的钢筋面。操作方式如图 4.1-20—图 4.1-24 所示。侧面各钢筋网片几何坐标值见表 4-2。

表 4-2　侧面钢筋网片几何坐标值

Grid4	(0.5, 0, 1.5),(2.95, 0, 1.5),(2.95, 0, 0),(0.5, 0, 0)
Grid5	(2.95, 0, 1.5),(5.4, 0, 1.5),(5.4, 0, 0),(2.95, 0, 0)
Grid6	(5.4, 3.5, 1.5),(5.4, 3.5, 0),(2.95, 3.5, 0),(2.95, 3.5, 1.5)
Grid7	(2.95, 3.5, 1.5),(2.95, 3.5, 0),(0.5, 3.5, 0),(0.5, 3.5, 1.5)

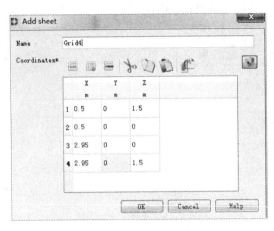

图 4.1-20　创建 Grid4　　　　　图 4.1-21　移动 Grid4,Grid5

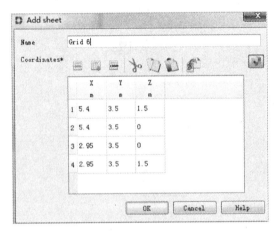

图 4.1-22　创建 Grid6　　　　　图 4.1-23　创建 Grid7

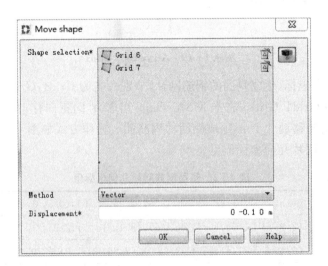

图 4.1-24　移动 Grid6,Grid7

赋予钢筋网片属性,选中 Grid1,Gird2,Grid3 命名为 Grid1 组,点击快捷图标

Reinforcement assignments 按钮赋予钢材的材料属性为 Von Mises 塑性屈服模型,其中弹性模量为 2.1×10^{11} kg/mday^2,泊松比为 0.3,钢材的密度为 7800 kg/m^3,热膨胀系数设置为 0.000 01,硬化函数为非硬化(No hardening),屈服强度为 4.4×10^8 kg/mday^2。如图 4.1-25—图 4.1-28 所示。

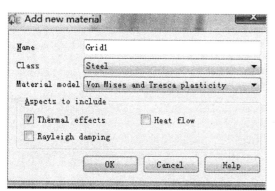

图 4.1-25 定义材料属性

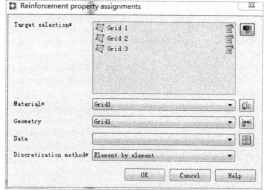

图 4.1-26 Grid1 组

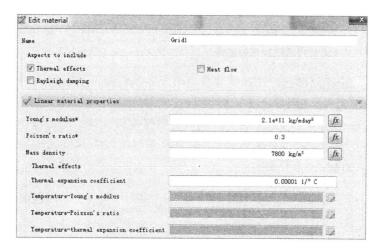

图 4.1-27 材料参数赋予

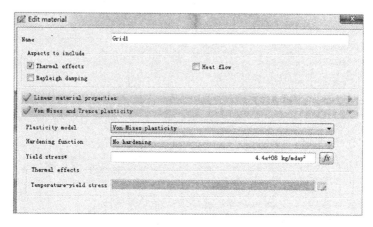

图 4.1-28 材料参数赋予

赋予 Grid1 组钢筋网片几何特性。采用直径和间距的方式定义钢筋网片厚度。其中 X 与 Y 方向的钢筋直径均定义为 32 mm，间距均为 100 mm。如图 4.1-29 所示。

将面 Grid4，Grid5，Grid6，Grid7 定义为 Grid2 钢筋网片组，钢筋网片的局部坐标 X 方向对应整体坐标系下的 Y 方向。再用同样的方式定义 Grid2 钢筋组材料属性和几何属性。

点击上方红色正方体图标的快捷键，创立边界单元，命名为 Boundary，如图 4.1-30 所示。

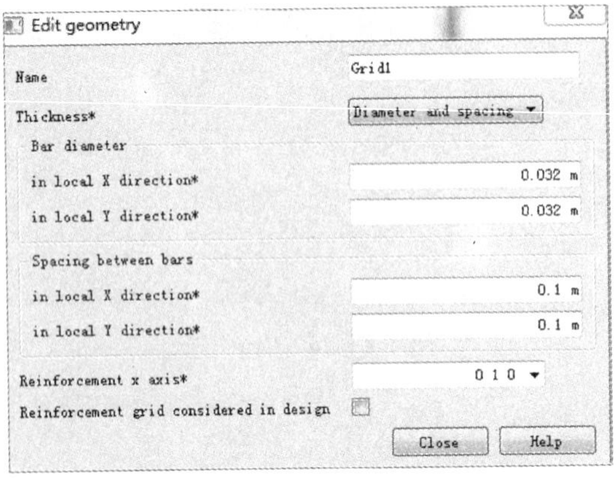

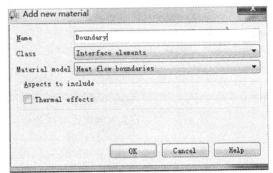

图 4.1-29 赋予 Grid1 组钢筋网片几何属性　　图 4.1-30 选用界面单元的热流边界条件

选择热对流。由于之前已经选择的混凝土单元是实体单元，因此不需要再赋予几何属性。在编辑热对流边界类型中，有仅仅考虑对流、仅仅考虑辐射、辐射和对流同时考虑、什么都不考虑四种类型，这里选择 Convection only，表明仅仅考虑热对流，且热对流系数为一个常数。与之前定义混凝土热传导函数相同，认为热传导率以及热容量均为常数，因此选择 No dependency。如图 4.1-31 所示。

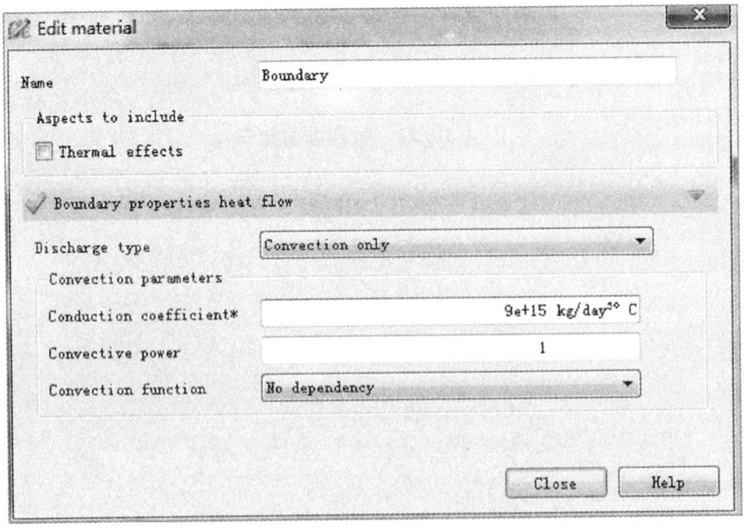

图 4.1-31 定义热对流边界属性

界面单元材料属性定义。选择垂直于 X 轴实体六面体的两个侧面及上顶面为热对流边界面,连接类型为边界界面方式,界面单元的单元类型为热流边界。如图4.1-32所示。

设置热流边界的边界条件。边界条件分组(Boundary condition set)选择边界(Boundary)。施加热流目标类型(Target type)为面类型(Face),边界条件类型为外部温度(External temperature),外部温度设置为35℃。如图4.1-33所示。

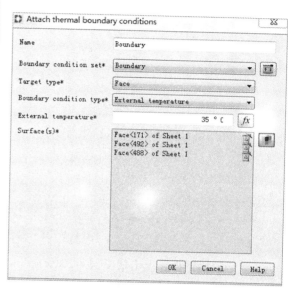

图 4.1-32 界面单元材料属性定义　　图 4.1-33 热对流边界的边界条件

点击 OK,生成如图4.1-34所示的热对流边界。绿色表示热对流边界已经成功定义。

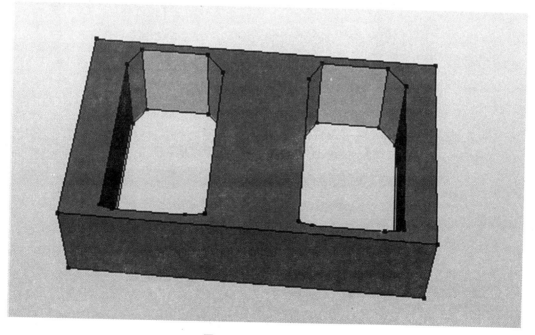

图 4.1-34 热对流边界

对边界条件进行时变参数定义,本案例中设定 60 天内影响因子全为 1,如图 4.1-35 所示。

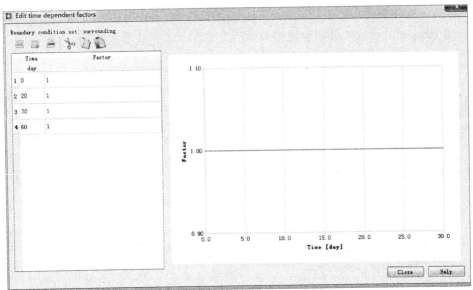

图 4.1-35 时变参数定义界面

定义初始温度。点击快捷键 Attach an initial field to shape/face/line/point,如图 4.1-36 所示,选中整个实体模型,命名为 initial,施加温度场类型为实体,温度场初始温度为 20℃,如图 4.1-37 所示。

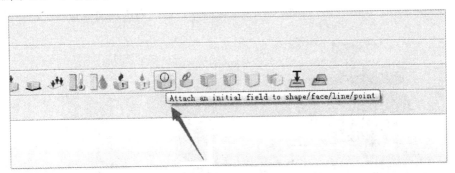

图 4.1-36 Attach an initial field 快捷键位置

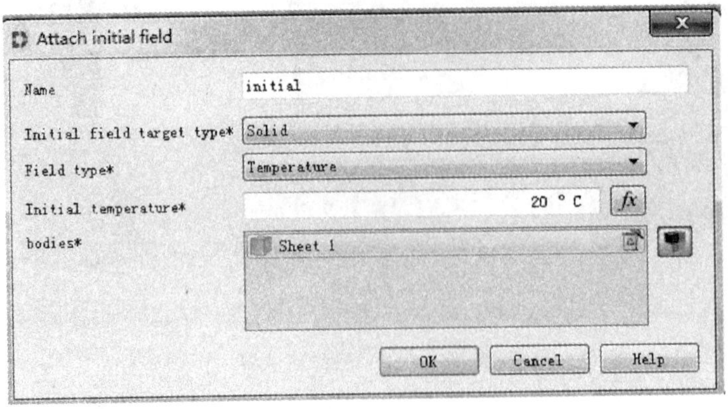

图 4.1-37 定义初始温度

待 Load 属性栏添加重力荷载之后,划分单元网格。选中整个实体几何模型 Sheet 1,右击选择设置网格属性,这里采用定义单元尺寸的方式进行网格划分,其中单元网格的尺寸为 0.1 m,划分类型采用六面体/四边形类型,如图 4.1-38 所示。

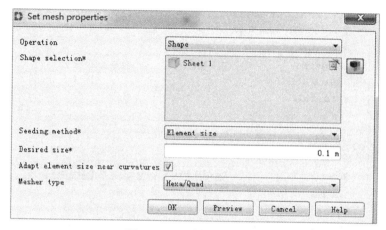

图 4.1-38 划分单元网格

点击快捷键图标 Generate mesh of a shape 按钮,生成如图 4.1-39 所示的单元网格。

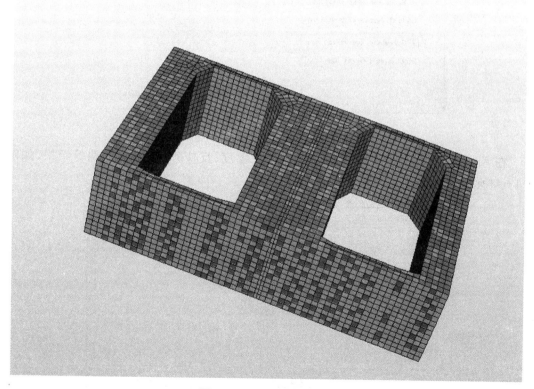

图 4.1-39 单元网格形成

设置分析工况。点击 Analysis,右击选择瞬态热传递,在初始瞬态分析中选择初始温度场,分析类型中选择非线性分析,起始时间(Start time)设置为从第 1 天(1day)开始。并且勾选下方的水化热分析,水化热反应程度设置为默认值 0.01,勾选下一栏混凝土水化反应计算等效龄期，等效龄期选择默认值 0 天。如图 4.1-40 所示。在荷载工况中的荷载步定义一栏将 35 天水化热反应时间步设置成 0.500000(20) 1.00000(5) 10.0000(2)。

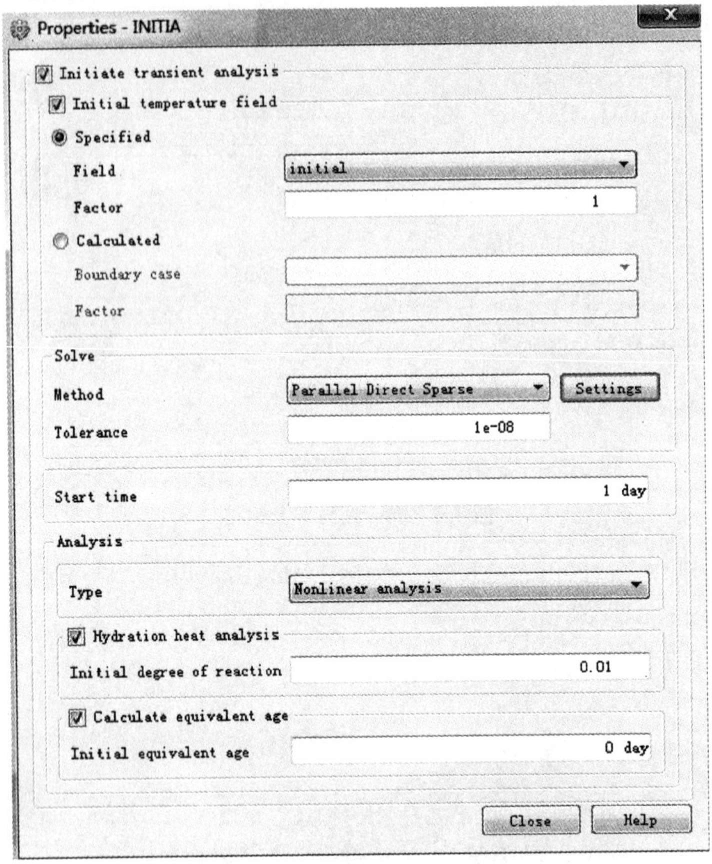

图 4.1-40

在 Analysis output 结果输出栏中选择初始温度、反应程度、等效龄期等结果，如图 4.1-41 所示。

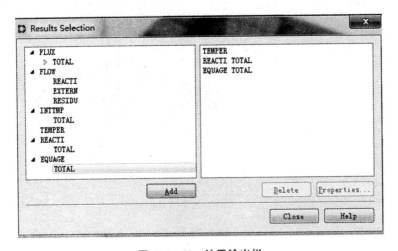

图 4.1-41 结果输出栏

点击 Run analysis 生成计算结果。如图 4.1-42 所示为水化热反应分析结束后的变形图。

图 4.1-42　水化热反应分析结束后的变形图

选取最后一个荷载步(第 27 荷载步)即对应第 36 天时水化热的反应结果,点击输出结果 Analysis output→Nodal results 下方的温度栏,显示温度云图。从云图 4.1-43 中可以看到,实体单元中没有进行热对流边界定义的区域在水化热反应后温度明显较高。

图 4.1-43　温度云图

等效龄期(成熟度)云图显示如图 4.1-44 所示。可见水化热反应放热温度较高区域,等效龄期也较大。

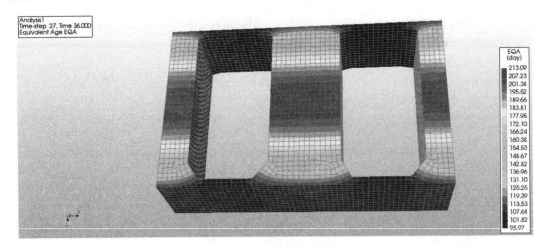

图 4.1-44 等效龄期(成熟度)云图

Python 语言命令流如下：

```
newProject( "guanlang", 100 )
setModelAnalysisAspects( [ "STRUCT", "HEATFL" ] )
setModelDimension( "3D" )
setDefaultMeshOrder( "LINEAR" )
setDefaultMesherType( "HEXQUAD" )
createSheet( "Sheet 1", [[ 0, 0, 0 ],[ 5.7, 0, 0 ],[ 5.7, 3.5, 0 ],[ 0, 3.5, 0 ]] )
createSheet( "Sheet 2", [[ 0.8, 0.15, 0 ],[ 0.6, 0.2, 0 ],[ 0.6, 3, 0 ],[ 0.8, 3.35, 0 ],[ 1.9, 3.35, 0 ],[ 2.2, 3, 0 ],[ 2.2, 0.2, 0 ],[ 1.9, 0.15, 0 ]] )
arrayCopy( [ "Sheet 2" ], [ 3, 0, 0 ], [ 0, 0, 0 ], [ 0, 0, 0 ], 1 )
saveProject(  )
subtract( "Sheet 1", [ "Sheet 3", "Sheet 2" ], False, True )
saveProject(  )
extrudeProfile( [ "Sheet 1" ], [ 0, 0, 1.5 ] )
addMaterial( "concrete", "CONCR", "TSCR", [ "HEATFL", "THERMA" ])
setParameter( "MATERIAL", "concrete", "LINEAR /ELASTI /YOUNG", 3.45e + 10 )
setParameter( "MATERIAL", "concrete", "LINEAR /ELASTI /POISON", 0.15 )
setParameter( "MATERIAL", "concrete", "LINEAR /ELASTI /POISON", 0.15 )
setParameter( "MATERIAL", "concrete", "LINEAR /MASS /DENSIT", 2500 )
setParameter( "MATERIAL", "concrete", "LINEAR /THERMA /THERMX", 1e - 05 )
setParameter( "MATERIAL", "concrete", "MODTYP /TOTCRK", "ROTATE" )
setParameter( "MATERIAL", "concrete", "TENSIL /TENSTR", 3700000 )
setParameter( "MATERIAL", "concrete", "TENSIL /TENSTR", 3700000 )
setParameter( "MATERIAL", "concrete", "TENSIL /TENCRV", "HORDYK" )
setParameter( "MATERIAL", "concrete", "TENSIL /GF1", 150 )
setParameter( "MATERIAL", "concrete", "TENSIL /RESTST", 0 )
setParameter( "MATERIAL", "concrete", "TENSIL /RESTST", 0 )
```

```
setParameter( "MATERIAL", "concrete", "COMPRS /COMCRV", "MC1990" )
setParameter( "MATERIAL", "concrete", "COMPRS /COMSTR", 58000000 )
setParameter( "MATERIAL", "concrete", "HEATFL /CONDUC", 1.73e + 15 )
setParameter( "MATERIAL", "concrete", "HEATFL /CAPACI", 2e + 16 )
setParameter( "MATERIAL", "concrete", "HEATFL /HEATHY /HYDRAT", "PREPRO" )
setParameter( "MATERIAL", "concrete", "HEATFL /HEATHY /ADIAB", [] )
setParameter( MATERIAL, "concrete", "HEATFL /HEATHY /ADIAB", [ 0, 20, 0.1,
24.98, 0.2, 29.47, 0.3, 33.51, 0.4, 37.15, 0.5, 40.42, 0.6, 43.37, 0.7, 46.02, 0.8,
48.41, 0.9, 50.57, 1, 52.5, 1.5, 59.65, 2, 63.88, 2.5, 66.38, 3, 67.86, 4, 69.25, 5,
69.73, 10, 70, 60, 70 ] )
setUnit( "TEMPER", "CELSIU" )
setUnit( "ANGLE", "DEGREE" )
setUnit( "TIME", "DAY" )
saveProject( )
addGeometry( "Element geometry 1", "SOLID", "STRSOL", [] )
rename( "GEOMET", "Element geometry 1", "concrete" )
clearReinforcementAspects( [ "Sheet 1" ] )
setElementClassType( "SHAPE", [ "Sheet 1" ], "STRSOL" )
assignMaterial( "concrete", "SHAPE", [ "Sheet 1" ] )
resetGeometry( "SHAPE", [ "Sheet 1" ] )
resetElementData( "SHAPE", [ "Sheet 1" ] )
createSheet( "Grid1", [[ 0.5, 0, 0 ], [ 0.5, 0, 1.5 ], [ 0.5, 3.5, 1.5 ], [ 0.5,
3.5, 0 ]] )
saveProject( )
arrayCopy( [ "Grid1" ], [ 2.45, 0, 0 ], [ 0, 0, 0 ], [ 0, 0, 0 ], 2 )
createSheet( "Grid5", [[ 0.5, 0, 1.5 ], [ 0.5, 0, 0 ], [ 2.95, 0, 0 ], [ 2.95, 0, 1.
5 ]] )
createSheet( "Grid6", [[ 2.95, 0, 1.5 ], [ 2.95, 0, 0 ], [ 5.4, 0, 0 ], [ 5.4, 0,
1.5 ]] )
renameShape( "Grid5", "Grid4" )
renameShape( "Grid6", "Grid5" )
renameShape( "Grid4", "Grid 4" )
renameShape( "Grid5", "Grid 5" )
renameShape( "Grid3", "Grid 3" )
renameShape( "Grid2", "Grid 2" )
renameShape( "Grid1", "Grid 1" )
translate( [ "Grid 4", "Grid 5" ], [ 0, 0.1, 0 ] )
saveProject( )
createSheet( "Grid 6", [[ 5.4, 3.5, 1.5 ], [ 5.4, 3.5, 0 ], [ 2.95, 3.5, 0 ], [
2.95, 3.5, 1.5 ]] )
```

```
createSheet( "Grid 7", [[ 2.95, 3.5, 1.5 ],[ 2.95, 3.5, 0 ],[ 0.5, 3.5, 0 ],[ 0.5, 3.5, 1.5 ]] )
translate( [ "Grid 6", "Grid 7" ], [ 0, -0.1, 0 ] )
saveProject( )
addMaterial( "Grid1", "MCSTEL", "TRESCA", [ "THERMA" ] )
setParameter( "MATERIAL", "Grid1", "LINEAR /ELASTI /YOUNG", 2.1e+11 )
setParameter( "MATERIAL", "Grid1", "LINEAR /ELASTI /POISON", 0.3 )
setParameter( "MATERIAL", "Grid1", "LINEAR /MASS /DENSIT", 7800 )
setParameter( "MATERIAL", "Grid1", "LINEAR /THERMA /THERMX", 1e-05 )
setParameter( "MATERIAL", "Grid1", "LINEAR /THERMA /THERMX", 1e-05 )
setParameter( "MATERIAL", "Grid1", "LINEAR /THERMA /THERMX", 1e-05 )
setParameter( "MATERIAL", "Grid1", "TREPLA /YLDSTR", 4.4e+08 )
setMaterialAspects( "Grid1", [ "THERMA", "HEATFL" ] )
setMaterialAspects( "Grid1", [ "THERMA" ] )
saveProject( )
addGeometry( "Element geometry 1", "RSHEET", "REGRID", [] )
rename( "GEOMET", "Element geometry 1", "Grid1" )
setParameter( "GEOMET", "Grid1", "PHI", [ 0.032, 0.032 ] )
setParameter( "GEOMET", "Grid1", "SPACIN", [ 0.1, 0.1 ] )
setParameter( "GEOMET", "Grid1", "XAXIS", [ 0, 1, 0 ] )
setParameter( "GEOMET", "Grid1", "XAXIS", [ 0, 1, 0 ] )
setParameter( "GEOMET", "Grid1", "XAXIS", [ 0, 1, 0 ] )
setParameter( "GEOMET", "Grid1", "XAXIS", [ 0, 1, 0 ] )
setParameter( "GEOMET", "Grid1", "XAXIS", [ 0, 1, 0 ] )
setReinforcementAspects( [ "Grid 1", "Grid 2", "Grid 3" ] )
assignMaterial( "Grid1", "SHAPE", [ "Grid 1", "Grid 2", "Grid 3" ] )
assignGeometry( "Grid1", "SHAPE", [ "Grid 1", "Grid 2", "Grid 3" ] )
resetElementData( "SHAPE", [ "Grid 1", "Grid 2", "Grid 3" ] )
setReinforcementDiscretization( [ "Grid 1", "Grid 2", "Grid 3" ], "ELEMENT" )
addMaterial( "Grid2", "MCSTEL", "TRESCA", [ "THERMA" ] )
setParameter( "MATERIAL", "Grid2", "LINEAR /ELASTI /YOUNG", 2.1e+11 )
setParameter( "MATERIAL", "Grid2", "LINEAR /ELASTI /POISON", 0.3 )
setParameter( "MATERIAL", "Grid2", "LINEAR /MASS /DENSIT", 7800 )
setParameter( "MATERIAL", "Grid2", "LINEAR /THERMA /THERMX", 1e-05 )
setParameter( "MATERIAL", "Grid2", "TREPLA /YLDSTR", 4.4e+08 )
addGeometry( "Element geometry 2", "RSHEET", "REGRID", [] )
rename( "GEOMET", "Element geometry 2", "Grid 2" )
setParameter( "GEOMET", "Grid 2", "PHI", [ 0.032, 0.032 ] )
setParameter( "GEOMET", "Grid 2", "SPACIN", [ 0.1, 0.1 ] )
setParameter( "GEOMET", "Grid 2", "XAXIS", [ 1, 0, 0 ] )
```

第四章 水化热反应

```
    setParameter( "GEOMET", "Grid 2", "XAXIS", [ 0, 1, 0 ] )
    setParameter( "GEOMET", "Grid 2", "XAXIS", [ 0, 1, 0 ] )
    saveProject(  )
    setReinforcementAspects( [ "Grid 4", "Grid 5", "Grid 6", "Grid 7" ] )
    assignMaterial( "Grid2", "SHAPE", [ "Grid 4", "Grid 5", "Grid 6", "Grid 7" ] )
    assignGeometry( "Grid 2", "SHAPE", [ "Grid 4", "Grid 5", "Grid 6", "Grid 7" ] )
    resetElementData( "SHAPE", [ "Grid 4", "Grid 5", "Grid 6", "Grid 7" ] )
    setReinforcementDiscretization( [ "Grid 4", "Grid 5", "Grid 6", "Grid 7" ],
"ELEMENT" )
    saveProject(  )
    addMaterial( "Boundary", "INTERF", "FLBOUN", [] )
    setParameter( "MATERIAL", "Boundary", "HTBOUN /CONPAR /CONVEC", 9e + 15 )
    createSurfaceConnection( "Boundary" )
    setParameter( "GEOMETRYCONNECTION", "Boundary", "CONTYP", "BOUNDA" )
    attachTo ( " GEOMETRYCONNECTION ", " Boundary ", " SOURCE ", " Sheet 1 ",
[[ 3.2693661, 3.5, 0.6396405 ], [ 2.4306339, 6.5822961e - 33, 0.8603595 ],
[ 0.2558562, 1.4924945, 1.5 ]] )
    setElementClassType( "GEOMETRYCONNECTION", "Boundary", "HEABOU" )
    assignMaterial( "Boundary", "GEOMETRYCONNECTION", "Boundary" )
    resetGeometry( "GEOMETRYCONNECTION", "Boundary" )
    resetElementData( "GEOMETRYCONNECTION", "Boundary" )
    addSet( "GEOMETRYBCSET", "surrounding" )
    createSurfaceBoundaryCondition( "THERMAL", "surrounding", "surrounding" )
    setParameter( "GEOMETRYBC", "surrounding", "BOUTYP", "EXTEMP" )
    setParameter( "GEOMETRYBC", "surrounding", "EXTEMP /VALUE", 35 )
    attach( "GEOMETRYBC", "surrounding", "Sheet 1", [[ 3.2693661, 3.5, 0.6396405
],[ 0.2558562, 1.4924945, 1.5 ],[ 2.4306339, 6.5822961e - 33, 0.8603595 ]] )
    saveProject(  )
    setTimeDependentLoadFactors( "GEOMETRYBCSET", "surrounding", [ 0, 20, 30, 60
], [ 1, 1, 1, 1 ] )
    createBodyInitialField( "initial" )
    setParameter( "GEOMETRYINIFIELD", "initial", "INITYP", "TEMPER" )
    setParameter( "GEOMETRYINIFIELD", "initial", "TEMPER /VALUE", 20 )
    attach( "GEOMETRYINIFIELD", "initial", [ "Sheet 1" ] )
    addSet( "GEOMETRYLOADSET", "gravity" )
    createModelLoad( "gravity", "gravity" )
    setElementSize( [ "Sheet 1" ], 0.1, -1, True )
    setMesherType( [ "Sheet 1" ], "HEXQUAD" )
    setParameter( "MATERIAL", "concrete", "LINEAR /ELASTI /YOUNG", 3.45e + 10 )
    setParameter( "MATERIAL", "concrete", "TENSIL /TENSTR", 3700000 )
```

```
    setParameter( "MATERIAL", "concrete", "TENSIL /GF1", 150 )
    setParameter( "MATERIAL", "concrete", "COMPRS /COMSTR", 58000000 )
    setParameter( "MATERIAL", "concrete", "HEATFL /CONDUC", 1.73e+15 )
    setParameter( "MATERIAL", "concrete", "HEATFL /CONDUC", 1.73e+15 )
    setParameter( "MATERIAL", "concrete", "HEATFL /CAPACI", 2e+16 )
    setParameter( "MATERIAL", "concrete", "HEATFL /CAPACI", 2e+16 )
    setParameter( MATERIAL, "concrete", "HEATFL /HEATHY /ADIAB", [ 0, 20, 0.1,
24.98, 0.2, 29.47, 0.3, 33.51, 0.4, 37.15, 0.5, 40.42, 0.6, 43.37, 0.7, 46.02, 0.8,
48.41, 0.9, 50.57, 1, 52.5, 1.5, 59.65, 2, 63.88, 2.5, 66.38, 3, 67.86, 4, 69.25, 5,
69.73, 10, 70, 60, 70 ] )
    saveProject( )
    generateMesh( [] )
    hideView( "GEOM" )
    showView( "MESH" )
    addAnalysis( "Analysis1" )
    addAnalysisCommand( "Analysis1", "HEATTR", "Transient heat transfer" )
    addAnalysisCommandDetail( "Analysis1", "Transient heat transfer", "INITIA /
TEMPER" )
    setAnalysisCommandDetail( "Analysis1", "Transient heat transfer", "INITIA /
TEMPER", True )
    addAnalysisCommandDetail( "Analysis1", "Transient heat transfer", "INITIA /
TIME0" )
    setAnalysisCommandDetail( "Analysis1", "Transient heat transfer", "INITIA /
TIME0", 1 )
    setAnalysisCommandDetail( "Analysis1", "Transient heat transfer", "INITIA /
ANATYP", "NONLIN" )
    addAnalysisCommandDetail( "Analysis1", "Transient heat transfer", "INITIA /
NONLIN /HYDRAT" )
    setAnalysisCommandDetail( "Analysis1", "Transient heat transfer", "INITIA /
NONLIN /HYDRAT", True )
    addAnalysisCommandDetail( "Analysis1", "Transient heat transfer", "INITIA /
NONLIN /EQUAGE" )
    setAnalysisCommandDetail( "Analysis1", "Transient heat transfer", "INITIA /
NONLIN /EQUAGE", True )
    saveProject( )
    setAnalysisCommandDetail( "Analysis1", "Transient heat transfer", "OUTPUT(1) /
SELTYP", "USER" )
    addAnalysisCommandDetail( "Analysis1", "Transient heat transfer", "OUTPUT(1) /
USER" )
    addAnalysisCommandDetail( "Analysis1", "Transient heat transfer", "OUTPUT(1) /
```

USER/TEMPER")
    addAnalysisCommandDetail("Analysis1", "Transient heat transfer", "OUTPUT(1) /USER/REACTI(1) /TOTAL")
    addAnalysisCommandDetail("Analysis1", "Transient heat transfer", "OUTPUT(1) /USER/EQUAGE(1) /TOTAL")
    setAnalysisCommandDetail("Analysis1", "Transient heat transfer", "EXECUT /SIZES", "0.500000(20) 1.00000(5) 10.0000 10.0000")
    runSolver("Analysis1")
    showView("RESULT")
    setResultCase(["Analysis1", "Analysis output", "Time - step 27, Time 36.000"])
    setResultPlot("contours", "Temperatures /node", "PTE")
    setResultPlot("contours", "Degrees of Reaction /node", "DGR")
    setResultPlot("contours", "Equivalent Age /node", "EQA")
    setResultPlot("contours", "Temperatures /node", "PTE")

4.2 案例二：大体积混凝土水化热开裂指数

如图 4.2-1 所示，本案例模型为一段作为配重块的素混凝土工程方桩。方桩长为 1 m，宽度和高度均为 0.6 m，混凝土强度为 C50。采用实体单元建模。其中热对流相关参数定义采用日本土木工程协会规范（JSCE）、欧洲 1990 规范（CEB-FIP1990）以及美国 AASHTO 规范三种不同规范进行模拟，并附上 Python 语句。

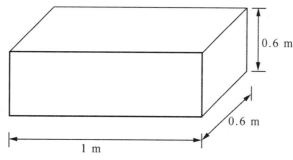

图 4.2-1 方桩几何尺寸

学习要义：
（1）学习直接建立立体几何模型生成实体单元几何形状和实体单元材料属性赋予。
（2）热对流边界条件的添加。
（3）初始温度的添加。
（4）各种规范中热对流和水化热参数定义。
（5）温度非线性计算荷载工况设置。
（6）查看水化热开裂指数云图分布。

注：作为 Diana10.1 的分析案例，本例题分析模型及分析条件全部是假定值，分析结果妥当性另当别论，望谅解。

首先,打开 Diana10.1 界面,点击 File→New,在 G 盘名称为例题的文件夹中创建新的.dpf 文件,命名为方桩水化热开裂指数。考虑到模拟水化热的影响,因此需要同时选择结构(Structural)和热流(Heat flow)两个模块。接下来要采用 Solid 实体单元进行模拟,因此模型的维数选择三维,模型的尺寸范围选择 10 m,默认的划分类型选择六面体/矩形单元,模型的划分阶数选择二次。中点位置选择在形状上(On shape)。如图 4.2-2 所示。

图 4.2-2 创立 New project 热流模块

设置单位。这里设置温度单位为摄氏度(celsius),时间单位为天(day),力单位自动换算为 kgm/day^2。如图 4.2-3 所示。

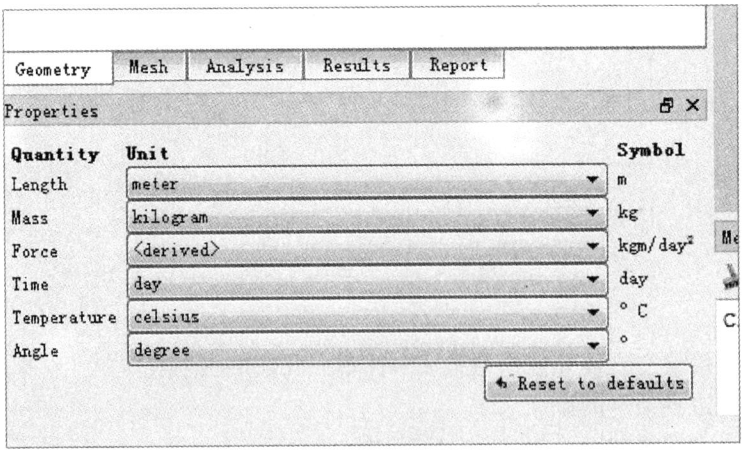

图 4.2-3 设置单位

创建混凝土模型。点击快捷键 Adds a block solid,选择实体单元的形状为砖块(Adds a block solid),如图 4.2-4 所示。输入起点坐标(0,0,0),长 1 m,宽 0.6 m,高 0.6 m,生成块体几何形状。如图 4.2-5 所示。

第四章 水化热反应

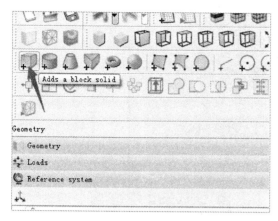

图 4.2-4 添加正六面体快捷工具栏操作

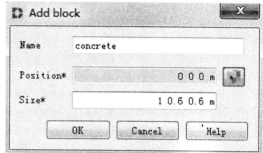

图 4.2-5 块体的坐标位置和尺寸

如图 4.2-5 所示，选择第一个点位置为 (0,0,0) 点，依次输入六面体的几何尺寸(Size)为长 1 m、宽 0.6 m、高 0.6 m，点击 OK，生成如图 4.2-6 所示的几何六面体单元立体模型，命名为 Block1。

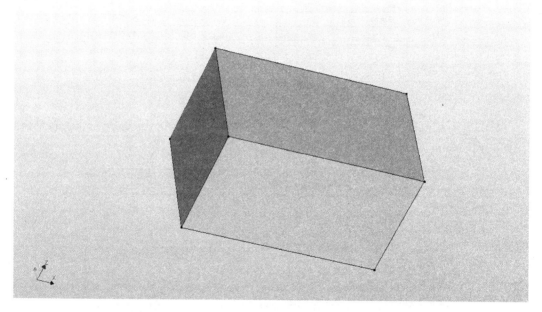

图 4.2-6 六面体单元几何模型

选中 Block1，右击选择 Property assignments 赋予材料属性。这里选择混凝土材料，单元类型选择 Structural Solid 实体单元类型，命名为 concrete。选择日本 JSCE 规范，勾选早强混凝土、开裂指数，以及热流三个模块，如图 4.2-7 所示。基本单位的设置也在这里做一下更改，长度为米，时间为天，力为牛顿，温度为摄氏度，角度为弧度，质量自动调整为 $Nday^2/m$，模型单位调整后如图 4.2-8 所示。

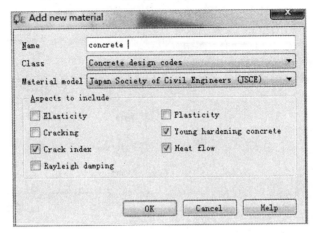

图 4.2-7 材料属性

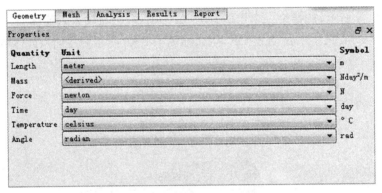

图 4.2-8 模型基本单位调整

设置 JSCE 规范中参数。其中 91 天抗压强度标准值为 $3.24\times10^7\,\mathrm{N/m^2}$,91 天弹性模量为 $2.7\times10^{10}\,\mathrm{N/m^2}$。如图 4.2-9 所示。

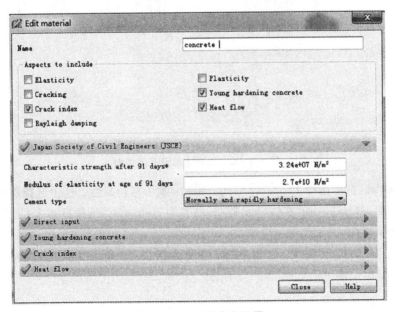

图 4.2-9 材料参数设置

一旦选中 JSCE 规范中的开裂指数,开裂指数就会自动选择符合 JSCE 规范规定。如图 4.2-10 所示,设定混凝土弹性模量为 $2.7 \times 10^{10} \text{N/m}^2$,泊松比为 0.2。注意混凝土的密度在这里已经不是 kg/m^3 了,而是在之前的基本单位调整中变成了 Nday^2/m,因此根据基本单位之间的换算,混凝土的密度换算为 $3.34898 \times 10^{-7} \text{Nday}^2/\text{m}$。在早强混凝土参数特性设置中,需要选择与模型能量准则(Power law model)相关的参数,这里统一采用软件中的默认设置值,如图 4.2-11 所示。

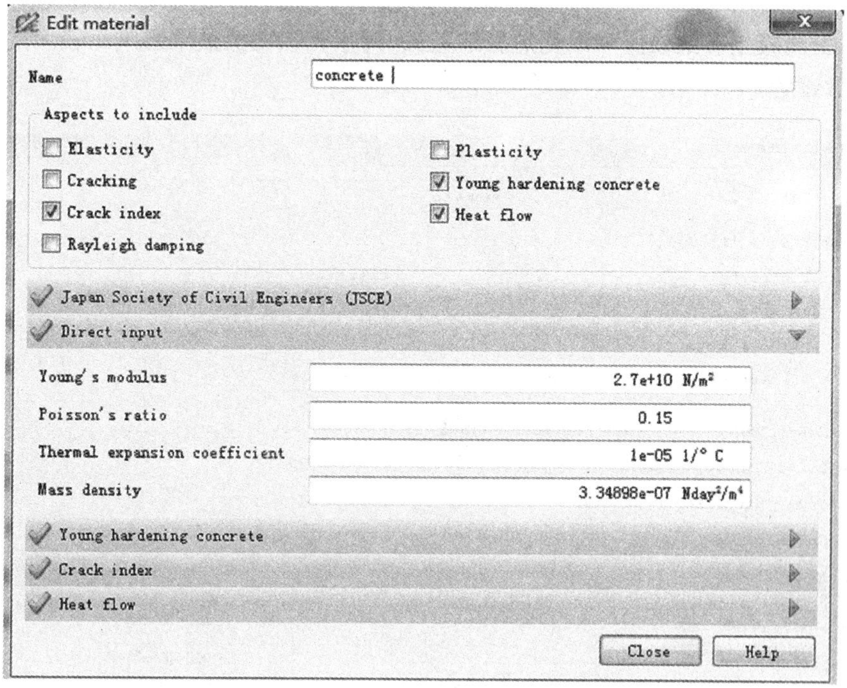

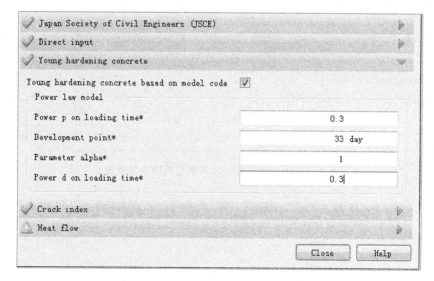

图 4.2-10 JSCE 规范直接输入基本参数

图 4.2-11 JSCE 规范中早强混凝土参数特性设置

采用 Preprocessing 预处理方式定义绝热升温曲线、热传导率、热容量。热传导率和热容量均可设置成与单元龄期、时间、温度等有关的函数，但这里认为热传导率以及热容量均为常数，因此选择 No dependency。在水化热方法一栏中，有预处理、直接输入以及二次开发三种方式，其中界面操作中提供了前两种方式，本案例选择热处理方法。热传导系数 (Conductivity) 在一开始设置为 320 N/day℃，热容量 (Capcity) 为 2660 J/m³℃，参考温度和阿列纽斯系数选择默认设定值。如图 4.2-12 所示。龄期时间—温度曲线定义如图 4.2-13 所示。

注：以上混凝土参数设置对水化热影响甚微，对水化热模拟效果影响比较大的是热传导模块的参数设置。

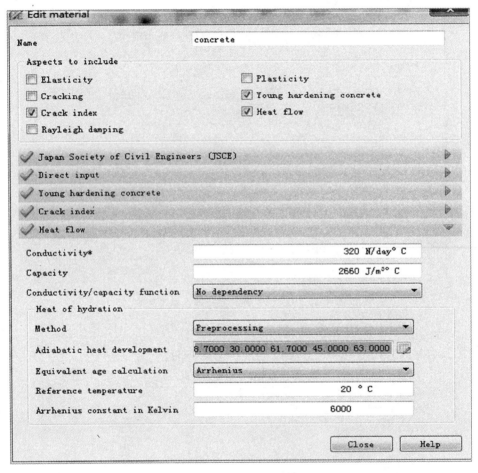

图 4.2-12 JSCE 规范本构模型热流参数设置

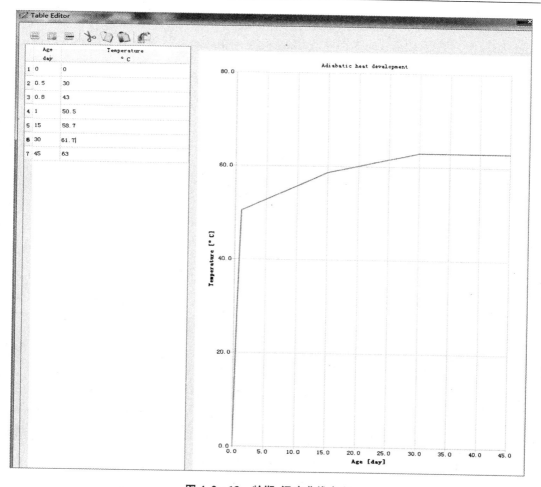

图 4.2-13　龄期-温度曲线定义

Solid 实体单元不需要定义截面几何特性,直接点击 OK,确认生成材料属性。如图 4.2-14 所示。

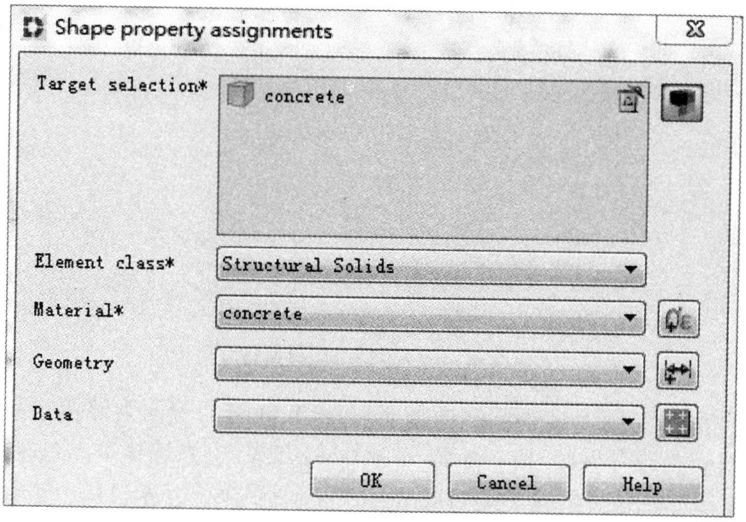

图 4.2-14　生成材料属性

如图 4.2-15 所示,点击红色正方体图标的边界单元快捷键,创立边界单元,命名为 Boundary,单元类型(Class)为界面单元(Interface elements),材料类型为热流边界(Heat flow boundaries)。

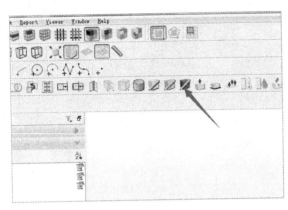

图 4.2-15 边界单元快捷工具栏位置

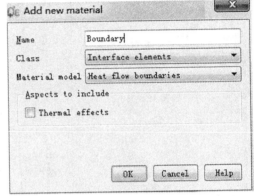

图 4.2-16 热对流边界操作界面

界面单元材料属性定义中,边界条件选择热流边界(Heat flow boundaries)。

选择热对流选项。由于之前已经选择的混凝土单元是实体单元,因此不需要再赋予截面几何特性。在编辑热对流边界类型中,有仅考虑对流、仅考虑辐射、辐射对流同时考虑、什么都不考虑四种类型,这里选择 Convetion only,表明仅考虑热对流,且热对流系数为一个常数。热传导系数(Conduction coefficient)为 9×10^{15} kg/day³℃。热传导函数的幂指数(Convective power)选择为 1 次。与之前定义混凝土热传导函数一致,认为热传导率以及热容量均为常数,因此选择 No dependency。如图 4.2-17 所示。

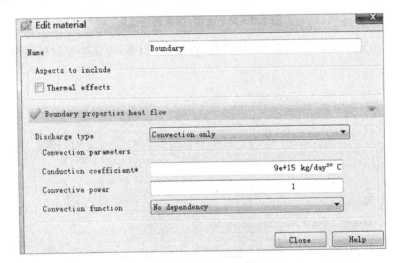

图 4.2-17 热流边界属性

选择垂直于 Y 轴实体六面体的侧面和平行于 Y 轴的上顶面为热流边界面,连接类型(Connection type)为边界界面(Boundary interface)方式,边界选择类型(Selection type)为边界面(Face),界面单元的单元类型(Element class)为热流边界(Heat Flow Boundary)。如图 4.2-18 所示。

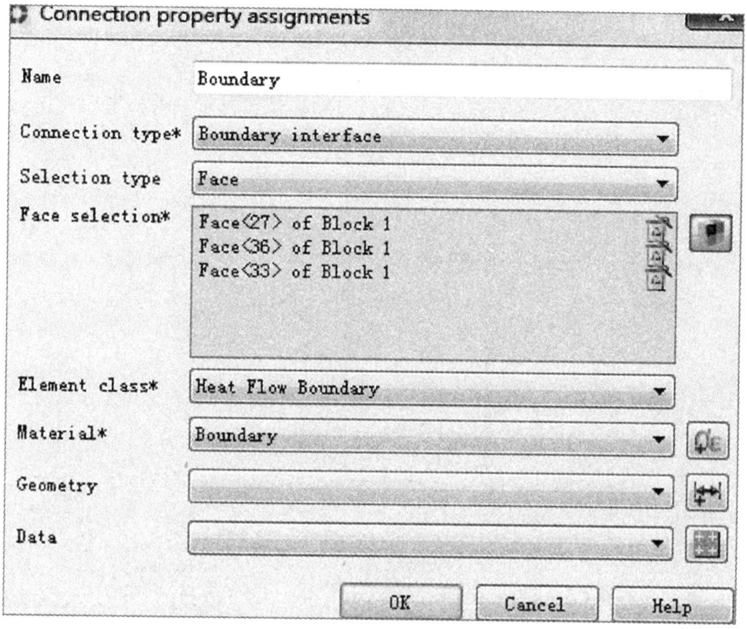

图 4.2-18 热对流边界的选择

如图 4.2-19 所示,定义热流边界属性。选择只考虑热对流方式(Convection only),热传导系数(Conduction coefficient)为 700 N/mday℃。热对流函数选择与时间有关的函数(Time dependent),编辑时间—热传导系数曲线,即时间为自变量,热传导系数为因变量,考虑时间与热传导系数之间的函数关系。这里将时间设定为 60 天。其中前两周(14 天)热传导系数值为 700 N/mday℃,从第 14 天开始上升到 2000 N/mday℃,直到第 60 天结束。如图 4.2-20 所示。

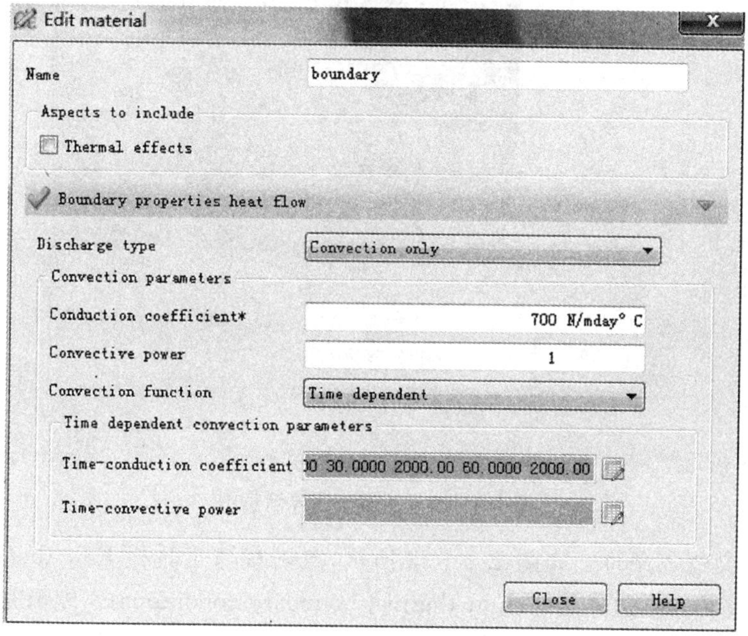

图 4.2-19 热流边界的属性定义

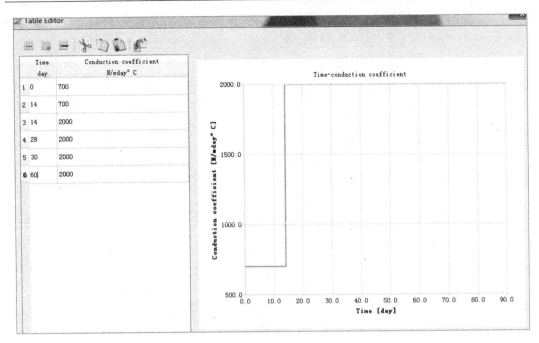

图 4.2-20　热对流边界热传导系数和时间的关系

点击 OK,生成如图 4.2-21 所示的热对流边界。绿色表示热对流边界已经成功定义。

图 4.2-21　热对流边界示意图

接下来定义边界条件。如图 4.2-22 所示,点击快捷工具栏 Edit thermal boundary conditions,编辑热力学边界条件(Edit thermal boundary conditions)。编辑完成后,再点击绿色时钟小图标 Time dependency 定义时间(60 天)与系数(系数恒定为 1)的关系。

第四章　水化热反应

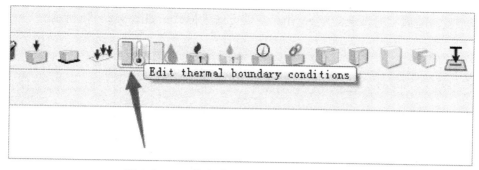

图 4.2-22　热力学边界条件快捷工具栏位置

选中定义的绿色的热对流边界面,定义外界环境温度。边界条件类型组(Boundary condition set)名称定义为 surronding,目标类型(Target type)为面(Face),边界条件类型(Boundary condition type)为外界温度(External temperature)。外界温度值定义为 35℃。热对流环境温度的定义方式如图 4.2-23 所示。

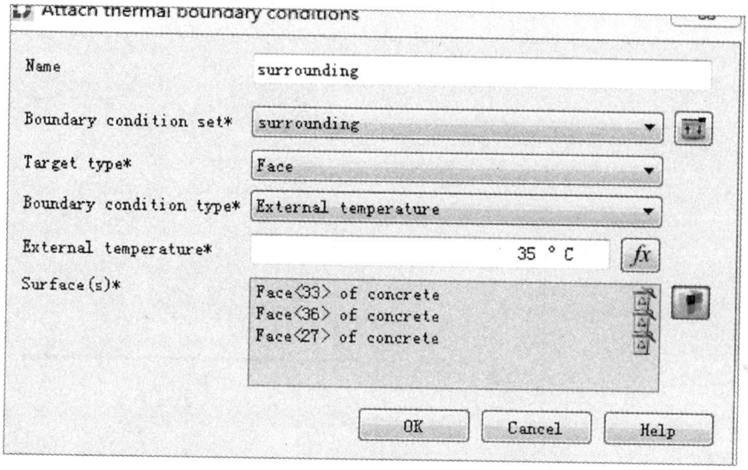

图 4.2-23　热对流环境温度定义界面

定义初始温度,对整个模型施加初始温度场。点击快捷键 Attach an initial field to shape/face/line/point,选中整个实体模型,命名为 initial。如图 4.2-24 所示,施加温度场类型为实体,温度场初始温度为 25℃。

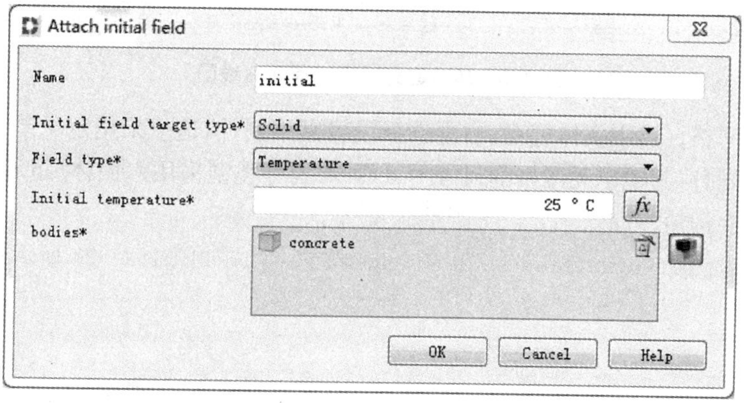

图 4.2-24　对整个模型施加初始温度场

添加重力荷载工况。定义为 gravity,荷载方式为恒载。如图 4.2-25 所示。

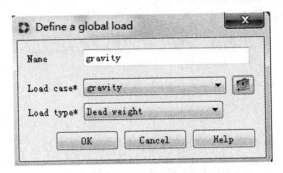

图 4.2-25 重力荷载的荷载工况定义

定义时间—荷载曲线图。这里选择 45 天内重力荷载不随时间变化。如图 4.2-26 所示。

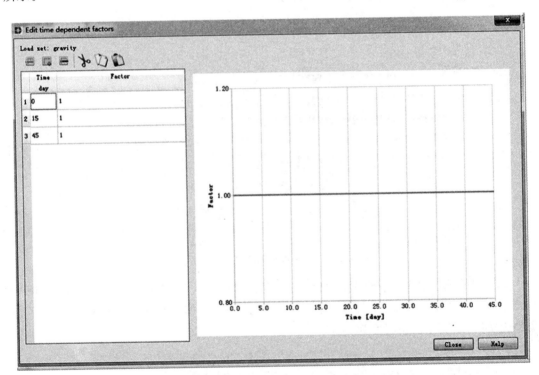

图 4.2-26 定义时间—荷载曲线图

划分单元网格。选中实体模型,右击选择设置网格属性,这里采用定义单元网格尺寸的方式进行网格划分,其中单元网格的尺寸为 0.1m,划分类型采用六面体/矩形类型,中点位置坐标采用线性插值。如图 4.2-27 所示。

点击快捷键图标 Generate mesh of a shape 按钮,生成如图 4.2-28 所示的单元网格。

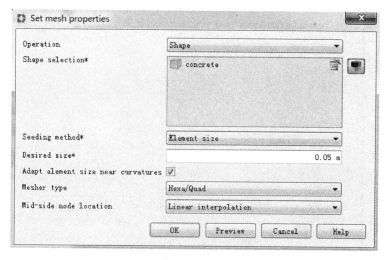

图 4.2-27　单元网格尺寸定义

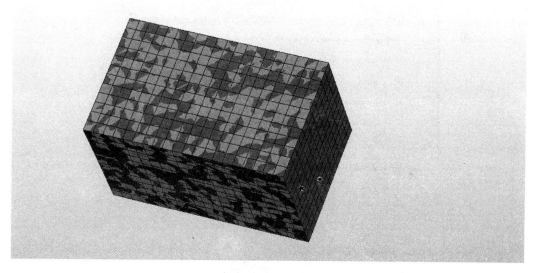

图 4.2-28　单元网格生成

设置分析工况。点击左下角带红色＋图标的快捷键,添加分析工况 Analysis。鼠标右键点击 Analysis 选择瞬态热转化分析模块(Transient heat transfer),在初始瞬态分析中(Initial conditions)选择初始温度场作为初始条件添加,分析类型中(Type)选择非线性分析(Nonlinear analysis),并且勾选下方的水化热分析(Hydration heat analysis),水化热初始反应程度(Initial degree of reaction)设置为默认值 0.01,勾选下一栏混凝土水化热反应自动计算等效龄期(Calculate equivalent age),其中初始的等效龄期(Initial equivalent age)选择默认值 0 天,如图 4.2-29 所示。在运算分析(Execute analysis)模块,输入荷载工况选取 35 天时间作为水化热反应的总时间,其中水化热反应时间荷载步设置成 0.500000(20) 1.00000(5) 10.0000(2),设置非线性计算的最大迭代步数(Maximum number of interations)为 5 步,收敛残差许可值为 1×10^{-6},迭代方法为常规的牛顿迭代法(Newton regular),如图 4.2-30 所示。

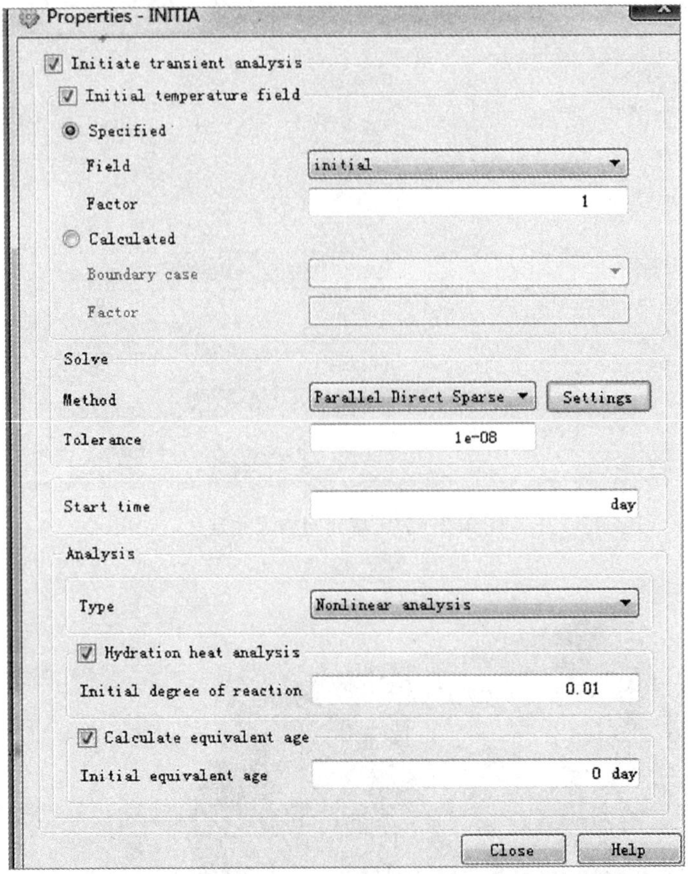

图 4.2‐29　初始温度场瞬态热分析非线性分析设置

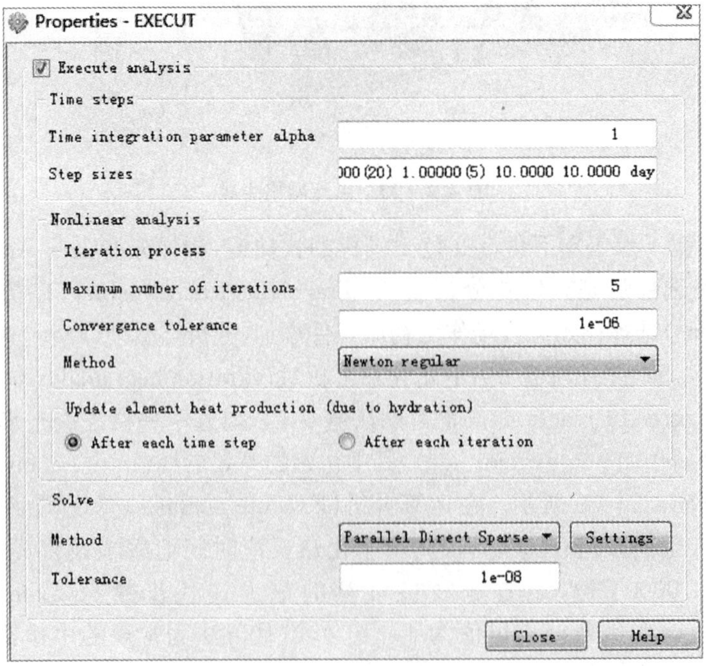

图 4.2‐30　运算分析模块设置

第四章 水化热反应

接下来设置结构非线性分析(Structural nonlinear)阶段分析模块。首先施加重力荷载工况,作为初始工况,最大迭代步数为 20 步,选择收敛准则为力和位移同时勾选的收敛准则。选择采用之前阶段的荷载(Use load of previous phase),荷载类型(Load set)为重力荷载。如图 4.2-31 所示。

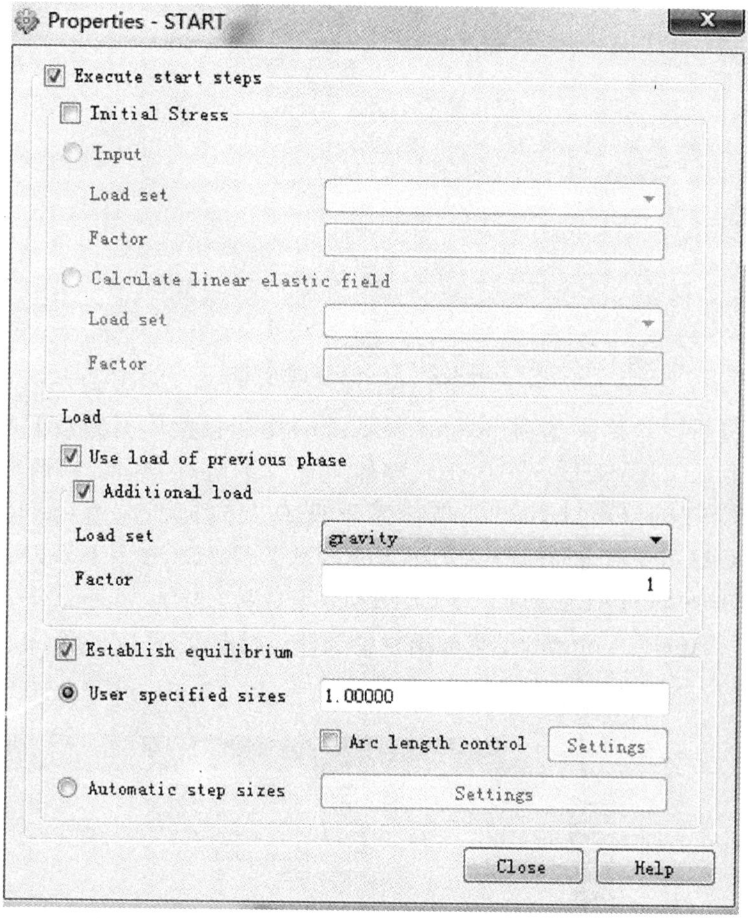

图 4.2-31 非线性分析初始重力荷载界面设置

设置 Time steps 时间荷载。荷载步数与之前瞬态热分析模块的设置方式一致。设置结果与之前的热分析荷载步一致。最大迭代步数设置为 50 次。如图 4.2-32 所示。

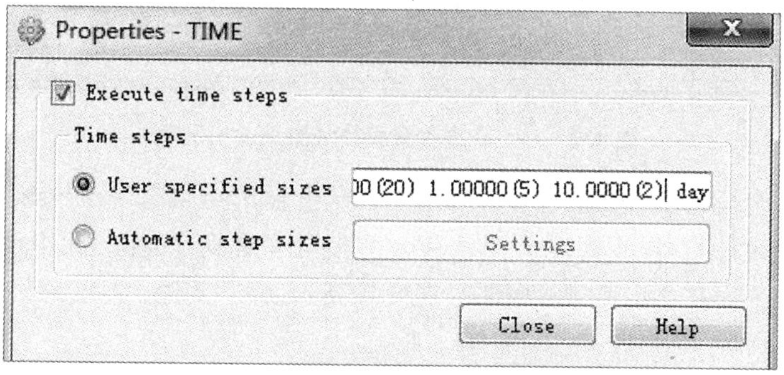

图 4.2-32 时间荷载的荷载步

在瞬态热分析模块的 OUTPUT 结果选项中选择用户选项(User selection),其中选择 INITMP TOTAL,TEMPER,REACTI TOTAL,EQUAGE TOTAL 等选项用以查看与温度和等效龄期有关的结果云图,如图 4.2-33 所示。

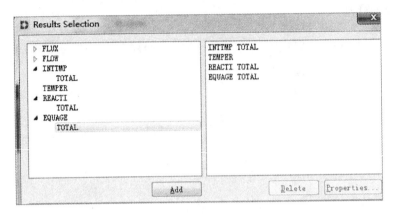

图 4.2-33　结果选项

在非线性分析模块结果选项(Result selection)中选择整体坐标系下所有方向位移(DISPLA TOTAL TRANSL GLOBAL)、整体坐标系下所有方向的柯西应力(STRESS TOTAL CAUCHY GLOBAL)、局部坐标系下所有方向柯西应力(STRESS TOTAL CAUCHY LOCAL)、所有方向柯西主应力(STRESS TOTAL CAUCHY PRINCI)、裂缝方向主应力(STRESS TOTAL CAUCHY CRKIND)等结果输出项,如图 4.2-34 所示。在结构的非线性分析 Analysis output 结果输出栏中选择初始温度、反应程度、等效龄期等结果输出项。

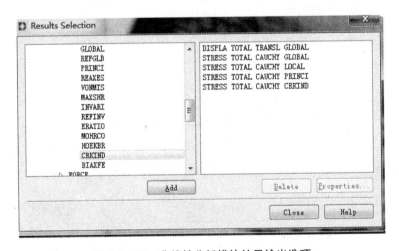

图 4.2-34　非线性分析模块结果输出选项

选取最后一个荷载步。如图 4.2-35 所示,点击输出结果栏 Analysis output→Nodal results 下方的温度一栏,显示温度云图。从云图中可以看到,实体单元中没有进行热对流边界定义的下方平面在水化热过程中明显温度较高。

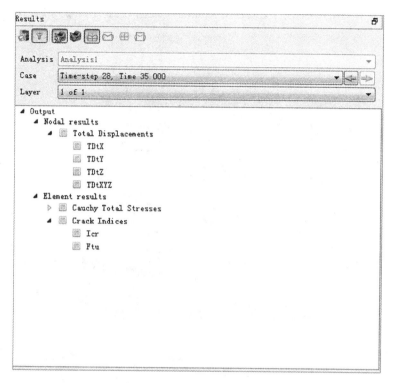

图 4.2-35　非线性输出结果栏

查看水化热反应结束后最终的 Z 向位移云图（TDtZ），如图 4.2-36 所示，从图中可看出，在水化热作用下，中间区域的 Z 向位移最大。

图 4.2-36　水化反应结束后 Z 向位移云图

不同时间作用下初始水化热反应程度特征的开裂指数云图如图 4.2-37—图 4.2-41 所示。

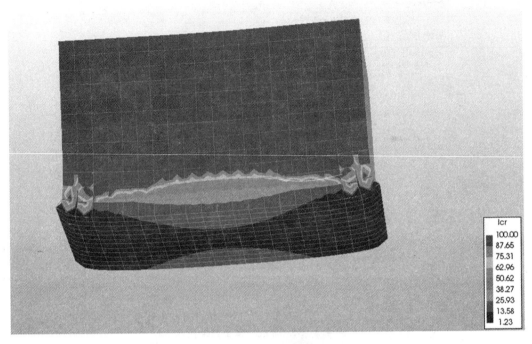

图 4.2-37　初始水化热反应开裂指数分布云图

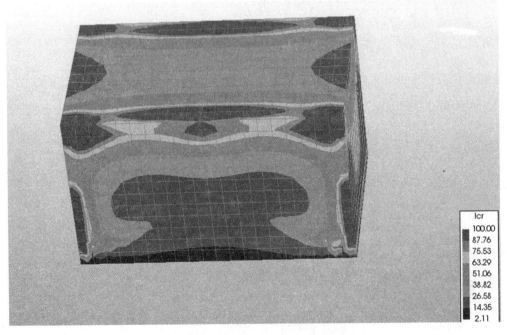

图 4.2-38　水化反应放热 2 天后的开裂指数分布云图

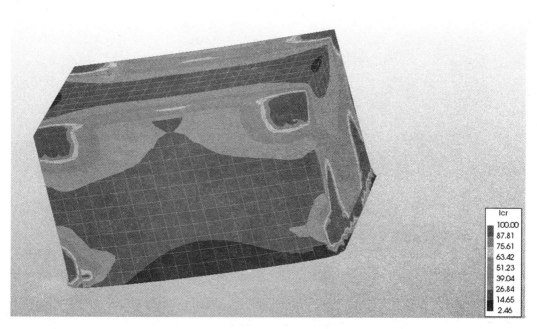

图 4.2-39 水化反应一周后开裂指数分布云图

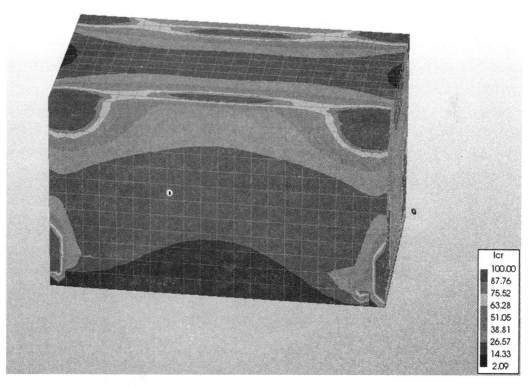

图 4.2-40 水化反应放热 4 天后的开裂指数分布云图

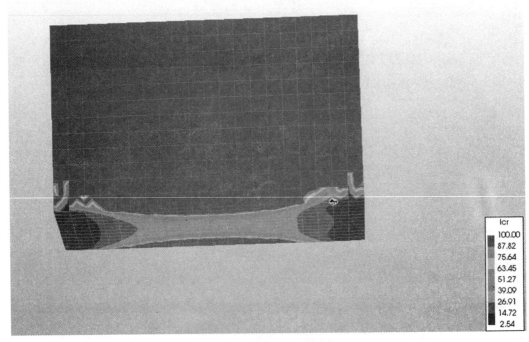

图 4.2-41　水化热反应达到 35 天后的开裂指数分布云图(底面展示)

由开裂指数分布云图可知,水化反应刚开始阶段,混凝土的热对流区域开裂可能性较大,随着水化热反应时间增加,热对流区域的开裂指数数值和开裂指数较高的区域面积不断增大,开裂可能性逐渐减少,至 15 天左右趋于稳定。水化热反应达到 35 天左右,底部开裂指数随着时间的增加而升高。

删除划分成功后的网格,重新定义材料属性,如图 4.2-42 所示,本构模型选用欧洲 CEB-FIP90 规范,同样选择早强混凝土、热流条件以及开裂指数三个方面。如图 4.2-43 所示,选择混凝土强度为 C50,温度 20℃,环境湿度 69%,圆柱体抗压强度平均值为 5.8×10^7 N/m²,其他条件不变,点击 Generace mesh of a shape 快捷图标按钮,待网格生成之后开始计算。

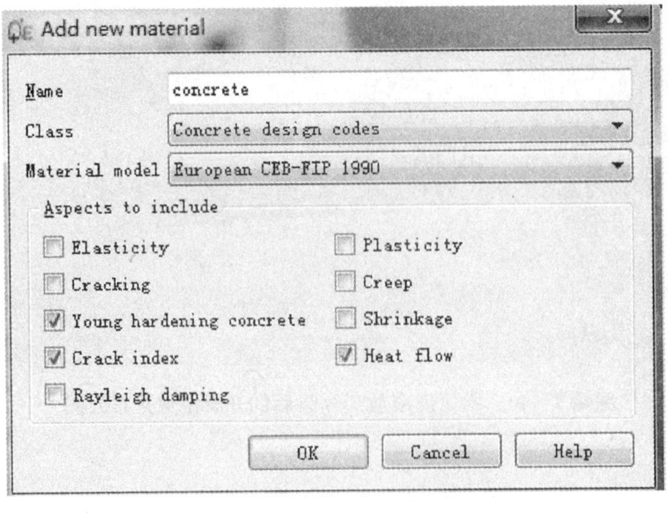

图 4.2-42　欧洲 CEB-FIP90 规范热对流本构设置

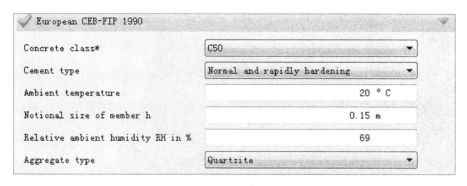

图 4.2-43 欧洲 CEB-FIP90 规范基本本构参数设置

计算完成,观察计算结果,Z 方向位移如图 4.2-44 所示。

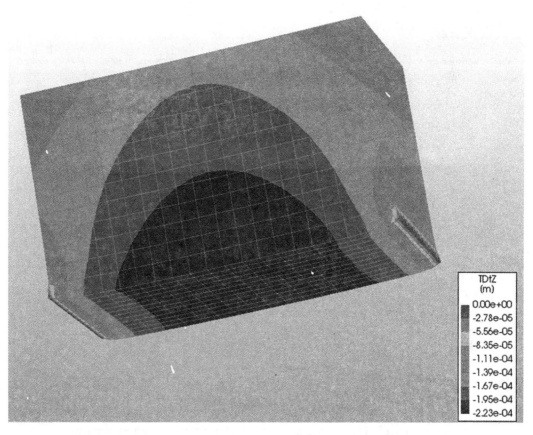

图 4.2-44 CEB-FIP90 规范本构下 Z 方向位移云图

选取不同时刻水化热反应的开裂指数云图分布如图 4.2-45—图 4.2-48 所示。

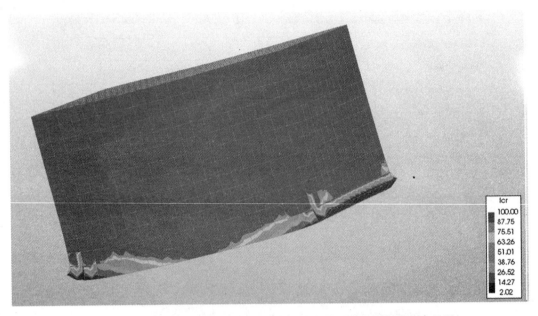

图 4.2-45　CEB-FIP90 规范本构下初始水化热反应开裂指数分布云图

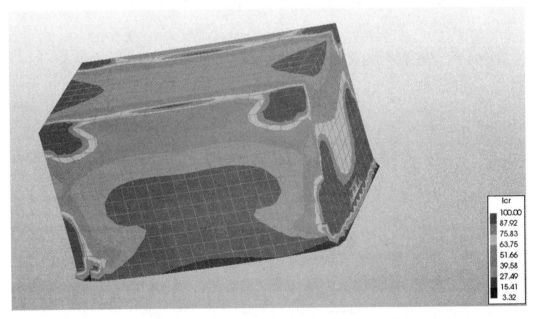

图 4.2-46　CEB-FIP90 规范本构下水化反应放热 5 天后的开裂指数分布云图

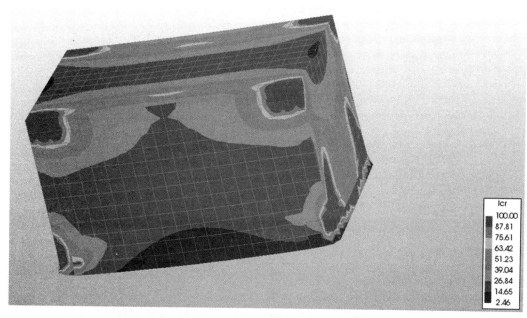

图 4.2-47 CEB-FIP90 规范本构下水化反应放热 10 天后的开裂指数分布云图

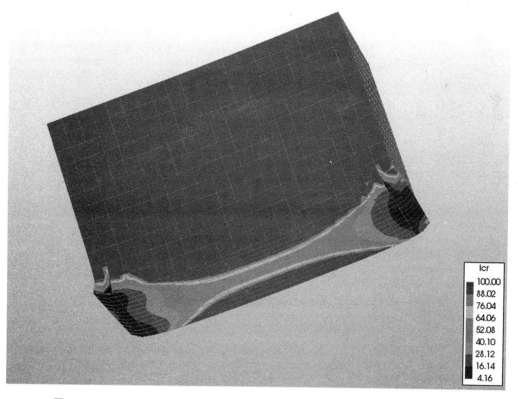

图 4.2-48 CEB-FIP90 规范本构下水化反应放热 35 天后的开裂指数分布云图

类似地，采用美国 AASHTO 规范，输入相应的混凝土本构参数，其余条件不变，如图 4.2-49所示，生成跨中位移的计算云图。

图 4.2-49 美国 AASHTO 规范基本本构参数设置

水化热反应结束后 Z 方向位移云图及不同水化热反应时刻开裂指数云图如图 4.2-50—图 4.2-54 所示。

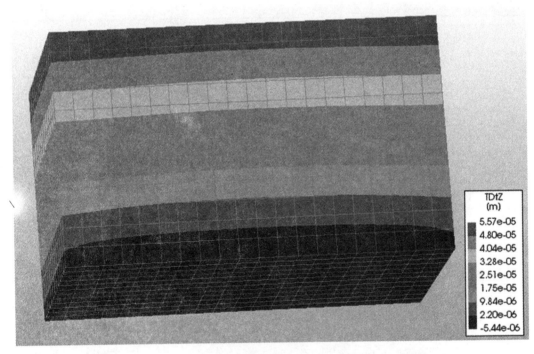

图 4.2-50 美国 AASHTO 规范下水化热反应结束后的 Z 向位移云图

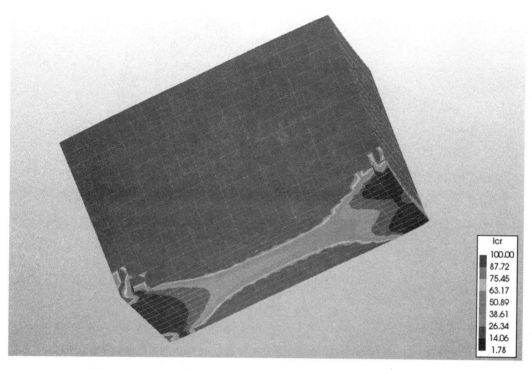

图 4.2-51　美国 AASHTO 规范下初始水化热反应开裂指数分布云图

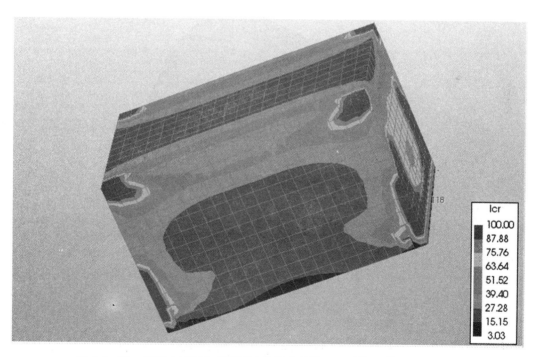

图 4.2-52　美国 AASHTO 规范下水化反应放热 2 天后的开裂指数分布云图

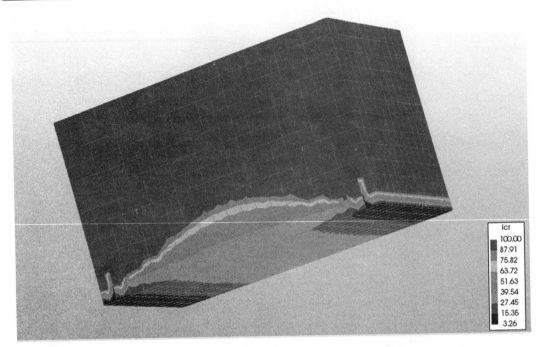

图 4.2-53　美国 AASHTO 规范下水化反应放热 10 天后的开裂指数分布云图

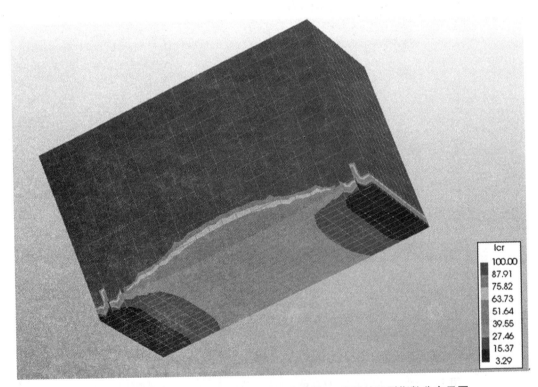

图 4.2-54　美国 AASHTO 规范下水化反应放热 35 天后的开裂指数分布云图

附:JSCE 规范下的水化热反应 Python 语言命令流。

```
newProject( "Pile", 10 )
setModelAnalysisAspects( [ "STRUCT", "HEATFL" ] )
```

```
setModelDimension( "3D" )
setDefaultMeshOrder( "QUADRATIC" )
setDefaultMesherType( "HEXQUAD" )
setDefaultMidSideNodeLocation( "ONSHAP" )
createBlock( "concrete", [ 0, 0, 0], [ 1, 0.6, 0.6 ] )
addMaterial( "concrete ", "CONCDC", "JSCE", [ "CRKIDX", "HEATFL", "YOUNGH" ] )
setParameter( "MATERIAL", "concrete ", "JSCE /YOUN91", 2.7e+10 )
setParameter( "MATERIAL", "concrete ", "JSCE /FCK91", 29000000 )
setParameter( "MATERIAL", "concrete ", "CONCDI /POISON", 0.15 )
setUnit( "TEMPER", "CELSIU" )
setParameter( "MATERIAL", "concrete ", "CONCDI /THERMX", 1e-05 )
setParameter( "MATERIAL", "concrete ", "JSCE /FCK91", 32400000 )
setParameter( "MATERIAL", "concrete ", "JSCE /YOUN91", 2.7e+10 )
setParameter( "MATERIAL", "concrete ", "CONCDI /YOUNG", 2.7e+10 )
setUnit( "TIME", "DAY" )
setUnit( "FORCE", "N" )
setParameter( "MATERIAL", "concrete ", "CONCDI /DENSIT", 3.34898e-07 )
setParameter( "MATERIAL", "concrete ", "CONCYH /POWER", [ 0.3, 33, 0, 0 ] )
setParameter( "MATERIAL", "concrete ", "CONCYH /POWER", [ 0.3, 33, 1, 0 ] )
setParameter( "MATERIAL", "concrete ", "CONCYH /POWER", [ 0.3, 33, 1, 0 ] )
setParameter( "MATERIAL", "concrete ", "CONCYH /POWER", [ 0.3, 33, 1, 0.3 ] )
setParameter( "MATERIAL", "concrete ", "CONCYH /POWER", [ 0.3, 33, 1, 0.3 ] )
setParameter( "MATERIAL", "concrete ", "CONCYH /POWER", [ 0.3, 33, 1, 0.3 ] )
setParameter( "MATERIAL", "concrete ", "HEATFL /CONDUC", 320 )
setParameter( "MATERIAL", "concrete ", "HEATFL /CAPACI", 2660 )
setParameter( "MATERIAL", "concrete ", "HEATFL /HEATHY /HYDRAT", "PREPRO" )
setParameter( "MATERIAL", "concrete ", "HEATFL /HEATHY /ADIAB", [] )
setParameter( "MATERIAL", "concrete ", "HEATFL /HEATHY /ADIAB", [ 0, 0, 0.5, 30, 0.8, 43, 1, 50.5, 15, 58.7, 30, 61.7, 45, 63 ] )
clearReinforcementAspects( [ "concrete" ] )
setElementClassType( "SHAPE", [ "concrete" ], "STRSOL" )
assignMaterial( "concrete ", "SHAPE", [ "concrete" ] )
resetGeometry( "SHAPE", [ "concrete" ] )
resetElementData( "SHAPE", [ "concrete" ] )
saveProject( )
addMaterial( "boundary", "INTERF", "FLBOUN", [] )
setParameter( "MATERIAL", "boundary", "HTBOUN /CONPAR /CONVEC", 700 )
setParameter( "MATERIAL", "boundary", "HTBOUN /CONPAR /CONVEC", 700 )
setParameter( "MATERIAL", "boundary", "HTBOUN /CONPAR /CVTYPE", "TIMDEP" )
```

```
setParameter( "MATERIAL", "boundary", "HTBOUN /CONPAR /TIMDEP /TIMCNV", [] )
setParameter( "MATERIAL", "boundary", "HTBOUN /CONPAR /TIMDEP /TIMCNV", [ 0,
700, 14, 700, 14.1, 2000, 28, 2000, 30, 2000, 60, 2000 ] )
createSurfaceConnection( "boundary" )
setParameter( "GEOMETRYCONNECTION", "boundary", "CONTYP", "BOUNDA" )
attachTo ( " GEOMETRYCONNECTION ", " boundary ", " SOURCE ", " concrete ",
[[ 0.573573, 0.3441438, 0.6 ], [ 0.573573, 0.6, 0.3441438 ], [ 0.573573, 0,
0.2558562 ]] )
setElementClassType( "GEOMETRYCONNECTION", "boundary", "HEABOU" )
assignMaterial( "boundary", "GEOMETRYCONNECTION", "boundary" )
resetGeometry( "GEOMETRYCONNECTION", "boundary" )
resetElementData( "GEOMETRYCONNECTION", "boundary" )
saveProject( )
addSet( "GEOMETRYBCSET", "surrounding" )
createSurfaceBoundaryCondition( "THERMAL", "surrounding", "surrounding" )
setParameter( "GEOMETRYBC", "surrounding", "BOUTYP", "EXTEMP" )
setParameter( "GEOMETRYBC", "surrounding", "EXTEMP /VALUE", 35 )
attach( "GEOMETRYBC", "surrounding", "concrete", [[ 0.573573, 0, 0.2558562 ],
[ 0.573573, 0.3441438, 0.6 ],[ 0.573573, 0.6, 0.3441438 ]] )
setTimeDependentLoadFactors( "GEOMETRYBCSET", " surrounding", [ 0, 15, 45 ],
[ 1, 1, 1 ])
createBodyInitialField( "initial" )
setParameter( "GEOMETRYINIFIELD", "initial", "INITYP", "TEMPER" )
setParameter( "GEOMETRYINIFIELD", "initial", "TEMPER /VALUE", 25 )
attach( "GEOMETRYINIFIELD", "initial", [ "concrete" ] )
saveProject( )
addSet( "GEOMETRYLOADSET", "gravity" )
createModelLoad( "gravity", "gravity" )
setTimeDependentLoadFactors( "GEOMETRYLOADSET", "gravity", [ 0, 15,45 ], [ 1,
1, 1 ] )
saveProject( )
addSet( "GEOMETRYSUPPORTSET", "co1" )
createLineSupport( "co1", "co1" )
setParameter( "GEOMETRYSUPPORT", "co1", "AXES", [ 1, 2 ] )
setParameter( "GEOMETRYSUPPORT", "co1", "TRANSL", [ 1, 0, 1 ] )
setParameter( "GEOMETRYSUPPORT", "co1", "ROTATI", [ 0, 0, 0 ] )
attach( "GEOMETRYSUPPORT", "co1", "concrete", [[ 0, 0.3, 0 ]] )
saveProject( )
addSet( "GEOMETRYSUPPORTSET", "Geometry support set 2" )
```

第四章 水化热反应

```
rename( "GEOMETRYSUPPORTSET", "Geometry support set 2", "co2" )
createLineSupport( "co2", "co2" )
setParameter( "GEOMETRYSUPPORT", "co2", "AXES", [ 1, 2 ] )
setParameter( "GEOMETRYSUPPORT", "co2", "TRANSL", [ 0, 0, 1 ] )
setParameter( "GEOMETRYSUPPORT", "co2", "ROTATI", [ 0, 0, 0 ] )
attach( "GEOMETRYSUPPORT", "co2", "concrete", [[ 1, 0.3, 0 ]] )
saveProject( )
setElementSize( [ "concrete" ], 0.05, -1, True )
setMesherType( [ "concrete" ], "HEXQUAD" )
setMidSideNodeLocation( [ "concrete" ], "LINEAR" )
saveProject( )
generateMesh( [] )
hideView( "GEOM" )
showView( "MESH" )
addAnalysis( "Analysis1" )
addAnalysisCommand( "Analysis1", "HEATTR", "Transient heat transfer" )
addAnalysisCommandDetail( "Analysis1", "Transient heat transfer", "INITIA/TEMPER" )
setAnalysisCommandDetail( "Analysis1", "Transient heat transfer", "INITIA/TEMPER", True )
setAnalysisCommandDetail( "Analysis1", "Transient heat transfer", "INITIA/ANATYP", "NONLIN" )
addAnalysisCommandDetail( "Analysis1", "Transient heat transfer", "INITIA/NONLIN/HYDRAT" )
setAnalysisCommandDetail( "Analysis1", "Transient heat transfer", "INITIA/NONLIN/HYDRAT", True )
addAnalysisCommandDetail( "Analysis1", "Transient heat transfer", "INITIA/NONLIN/EQUAGE" )
setAnalysisCommandDetail( "Analysis1", "Transient heat transfer", "INITIA/NONLIN/EQUAGE", True )
setAnalysisCommandDetail( "Analysis1", "Transient heat transfer", "EXECUT/SIZES", "0.500000(20) 1.00000(5) 10.0000 10.0000" )
addAnalysisCommand( "Analysis1", "NONLIN", "Structural nonlinear" )
renameAnalysisCommandDetail( "Analysis1", "Structural nonlinear", "EXECUT(1)", "gravity" )
renameAnalysisCommandDetail( "Analysis1", "Structural nonlinear", "EXECUT(1)", "gravity" )
removeAnalysisCommandDetail( "Analysis1", "Structural nonlinear", "EXECUT(1)" )
```

```
setAnalysisCommandDetail( "Analysis1", "Structural nonlinear", "EXECUT /EXETYP", "START" )
renameAnalysisCommandDetail( "Analysis1", "Structural nonlinear", "EXECUT (1)", "gravity" )
addAnalysisCommandDetail( "Analysis1", "Structural nonlinear", "EXECUT(1) /START /LOAD /ADD" )
setAnalysisCommandDetail( "Analysis1", "Structural nonlinear", "EXECUT(1) /START /LOAD /ADD", True )
setAnalysisCommandDetail( "Analysis1", "Structural nonlinear", "EXECUT(1) /ITERAT /MAXITE", 20 )
saveProject( )
setAnalysisCommandDetail( "Analysis1", "Structural nonlinear", "EXECUT /EXETYP", "TIME" )
setAnalysisCommandDetail( "Analysis1", "Structural nonlinear", "EXECUT(2) /TIME /STEPS /EXPLIC /SIZES", "0.500000(20) 1.00000(5) 10.0000(2)" )
setAnalysisCommandDetail( "Analysis1", "Structural nonlinear", "EXECUT(2) /ITERAT /MAXITE", 50 )
saveProject( )
setAnalysisCommandDetail( "Analysis1", "Structural nonlinear", "OUTPUT(1) /SELTYP", "USER" )
addAnalysisCommandDetail( "Analysis1", "Structural nonlinear", "OUTPUT(1) /USER" )
addAnalysisCommandDetail( "Analysis1", "Structural nonlinear", "OUTPUT(1) /USER /DISPLA(1) /TOTAL /TRANSL /GLOBAL" )
addAnalysisCommandDetail( "Analysis1", "Structural nonlinear", "OUTPUT(1) /USER /STRESS(1) /TOTAL /CAUCHY /GLOBAL" )
addAnalysisCommandDetail( "Analysis1", "Structural nonlinear", "OUTPUT(1) /USER /STRESS(2) /TOTAL /CAUCHY /LOCAL" )
addAnalysisCommandDetail( "Analysis1", "Structural nonlinear", "OUTPUT(1) /USER /STRESS(3) /TOTAL /CAUCHY /PRINCI" )
addAnalysisCommandDetail( "Analysis1", "Structural nonlinear", "OUTPUT(1) /USER /STRESS(4) /TOTAL /CAUCHY /CRKIND" )
setAnalysisCommandDetail( "Analysis1", "Structural nonlinear", "EXECUT(2) /TIME /STEPS /EXPLIC /SIZES", "0.500000(20) 1.00000(5) 10.0000 10.0000" )
saveProject( )
setAnalysisCommandDetail( "Analysis1", "Structural nonlinear", "OUTPUT(1) /SELTYP", "USER" )
addAnalysisCommandDetail( "Analysis1", "Structural nonlinear", "OUTPUT(1) /USER" )
```

第四章 水化热反应

```
    addAnalysisCommandDetail( "Analysis1", "Structural nonlinear", "OUTPUT(1) /
USER /DISPLA(1) /TOTAL /TRANSL /GLOBAL" )
    addAnalysisCommandDetail( "Analysis1", "Structural nonlinear", "OUTPUT(1) /
USER /STRESS(1) /TOTAL /CAUCHY /GLOBAL" )
    addAnalysisCommandDetail( "Analysis1", "Structural nonlinear", "OUTPUT(1) /
USER /STRESS(2) /TOTAL /CAUCHY /LOCAL" )
    addAnalysisCommandDetail( "Analysis1", "Structural nonlinear", "OUTPUT(1) /
USER /STRESS(3) /TOTAL /CAUCHY /PRINCI" )
    addAnalysisCommandDetail( "Analysis1", "Structural nonlinear", "OUTPUT(1) /
USER /STRESS(4) /TOTAL /CAUCHY /CRKIND" )
    runSolver( "Analysis1" )
    showView( "RESULT" )
    setResultCase( [ "Analysis1", "Output", "Time-step 28, Time 35.000" ] )
    setResultPlot( "contours", "Total Displacements /node", "TDtZ" )
    setResultPlot( "contours", "Total Displacements /node", "TDtY" )
    setResultPlot( "contours", "Total Displacements /node", "TDtX" )
    setResultPlot( "contours", "Crack Indices /node", "Icr" )
    setResultCase( [ "Analysis1", "Output", "Time-step 27, Time 25.000" ] )
    setResultCase( [ "Analysis1", "Output", "Time-step 26, Time 15.000" ] )
    setResultCase( [ "Analysis1", "Output", "Time-step 25, Time 14.000" ] )
    setResultCase( [ "Analysis1", "Output", "Time-step 24, Time 13.000" ] )
    setResultCase( [ "Analysis1", "Output", "Time-step 23, Time 12.000" ] )
    setResultCase( [ "Analysis1", "Output", "Time-step 22, Time 11.000" ] )
    setResultCase( [ "Analysis1", "Output", "Time-step 21, Time 10.000" ] )
    setResultCase( [ "Analysis1", "Output", "Time-step 20, Time 9.5000" ] )
    setResultCase( [ "Analysis1", "Output", "Time-step 19, Time 9.0000" ] )
    setResultCase( [ "Analysis1", "Output", "Time-step 18, Time 8.5000" ] )
    setResultCase( [ "Analysis1", "Output", "Time-step 17, Time 8.0000" ] )
    setResultCase( [ "Analysis1", "Output", "Time-step 16, Time 7.5000" ] )
    setResultCase( [ "Analysis1", "Output", "Time-step 15, Time 7.0000" ] )
    setResultCase( [ "Analysis1", "Output", "Time-step 14, Time 6.5000" ] )
    setResultCase( [ "Analysis1", "Output", "Time-step 13, Time 6.0000" ] )
    setResultCase( [ "Analysis1", "Output", "Time-step 12, Time 5.5000" ] )
    setResultCase( [ "Analysis1", "Output", "Time-step 11, Time 5.0000" ] )
    setResultCase( [ "Analysis1", "Output", "Time-step 10, Time 4.5000" ] )
    setResultCase( [ "Analysis1", "Output", "Time-step 9, Time 4.0000" ] )
    setResultCase( [ "Analysis1", "Output", "Time-step 8, Time 3.5000" ] )
    setResultCase( [ "Analysis1", "Output", "Time-step 7, Time 3.0000" ] )
    setResultCase( [ "Analysis1", "Output", "Time-step 6, Time 2.5000" ] )
```

```
setResultCase( [ "Analysis1", "Output", "Time-step 5, Time 2.0000" ] )
setResultCase( [ "Analysis1", "Output", "Time-step 4, Time 1.5000" ] )
setResultCase( [ "Analysis1", "Output", "Time-step 3, Time 1.0000" ] )
setResultCase( [ "Analysis1", "Output", "Time-step 2, Time 0.50000" ] )
setResultCase( [ "Analysis1", "Output", "Start-step 1, Load-factor 1.0000" ] )
setResultCase( [ "Analysis1", "Output", "Time-step 2, Time 0.50000" ] )
setResultCase( [ "Analysis1", "Output", "Time-step 3, Time 1.0000" ] )
setResultCase( [ "Analysis1", "Output", "Time-step 4, Time 1.5000" ] )
setResultCase( [ "Analysis1", "Output", "Time-step 5, Time 2.0000" ] )
setResultCase( [ "Analysis1", "Output", "Time-step 6, Time 2.5000" ] )
setResultCase( [ "Analysis1", "Output", "Time-step 7, Time 3.0000" ] )
setResultCase( [ "Analysis1", "Output", "Time-step 8, Time 3.5000" ] )
setResultCase( [ "Analysis1", "Output", "Time-step 9, Time 4.0000" ] )
setResultCase( [ "Analysis1", "Output", "Time-step 10, Time 4.5000" ] )
setResultCase( [ "Analysis1", "Output", "Time-step 11, Time 5.0000" ] )
setResultCase( [ "Analysis1", "Output", "Time-step 12, Time 5.5000" ] )
setResultCase( [ "Analysis1", "Output", "Time-step 13, Time 6.0000" ] )
setResultCase( [ "Analysis1", "Output", "Time-step 14, Time 6.5000" ] )
setResultCase( [ "Analysis1", "Output", "Time-step 15, Time 7.0000" ] )
setResultCase( [ "Analysis1", "Output", "Time-step 16, Time 7.5000" ] )
setResultCase( [ "Analysis1", "Output", "Time-step 17, Time 8.0000" ] )
setResultCase( [ "Analysis1", "Output", "Time-step 18, Time 8.5000" ] )
setResultCase( [ "Analysis1", "Output", "Time-step 19, Time 9.0000" ] )
setResultCase( [ "Analysis1", "Output", "Time-step 20, Time 9.5000" ] )
setResultCase( [ "Analysis1", "Output", "Time-step 21, Time 10.000" ] )
setResultCase( [ "Analysis1", "Output", "Time-step 22, Time 11.000" ] )
setResultCase( [ "Analysis1", "Output", "Time-step 23, Time 12.000" ] )
setResultCase( [ "Analysis1", "Output", "Time-step 24, Time 13.000" ] )
setResultCase( [ "Analysis1", "Output", "Time-step 25, Time 14.000" ] )
setResultCase( [ "Analysis1", "Output", "Time-step 26, Time 15.000" ] )
setResultCase( [ "Analysis1", "Output", "Time-step 27, Time 25.000" ] )
setResultCase( [ "Analysis1", "Output", "Time-step 28, Time 35.000" ] )
```

采用 CEB-FIP90 欧洲规范水化热反应 Python 语言命令流（**注：以下命令流仅列出采用 CEB-FIP 规范替换 JSCE 规范中定义本构模型部分的 Python 语言，其余部分完全相同，请用户自行替换**）

```
addMaterial( "concrete", "CONCDC", "MC1990", [ "CRKIDX", "HEATFL", "YOUNGH" ] )
setParameter( "MATERIAL", "concrete", "MC90CO/GRADE", "C50" )
setParameter( "MATERIAL", "concrete", "MC90CO/RH", 69 )
setParameter( "MATERIAL", "concrete", "MC90CO/RH", 69 )
```

第四章 水化热反应

```
    setParameter( "MATERIAL", "concrete", "CONCDI/YOUNG", 3.8926e+10 )
    setParameter( "MATERIAL", "concrete", "CONCDI/YOUN28", 3.45e+10 )
    setParameter( "MATERIAL", "concrete", "CONCDI/POISON", 0.15 )
    setParameter( "MATERIAL", "concrete", "CONCDI/THERMX", 1e-05 )
    setParameter( "MATERIAL", "concrete", "CONCDI/DENSIT", 2500 )
    setParameter( "MATERIAL", "concrete", "CONCDI/FCK28", 50000000 )
    setParameter( "MATERIAL", "concrete", "CONCDI/FCM28", 58000000 )
    setParameter( "MATERIAL", "concrete", "CONCDI/DENSIT", 3.34898e-07 )
    setParameter( "MATERIAL", "concrete", "CONCYH/POWER", [ 0.3, 33, 1, 0.3 ] )
    setParameter( "MATERIAL", "concrete", "HEATFL/CONDUC", 320 )
    setParameter( "MATERIAL", "concrete", "HEATFL/CAPACI", 2660 )
    setParameter( "MATERIAL", "concrete", "HEATFL/CNDTYP", "TIMDEP" )
    setParameter( "MATERIAL", "concrete", "HEATFL/CNDTYP", "NONE" )
    setParameter( "MATERIAL", "concrete", "HEATFL/CNDTYP", "NONE" )
    setParameter( "MATERIAL", "concrete", "HEATFL/HEATHY/HYDRAT", "PREPRO" )
    setParameter( "MATERIAL", "concrete", "HEATFL/HEATHY/ADIAB", [] )
    setParameter( "MATERIAL", "concrete", "HEATFL/HEATHY/ADIAB", [ 0, 0, 0.5, 30,
0.8, 43, 1, 50.5, 15, 58.7, 30, 61.7, 45, 63 ] )
    saveProject( )
    clearReinforcementAspects( [ "concrete" ] )
    setElementClassType( "SHAPE", [ "concrete" ], "STRSOL" )
    assignMaterial( "concrete", "SHAPE", [ "concrete" ] )
    resetGeometry( "SHAPE", [ "concrete" ] )
    resetElementData( "SHAPE", [ "concrete" ] )
    saveProject( )
    generateMesh( [] )
    hideView( "GEOM" )
    showView( "MESH" )
    runSolver( "Analysis1" )
    showView( "RESULT" )
    setResultCase( [ "Analysis1", "Output", "Time-step 9, Time 4.0000" ] )
    setResultPlot( "contours", "Total Displacements/node", "TDtZ" )
    setResultCase( [ "Analysis1", "Output", "Time-step 28, Time 35.000" ] )
    setResultCase( [ "Analysis1", "Output", "Time-step 9, Time 4.0000" ] )
    setResultCase( [ "Analysis1", "Output", "Time-step 28, Time 35.000" ] )
    setParameter( "MATERIAL", "concrete", "MC90CO/CEMTYP", "RS" )
    setParameter( "MATERIAL", "concrete", "MC90CO/CEMTYP", "NR" )
    setResultPlot( "contours", "Crack Indices/node", "Icr" )
    setResultCase( [ "Analysis1", "Output", "Start-step 1, Load-factor 1.0000" ] )
```

```
setResultCase( [ "Analysis1", "Output", "Time-step 2, Time 0.50000" ] )
setResultCase( [ "Analysis1", "Output", "Time-step 10, Time 4.5000" ] )
setResultCase( [ "Analysis1", "Output", "Time-step 9, Time 4.0000" ] )
setResultCase( [ "Analysis1", "Output", "Time-step 11, Time 5.0000" ] )
setResultCase( [ "Analysis1", "Output", "Time-step 5, Time 2.0000" ] )
setResultCase( [ "Analysis1", "Output", "Time-step 9, Time 4.0000" ] )
setResultCase( [ "Analysis1", "Output", "Time-step 8, Time 3.5000" ] )
setResultCase( [ "Analysis1", "Output", "Time-step 10, Time 4.5000" ] )
setResultCase( [ "Analysis1", "Output", "Time-step 13, Time 6.0000" ] )
setResultCase( [ "Analysis1", "Output", "Time-step 28, Time 35.000" ] )
setResultCase( [ "Analysis1", "Output", "Time-step 27, Time 25.000" ] )
setResultCase( [ "Analysis1", "Output", "Time-step 26, Time 15.000" ] )
setResultCase( [ "Analysis1", "Output", "Start-step 1, Load-factor 1.0000" ] )
setResultCase( [ "Analysis1", "Output", "Time-step 2, Time 0.50000" ] )
setResultCase( [ "Analysis1", "Output", "Start-step 1, Load-factor 1.0000" ] )
setResultCase( [ "Analysis1", "Output", "Time-step 2, Time 0.50000" ] )
setResultCase( [ "Analysis1", "Output", "Time-step 9, Time 4.0000" ] )
setResultCase( [ "Analysis1", "Output", "Time-step 26, Time 15.000" ] )
setResultCase( [ "Analysis1", "Output", "Time-step 16, Time 7.5000" ] )
setResultCase( [ "Analysis1", "Output", "Time-step 11, Time 5.0000" ] )
```

采用美国 AASHTO 规范水化热反应 Python 语言命令流(**注:以下命令流仅列出采用 AASHTO 规范替换 JSCE 规范中定义本构模型部分的 Python 语言,其余部分完全相同,请用户自行替换**)。

```
addMaterial( "concrete", "CONCDC", "AASHTO", [ "CRKIDX", "HEATFL" ] )
setParameter( "MATERIAL", "CONCRETE", "HEATFL /HEATHY /HYDRAT", "PREPRO" )
setParameter( "MATERIAL", "CONCRETE", "HEATFL /HEATHY /ADIAB", [] )
setParameter( "MATERIAL", "CONCRETE", "HEATFL /HEATHY /ADIAB", [ 0, 0, 0.5, 30, 0.8, 43, 1, 50.5, 15, 58.7, 30, 61.7, 45, 63 ] )
setParameter( "MATERIAL", "CONCRETE", "HEATFL /CONDUC", 320 )
setParameter( "MATERIAL", "CONCRETE", "HEATFL /CONDUC", 320 )
saveProject(  )
setParameter( "MATERIAL", "CONCRETE", "HEATFL /CAPACI", 2660 )
setParameter( "MATERIAL", "CONCRETE", "HEATFL /CAPACI", 2660 )
rename( "MATERIAL", "CONCRETE", "concrete" )
setParameter( "MATERIAL", "concrete", "CONCDI /YOUNG", 0.3 )
removeParameter( "MATERIAL", "concrete", "CONCDI /YOUNG" )
setParameter( "MATERIAL", "concrete", "CONCDI /YOUNG", 3.45e+10 )
setParameter( "MATERIAL", "concrete", "CONCDI /POISON", 0.15 )
setParameter( "MATERIAL", "concrete", "CONCDI /THERMX", 1e-05 )
```

第四章 水化热反应

```
setParameter( "MATERIAL", "concrete", "CONCDI /DENSIT", 3.34898e-07 )
saveProject( )
setParameter( "MATERIAL", "concrete", "MCAASH /FT28", 2640000 )
saveProject( )
setParameter( "MATERIAL", "concrete", "MCAASH /FT28", 2640000 )
generateMesh( [] )
hideView( "GEOM" )
showView( "MESH" )
runSolver( "Analysis1" )
showView( "RESULT" )
setResultPlot( "contours", "Total Displacements /node", "TDtZ" )
setResultCase( [ "Analysis1", "Output", "Time-step 28, Time 35.000" ] )
setResultPlot( "contours", "Crack Indices /node", "Icr" )
setResultCase( [ "Analysis1", "Output", "Time-step 2, Time 0.50000" ] )
setResultCase( [ "Analysis1", "Output", "Start-step 1, Load-factor 1.0000" ] )
setResultCase( [ "Analysis1", "Output", "Time-step 2, Time 0.50000" ] )
setResultPlot( "contours", "Total Displacements /node", "TDtZ" )
setParameter( "MATERIAL", "concrete", "MCAASH /RH", 69 )
saveProject( )
runSolver( "Analysis1" )
showView( "RESULT" )
setResultCase( [ "Analysis1", "Output", "Time-step 28, Time 35.000" ] )
setResultPlot( "contours", "Total Displacements /node", "TDtZ" )
show( "GEOMETRYSUPPORTSET", [ "co1" ] )
show( "GEOMETRYSUPPORTSET", [ "co2" ] )
setResultPlot( "contours", "Crack Indices /node", "Icr" )
setResultCase( [ "Analysis1", "Output", "Time-step 2, Time 0.50000" ] )
showIds( "NODE", [ 814, 118, 2910 ] )
setResultCase( [ "Analysis1", "Output", "Time-step 10, Time 4.5000" ] )
setResultCase( [ "Analysis1", "Output", "Time-step 5, Time 2.0000" ] )
setResultCase( [ "Analysis1", "Output", "Time-step 11, Time 5.0000" ] )
setResultCase( [ "Analysis1", "Output", "Time-step 21, Time 10.000" ] )
setResultCase( [ "Analysis1", "Output", "Time-step 28, Time 35.000" ] )
setResultCase( [ "Analysis1", "Output", "Time-step 21, Time 10.000" ] )
setResultCase( [ "Analysis1", "Output", "Time-step 28, Time 35.000" ] )
setResultCase( [ "Analysis1", "Output", "Time-step 11, Time 5.0000" ] )
setResultCase( [ "Analysis1", "Output", "Time-step 12, Time 5.5000" ] )
setResultCase( [ "Analysis1", "Output", "Time-step 28, Time 35.000" ] )
setResultCase( [ "Analysis1", "Output", "Time-step 5, Time 2.0000" ] )
```

第五章 预制节段拼装构件非线性数值分析

5.1 案例一:预制拼装混凝土块键齿受剪破坏

本模型参考自东南大学宋守坛博士论文《高速铁路预制拼装箱梁桥抗弯及接缝抗剪试验研究与理论分析》中的 ABAQUS 有限元模型,本书另用 Diana10.1 进行建模,其中平面接触部分的高度为 0.2 m。整个键齿总高度为 0.42 m,构件厚度为 0.25 m,剪力键接触部分高为 0.1 m,其中剪力键端部尺寸为 0.05 m,键齿突出部分长度为 0.05 m,剪力键厚度为 0.25 m。构件平面尺寸如图 5.1-1 所示。划分单元的单元尺寸为 0.01 m,采用平面二维单元建模。其中拉伸软化模型为常用的 Hordijk 模型。整个模型承受竖向集中荷载和水平方向压力,分别考察基于多向固定裂缝模型和总应变固定裂缝模型下的竖向位移和裂缝分布云图,并绘制位移-荷载曲线图加以说明。

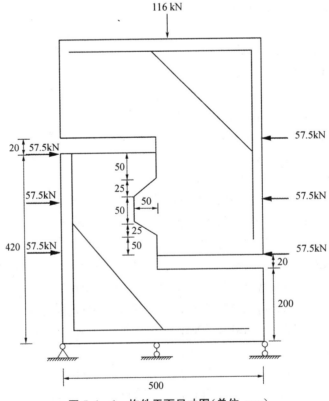

图 5.1-1 构件平面尺寸图(单位:mm)

混凝土采用弥散开裂模型,混凝土及钢筋的材料参数设定见表5-1、表5-2。

表5-1 混凝土材料参数

弹性模量	$3.45 \times 10^{10}\,\text{N/m}^2$
抗压强度标准值	$3.24 \times 10^{7}\,\text{N/m}^2$
抗拉强度标准值	$3.7 \times 10^{6}\,\text{N/m}^2$
泊松比	0.15
拉伸断裂能	200 N/m
密度	2500 kg/m³
开裂模型	多向固定裂缝模型

表5-2 钢筋材料参数

钢筋屈服应力	400 MPa
弹性模量	210 GPa
泊松比	0.33
横截面积	226 mm²

注:作为 Diana10.1 算例例题,本书例题分析条件均为假定值,关于分析结果妥当性另当别论,望谅解。

学习要义:

(1)掌握总应变开裂模型下材料各参数的输入。
(2)线对线界面单元的建立。
(3)库仑摩擦本构的设置。
(4)设置和查看非线性开裂计算的输出结果。

首先,启动 DianaIE,选择 2D 建模平面,平面范围长度设置为 10 m,键齿构件建模依次输入的坐标点见表5-3。

表5-3 键齿构件建模坐标点

1	(0,0)
2	(0.5,0)
3	(0.5,0.2)
4	(0.25,0.2)
5	(0.25,0.22)
6	(0.25,0.27)
7	(0.2,0.295)
8	(0.2,0.345)
9	(0.25,0.37)
10	(0.25,0.42)

(续表)

11	(0,0.42)
12	(0,0.22)
13	(0.5,0.22)
14	(0.5,0.62)
15	(0,0.62)
16	(0,0.44)
17	(0.25,0.44)
18	(0.25,0.42)
19	(0.25,0.37)
20	(0.2,0.345)
21	(0.2,0.295)
22	(0.25,0.27)

输入完各点坐标后,键齿几何模型如图 5.1-2 所示。

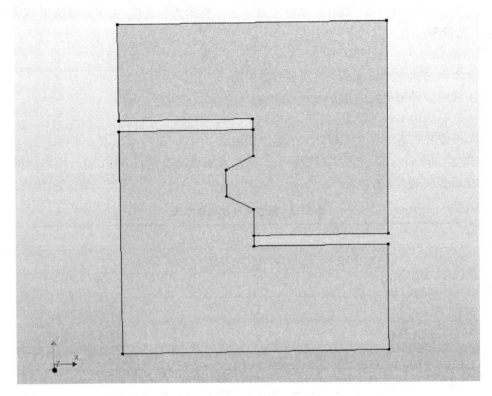

图 5.1-2 键齿几何模型

如图 5.1-3 所示,点击快捷工具栏图标 Add a poly line 创建多点直线。取消 Closed 闭合功能,下部试块配筋命名为 bar1,采用折线的添加方式,上部键齿试块钢筋折线部分各点坐标值输入如图 5.1-4 所示。

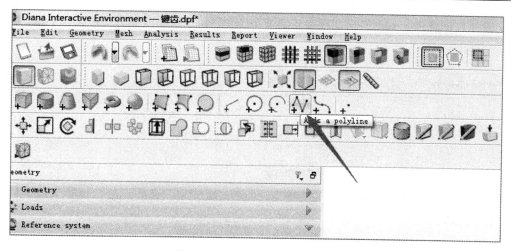

图 5.1-3 Add a polyline 图标

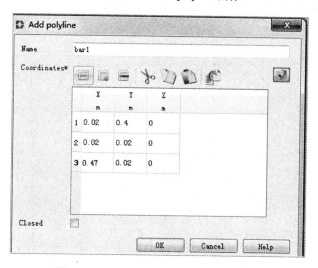

图 5.1-4 bar1 钢筋折线部分各类坐标

上部键齿试块折线部分的配筋命名为 bar2,采用同样的建模方式,各点坐标值如图 5.1-5 所示。

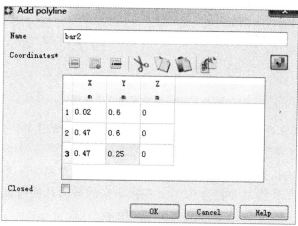

图 5.1-5 bar2 钢筋折线部分各类坐标

再采用同样的方式用直线建模方式建立斜直线钢筋,分别命名为 bar3,bar4,坐标如图 5.1-6 和图 5.1-7 所示。

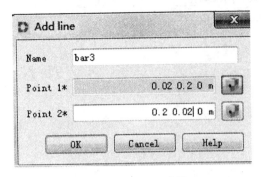

图 5.1-6 bar3 坐标

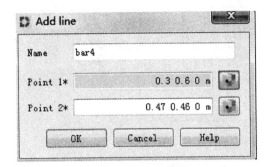

图 5.1-7 bar4 坐标

几何建模完成后,建立的钢筋混凝土键齿试块如图 5.1-8 所示。

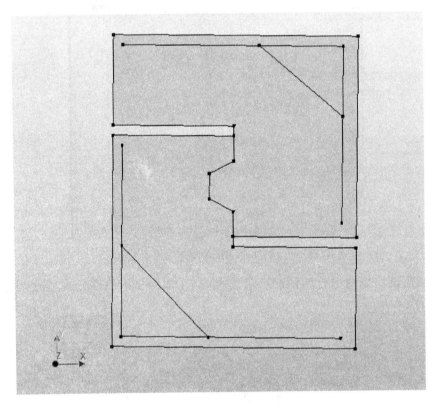

图 5.1-8 键齿配筋几何模型

接下来赋予材料属性。混凝土单元采用平面应力单元,裂缝本构模型采用弥散开裂模式下的多向固定裂缝模型,拉伸软化曲线类型为多向固定裂缝模型下的 Hordijk 拉伸软化曲线,拉伸软化曲线的抗拉强度为 $3.7 \times 10^6 \text{ N/m}^2$,断裂能为 200 N/m,剪力滞留类型选择恒定剪力滞留,其中剪力滞留常数为 0.01。在接下来的截面几何特征定义中,定义键齿的单元厚度为 0.25 m。如图 5.1-9 和图 5.1-10 所示。

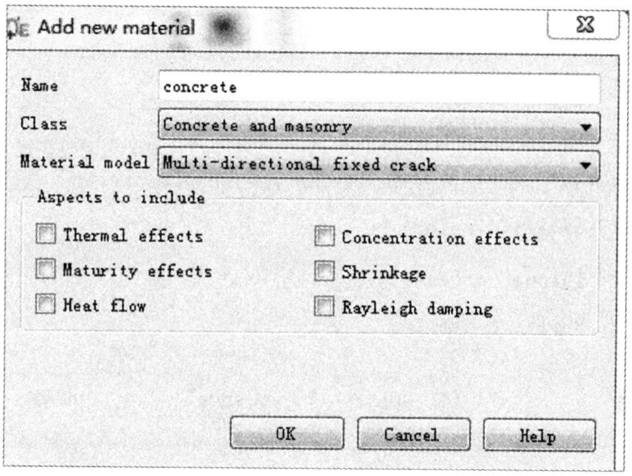

图 5.1-9　材料属性

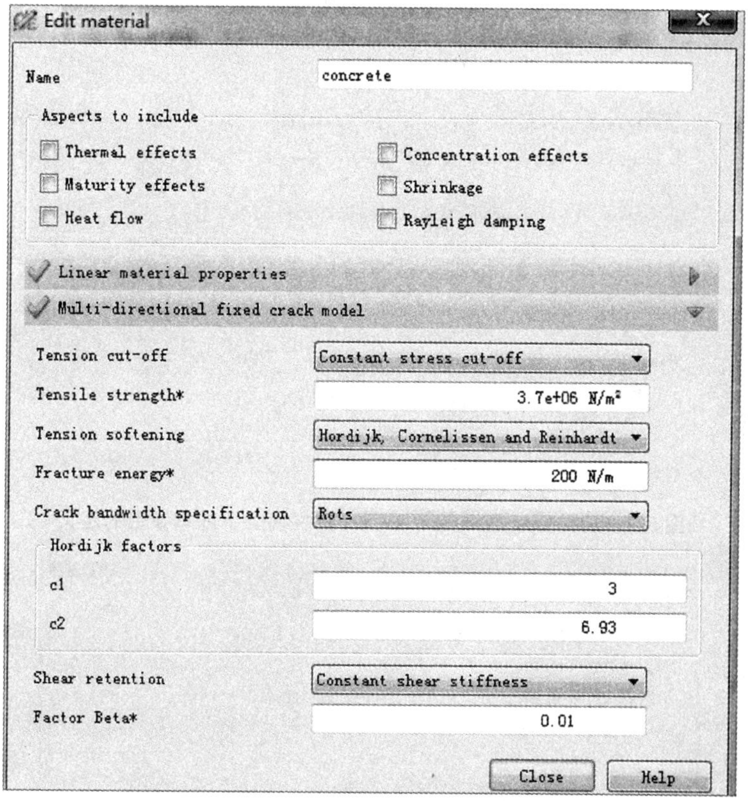

图 5.1-10　材料属性

选中 Geometry 模型树下的所有钢筋组(Bar1~Bar4)，右击选择钢筋材料，钢筋选择钢材类别，材料模型选择 Von Mises and Tresca 塑性模型(Von Mises and Tresca plasticity)。钢筋弹性模量为 $2.1×10^{11}$ N/m^2，钢材的泊松比为 0.33，密度为 7800 kg/m^3。Von Mises and Tresca 塑性模型下塑形类型选择为 Von Mises 塑性模型(Von Mises plasticity)，硬化类型为非硬化(Non hardening)，屈服强度为 $4×10^8$ N/m^2。如图 5.1-11—图 5.1-13 所示。

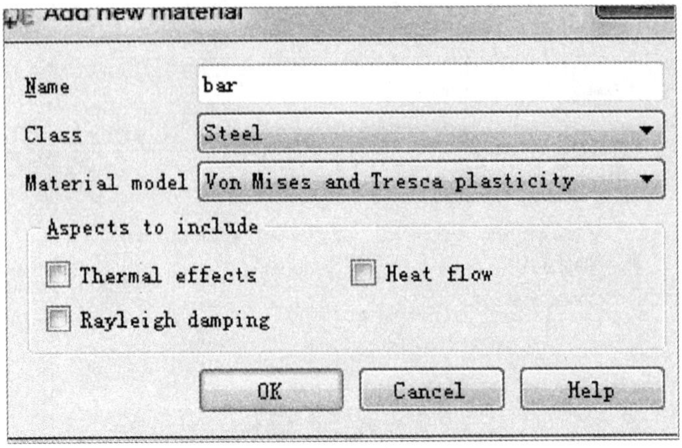

图 5.1-11　Von Mises and Tresca 塑性模型

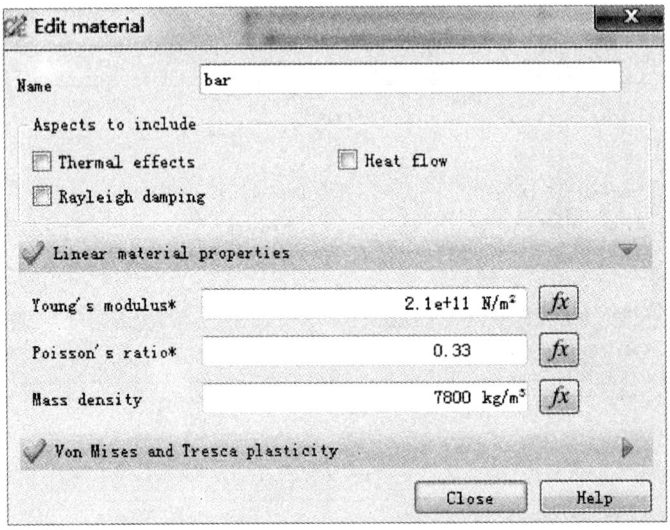

图 5.1-12　Von Mises and Tresca 塑性模型下基本参数的定义

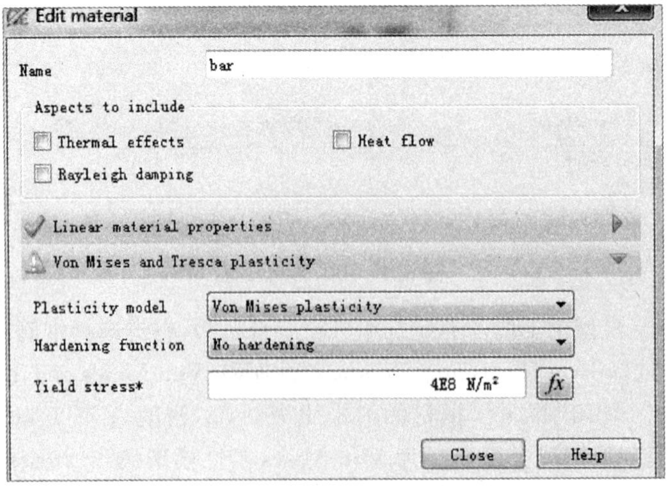

图 5.1-13　Von Mises and Tresca 塑性模型下塑形类型、硬化类型和屈服强度定义

材料属性定义完毕后,点击 OK,继续选择截面属性定义,钢筋采用 2Φ12,经面积折算后为 226 mm²,如图 5.1-14 所示。

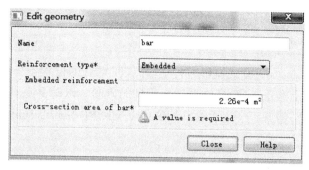

图 5.1-14 截面属性定义

添加界面单元。如图 5.1-15 所示,命名为 int,界面单元类型为二维平面线对线接触单元(2D line interface),考虑到实际受力时键齿之间会产生相对错动和相互挤压,并且会产生切向摩擦力,界面单元的本构类型为库仑摩擦类型(Coulomb friction),法向刚度为 $1\times 10^{16}\,\text{N/m}^3$,切向刚度为 $1\times 10^{12}\,\text{N/m}^3$,界面之间的接触选择库仑摩擦本构模型,其中摩擦角为 20 度。界面单元厚度在几何属性定义中也为 0.25 m。

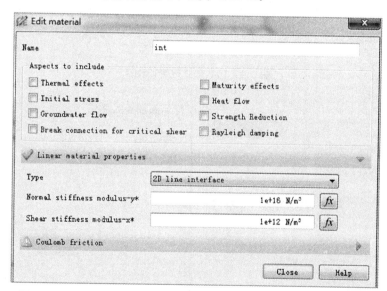

图 5.1-15 2D 线对线界面单元库仑摩擦本构下各方向刚度定义界面

接下来为了更好地施加荷载,同时也为了网格划分能够成功,采用先在键齿细部构造上方建立点再以投影(Projection)印刻(Imprint)的方式将点投影到模型正确的位置上,生成荷载和支座要作用的点。各点坐标见表 5-4。

表 5-4 荷载和支座作用点坐标

1	(-0.1, 0.32)
2	(0.6, 0.32)
3	(-0.1, 0.22)

（续表）

4	(0.6,0.42)
5	(0.25,0.62)
6	(0.25,-0.1)

采用 Projection imprint 方式投影印刻这些点，注意点的位置坐标既有向 X 轴投影，也有向 Y 轴投影，方向上有正有负，需注意投影印刻的方向。以第 2 点和第 4 点为例，投影印刻过程如图 5.1-16—图 5.1-18 所示。其余各点按照相同的方法进行投影印刻。

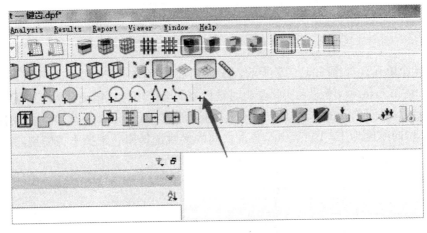

图 5.1-16　创立投影印刻点

图 5.1-17　第 2 点投影印刻坐标

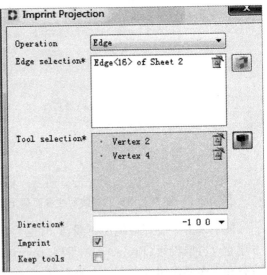

图 5.1-18　第 2 点和第 4 点投影印刻

投影印刻点完成后如图 5.1-19 所示。这样操作可以避免施加荷载和划分后因为单独添加点导致网格划分不成功。

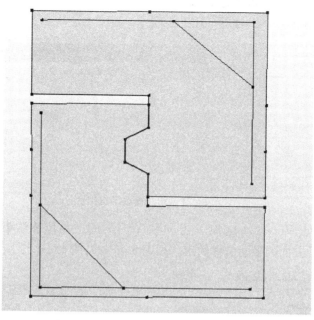

图 5.1-19 投影印刻完成后的几何图形

点击 Geomentry 模块下方 load 快捷键,创立荷载工况 1,命名为 lo1。右击选择 Attach load,选择施加荷载,大小为 10 kN,方向为 Y 轴负向,施加在已经印刻的点 Vertex 5 上,作为等荷载步的基准数值。再通过后面结构非线性分析中的荷载步数控制,逐级加载至试件破坏。然后分别对已经印刻的节点和原有的节点施加 X 方向的水平荷载,各点大小均为 57.5 kN。左边荷载沿 X 正向施加,右边荷载沿 X 负向施加。添加好的荷载如图 5.1-20 所示。

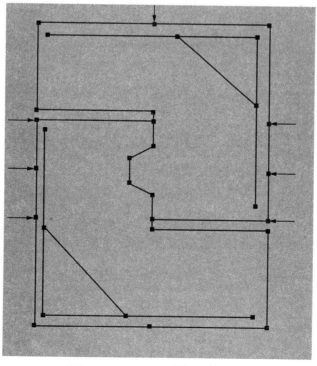

图 5.1-20 添加好荷载后的图形

创立荷载工况组合。将水平力 lo2,lo3 作为初始作用力添加,命名为荷载组合 1;将竖向荷载 lo1 命名为荷载组合 2,如图 5.1-21 所示。

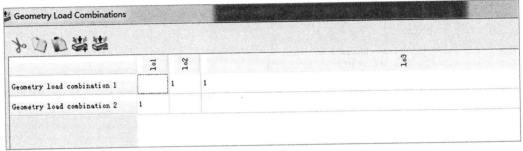

图 5.1-21 创立荷载工况组合

添加约束。对点 Vertex11 施加 X 方向和 Y 方向的平动位移约束,对点 Vertex70 和 Vertex15 仅施加 Y 方向的平动位移约束。如图 5.1-22、图 5.1-23 所示。

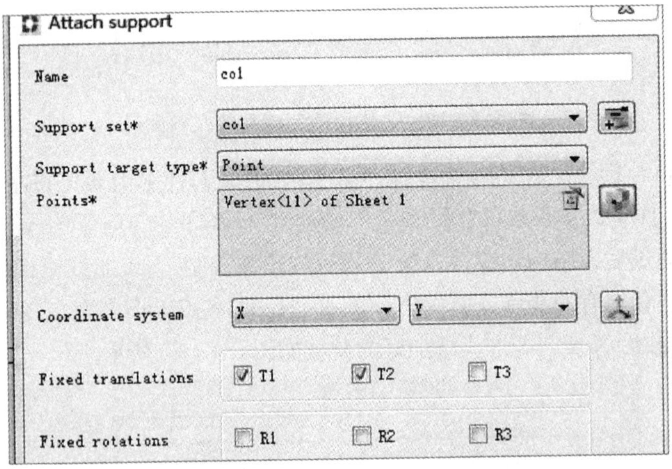

图 5.1-22 Vertex11 处 X 和 Y 向位移约束

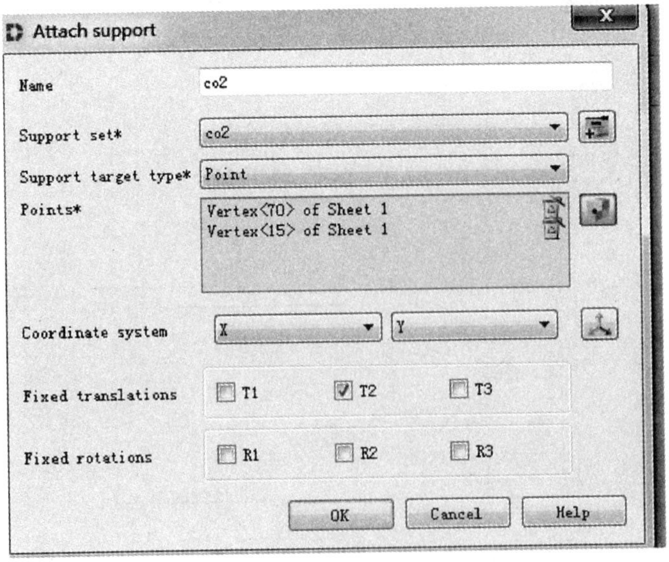

图 5.1-23 Vertex70 和 Vertex15 处 Y 向位移约束

约束施加完毕后如图 5.1-24 所示。

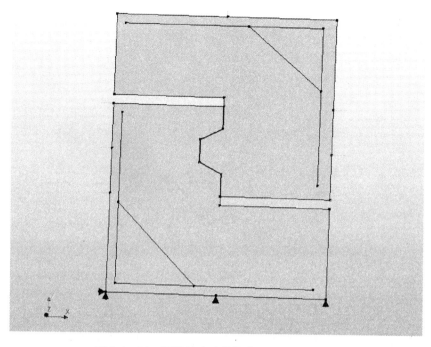

图 5.1-24　细部键齿试块约束施加完毕后图示

采用单元尺寸(Element size)的网格划分方式,混凝土试块及键齿部分网格单元尺寸大小为 0.01 m,划分类型为六面体/矩形单元,中间节点位置确定方式采用线性插值方式。如图 5.1-25、图 5.1-26 所示。

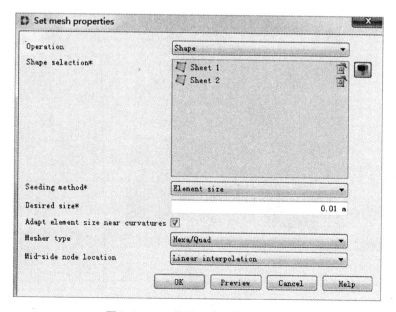

图 5.1-25　单元尺寸网格划分界面

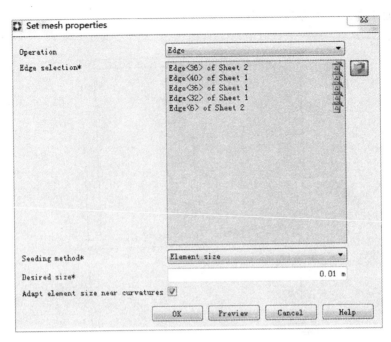

图 5.1-26

点击快捷键图标 Generate mesh of a shape 按钮，生成的网格如图 5.1-27 所示。

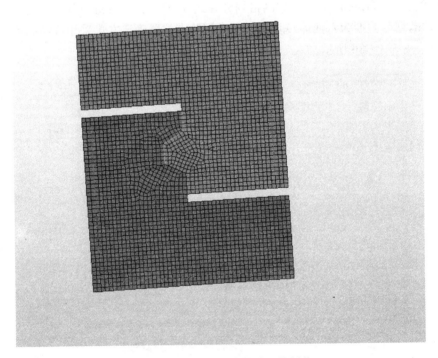

图 5.1-27 细部键齿构造网格划分

添加计算模块（Add an analysis），生成分析模块 1。右击选择 Add command 添加计算控制指令，在控制指令中选择结构非线性分析，在结构非线性分析中选择添加荷载步，选择荷载工况组合 1,设置计算步数为 1,总步长因子为 1,选择最大迭代次数 20 次,采用位移或力收敛的收敛准则。

施加外荷载所在的荷载工况组合 2。考虑到计算类型为非线性计算,故前面荷载步子步长因子设置得较大,后面荷载步中的子步长因子设置得较小。以 10 kN 为基准逐级加载,设置成 1(10) 0.4(5)。如图 5.1-28 所示。

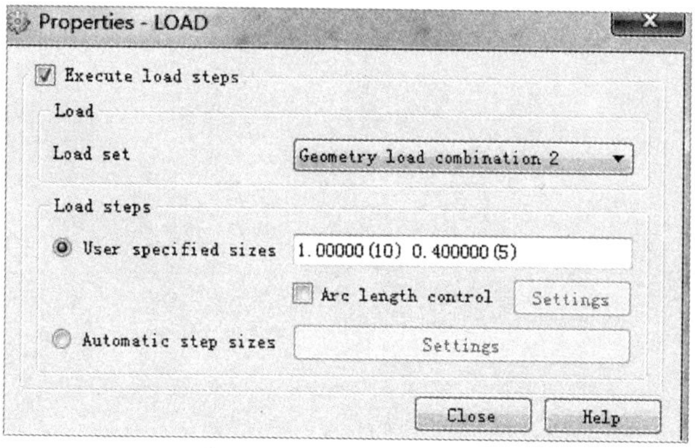

图 5.1-28 荷载工况 2 的荷载步分级加载

计算结果生成后,点击 Output→Total displacement→TDtY 查看 Y 方向的位移约为 0.28 mm。裂缝宽度云图如图 5.1-29 所示。X 方向的裂缝宽度(EcwXX)如图 5.1-29 所示。Y 方向的裂缝宽度(EcwYY)如图 5.1-30 所示。查看裂缝分布位置如图 5.1-31 所示。

得到极限状态时的荷载大小为 116 kN,其中裂缝宽度和裂缝应变分布云图如下图所示:

图 5.1-29 X 方向的裂缝宽度分布云图(EcwXX)

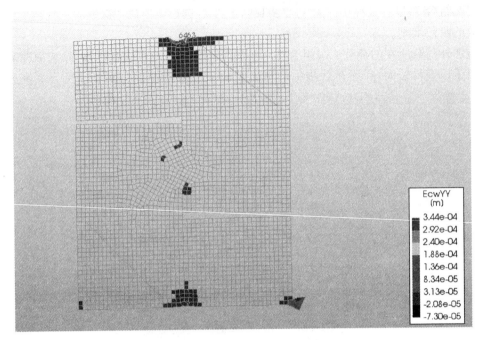

图 5.1-30 Y 方向的裂缝宽度分布云图

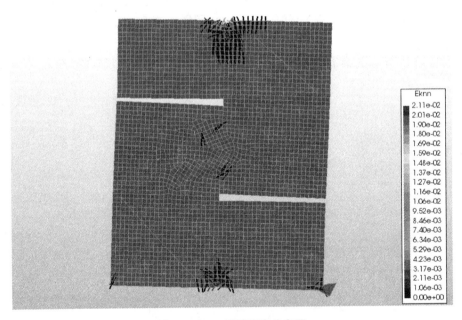

图 5.1-31 裂缝应变分布图

手动删除划分好的网格,将材料属性中的多向固定裂缝模型改为总应变固定裂缝模型,其他条件不变,重新划分网格计算。得到极限荷载大小约为 116 kN,同一节点位置处的 Y 方向位移约为 0.311 mm,整体坐标系下 X 方向裂缝宽度云图和裂缝位置分布图如图 5.1-32—图 5.1-33 所示。

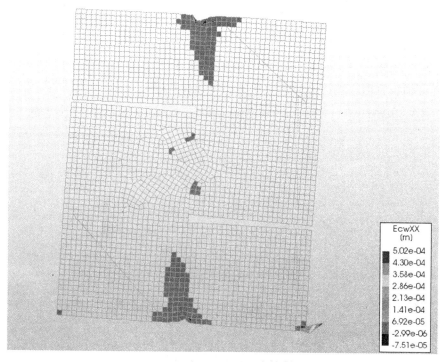

图 5.1-32　总应变固定裂缝模型 X 方向裂缝宽度云图

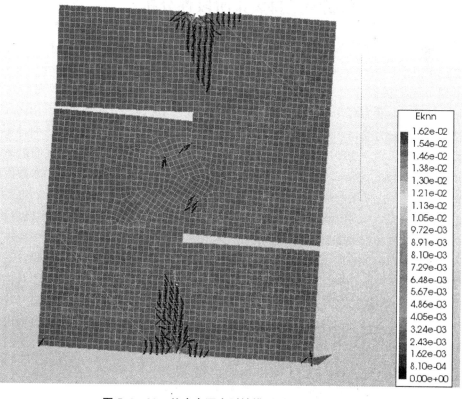

图 5.1-33　总应变固定裂缝模型裂缝应变分布图

为了更好地描述力与位移的关系，这时采用 Viewer-node selection→选中某个节点→右击选择 Show ids 查看节点→选择 Output-Total displacement-TDtY→右击选择 Show table 查看，在这里我们选择第 6463 号节点，提取从 0 至 116 kN 的过程中各荷载步的力和位移的数值，以位移为横坐标，荷载为纵坐标绘制曲线，比较在两种裂缝模型下该节点处外荷载与 Y 方向位移的关系。用 Origin 绘图软件简易处理如图 5.1-34 所示。

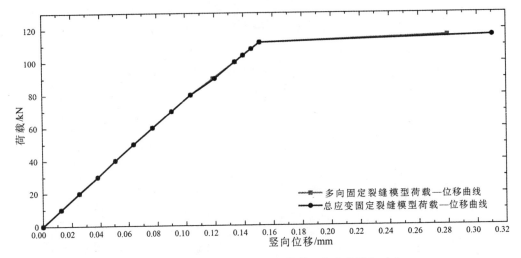

图 5.1-34　两种裂缝模型下的荷载—位移曲线（Y 向）

5.2　案例二：预制节段拼装梁阶段性分析

案例模型为两段简支节段拼装混凝土节段梁半结构模型。基于阶段性分析技术考察备用预应力筋张拉前后两个不同阶段作用下的跨中挠度值。采用二次普通曲壳单元（Regular Curved Elements）模拟混凝土，混凝土各节段长度为 0.8 m，高度为 0.3 m，壳单元厚度为 0.09 m。混凝土强度为 C50，体内体外预应力筋均为直径 15.24 mm 的钢绞线。用嵌入式钢筋单元模拟两种预应力筋，其中一种为体内折线筋，另一种为体内直线预应力筋。两种预应力筋的弹性模量均为 1.95×10^{11} N/m^2，屈服强度为 1860 MPa。折线预应力筋距底面最高高度为 0.22 m，最低高度为 0.06 m，直线预应力筋距底面距离为 0.05 m。预应力筋的净截面面积均为 139 mm^2。在本操作案例中采用两阶段张拉，即在初始条件下先张拉体内折线预应力筋，再张拉体内直线预应力筋，初始张拉力大小均为 180 kN。假设一年以后达到控制挠度许可值，考察一年收缩徐变和松弛多重时变效应作用下运用阶段性分析（Phased Analysis）模拟两次张拉体内预应力筋。整个节段拼装混凝土梁采用不带键齿的平接接触方式，整根梁在三等分点处对称作用 10 kN 集中荷载。节段混凝土梁的混凝土材料参数见表 5-5。由于对称性的存在，本例采用取半结构方式，正对称半结构示意图如图 5.2-1 所示。为说明网格尺寸和预应力筋回缩长度在本节非线性计算案例操作中对计算结果的重要性，本节采用对比验证的方式，即在初始预应力筋等条件相同的情况下，对比单元网格尺寸为 0.1 m 和预应力筋回缩长度（Retention Length）为 0.01 m、单元网格尺寸为 0.05 m 和预应力筋回缩长度（Retention Length）为 0.001 m 两种情况下初始位移云图反拱效果和钢筋拉力云图，说明这两个因素在本节案例操作中对模型最终计算结果的影响，从而说明其在非线性计算中的重要性。

第五章 预制节段拼装构件非线性数值分析

本节模型案例不仅介绍了如何进行常规的收缩徐变非线性计算和非线性阶段性分析，更重要的是为读者提供了解决数值模拟失真、数值模拟结果不够精确和非线性计算不收敛等一系列非线性计算中常见且无法避免的问题的一种解决思路。

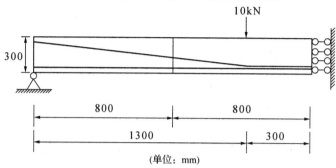

图 5.2-1 模型半结构尺寸图

注：作为 Diana10.1 算例例题，本书例题分析条件均为假定值，关于分析结果妥当性另当别论，望谅解。

学习要义：
(1) 预应力荷载的施加。
(2) 半结构约束条件的选取。
(3) 后处理查看开裂指数和开裂云图。
(4) 预应力回缩长度的设置以及后处理拉力查看。
(5) 张拉预应力筋荷载阶段性分析中荷载工况设置和单元激活。

首先，启动 DianaIE，D 盘中创立 .dpf 文件，模型名称输入 PSC，选择三维分析，模型最大尺寸范围(Model size)为 10 m，网格形状为六面体/四边形(Hexa/Quad)。网格划分阶数选择为 2 次单元(Quadratic)，中间节点位置选择 On shape 以便结合实际结构尺寸情况进行划分。如图 5.2-2所示。

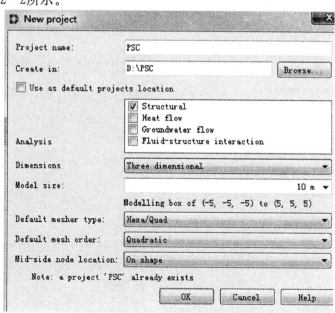

图 5.2-2 模型属性赋予

选定模型使用单位。采用国际单位制(米、千克、牛顿、秒、摄氏度、弧度角),如图 5.2-3 所示。

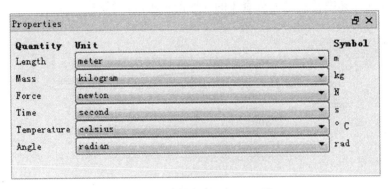

图 5.2-3 模型基本单位设置

创立左半部分的混凝土面,输入各点坐标,如图 5.2-4 所示。

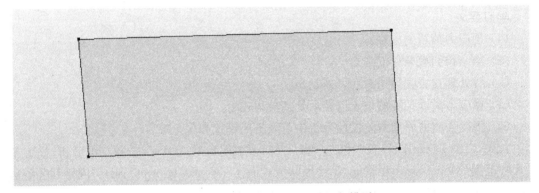

图 5.2-4 左半节段混凝土面几何模型

鼠标左键拾取截面 Sheet 1,右击选择 Array copy,沿着 Y 方向平移 0.8 m,份数为 1 份,得到右半部分平面,命名为 Sheet 2。如图 5.2-5 和图 5.2-6 所示。

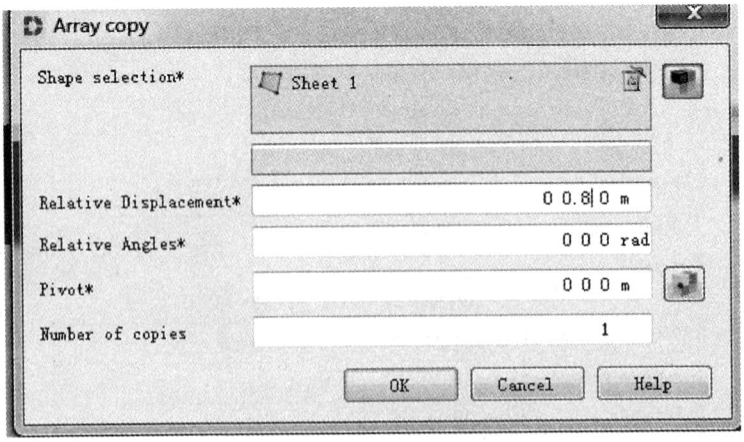

图 5.2-5 Array copy Sheet 1

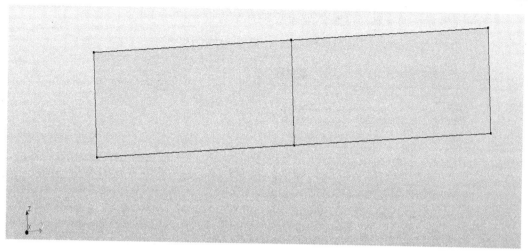

图 5.2-6　Sheet 2 平面

创建折线和直线预应力筋几何模型,分别命名为 tenin 和 teninphase,输入各点位置坐标值,如图 5.2-7、图 5.2-8 所示。

由于体内同时存在折线预应力筋和直线预应力筋,因此针对折线预应力筋采用多段直线(Polyline)建模,对于直线预应力筋采用直线(Line)方式建模。各点坐标如图 5.2-7 和图 5.2-8 所示。

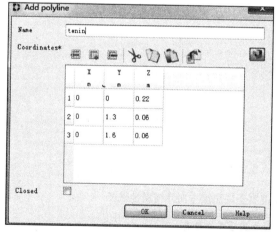

图 5.2-7　创建直线筋

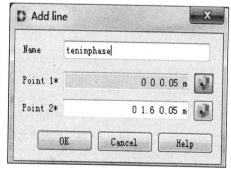

图 5.2-8　创建折线筋

赋予混凝土材料属性。创建 Concrete 组,其中混凝土收缩徐变采用欧洲 1990 规范(CEB-FIP1990),选用常规的曲壳单元(Regular Curved Shells),勾选徐变(Creep)和收缩(Shrinkage)模式,同时勾选开裂指数(Crack index)选项,如图 5.2-9 和图 5.2-10 所示。混凝土强度为 C50,混凝土的弹性模量为 $3.8629 \times 10^{10} \text{N/m}^2$。根据欧洲 CEB-FIP 1990 规范中对于抗压强度的设置,混凝土的 28 天抗压强度标准值为 50 MPa,抗压强度平均值为 58 MPa。对于松弛模型则采用不考虑混凝土老化并且以钢筋松弛为主的 Maxwell 弹簧单元链,混凝土加载龄期设置为 28 天(2419200 s),对收缩徐变影响较大的外界环境因素采用环境温度为 20℃。相对环境湿度设定为 69%。收缩固化曲线终点处对应混凝土龄期设置为 86400 s(1 天)。具体参数信息参见表 5-5。赋予混凝土厚度为 0.09 m,如图 5.2-11 所示。局部单元坐标系的 x 轴对应整体坐标系下的 Y 方向。

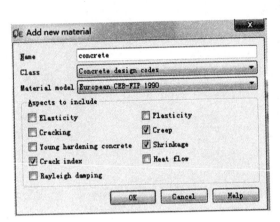

图 5.2-9　CEB-FIP 1990 单元类型界面

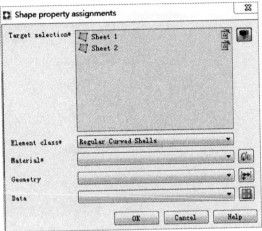

图 5.2-10　普通曲壳单元

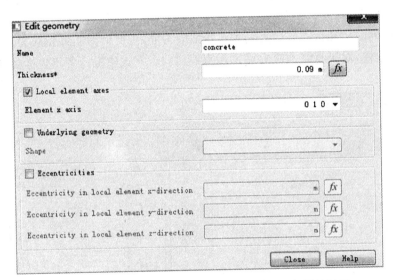

图 5.2-11　曲壳单元模拟混凝土块厚度方向几何属性赋予

表 5-5　混凝土建模参数信息

混凝土参数名称	参数值	单　位
强度等级	C50	—
弹性模量	3.8629×10^{10}	N/m²
28 天弹性模量	3.45×10^{10}	N/m²
泊松比	0.15	—
热膨胀系数	0.00001	1/℃
密度	2500	kg/m³
28 天抗压强度标准值	50	MPa
抗压强度平均值	58	MPa

(续表)

混凝土参数名称	参数值	单 位
徐变曲线类型	Non-Aging	—
混凝土单元龄期	2.419 2e+06	s
固化曲线结束混凝土龄期	86400	s
环境温度	20	℃
环境湿度	69%	—

混凝土属性赋予完成后,接着赋予预应力筋材料属性。由于预应力筋同时存在折线筋和直线筋类型,且二者都为有粘结,因此分别赋予折线预应力筋及直线预应力筋属性。

选中 tenin,右击选择 Reinforcement property assignment,点击 Add new material 设置。如图 5.2-12 所示,选择 Von mises plasticity,不勾选 Bonding 表示预应力筋与母体单元相连用以模拟有粘结预应力筋。如图 5.2-12 所示,弹性模量为 1.95×10^{11} Pa,预应力筋屈服强度为 1860 MPa。如图 5.2-14 所示,钢绞线截面几何属性赋值,选择嵌入混凝土单元方式,截面面积 0.000139 m^2。离散方式选用界面智能 Section wise。用同样的方式和同样参数给充当备用预应力筋的 teninphase 赋予类似的属性。

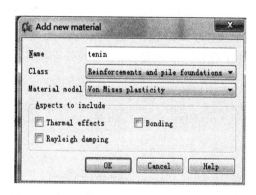

图 5.2-12 预应力筋 Von Mises 塑性屈服模型

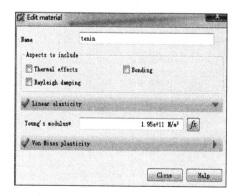

图 5.2-13 钢绞线弹性模量定义

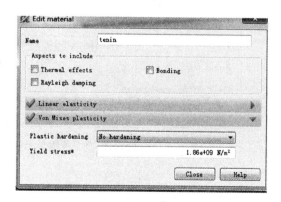

图 5.2-14 钢绞线塑性硬化类型和屈服强度定义

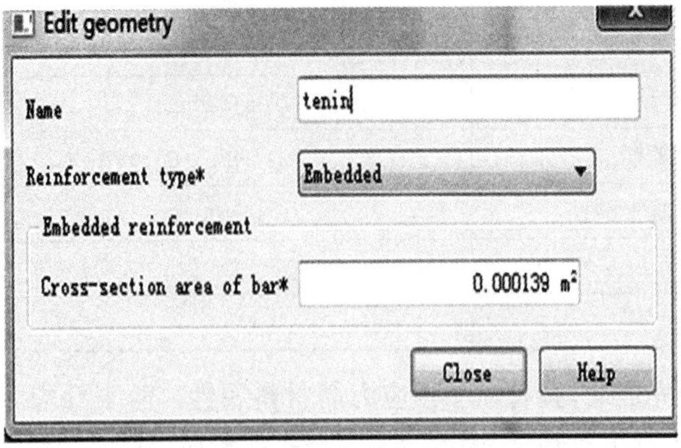

图 5.2-15　钢绞线截面几何特性定义

创建界面单元。点击界面上方快捷工具栏红色图标 Edit connection property assignments，创立界面单元，命名为 int 单元，选择类型为 Edge，界面单元的单元类型选择线对线接触的结构壳类单元界面(Structural Shell Interface)。赋予材料属性和厚度方向几何属性，其中界面单元几何材料属性设置为非线性弹性(Nonlinear elasticity)接触，线弹性材料，界面单元接触方式为 3D 曲壳单元之间的线对线接触(3D line Interface between shells)，非线性弹性属性选择剪切刚度为定值，抗拉刚度为零(No-tension with constant shear stiffness)，界面单元的厚度为 0.09 m，界面单元的单元方向平行于曲壳平面，平行于壳平面的方向向量为整体坐标系下的 Y 轴负方向，非线弹性抗压和抗剪刚度参数设置如图 5.2-16—图 5.2-17 所示。

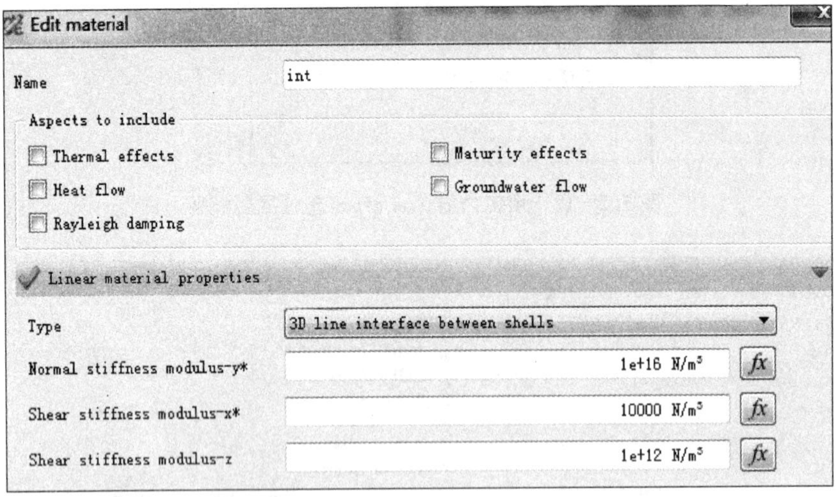

图 5.2-16　界面单元各坐标轴刚度定义

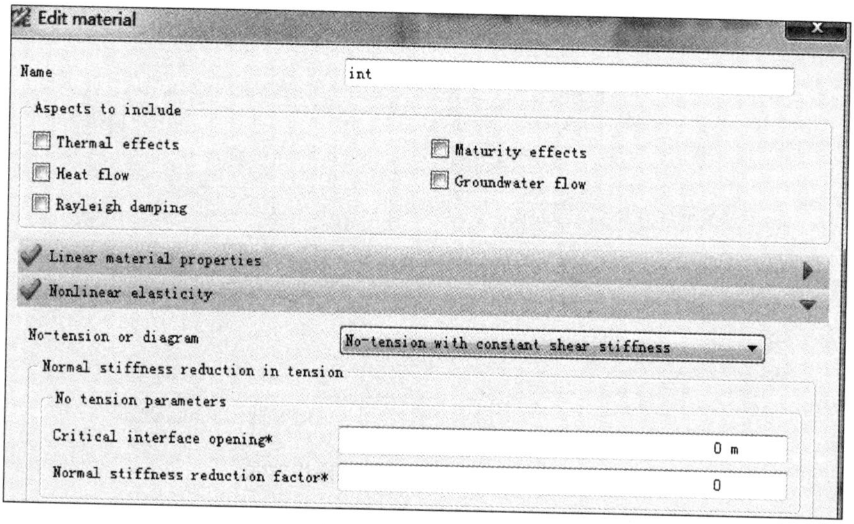

图 5.2-17　界面单元零拉力本构定义

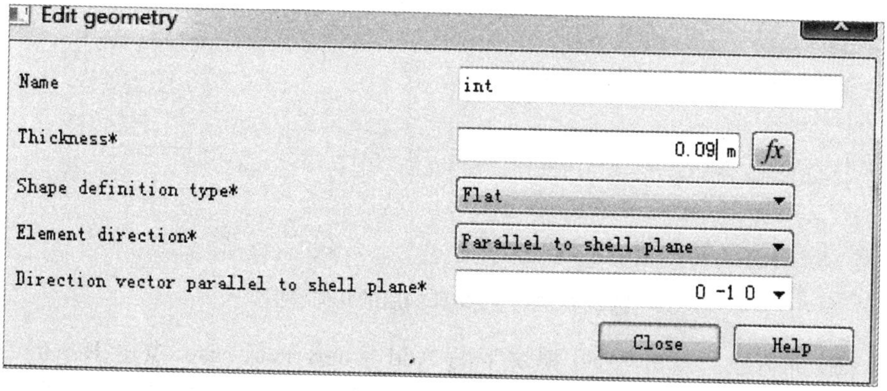

图 5.2-18　界面单元厚度方向积和属性设置

创建荷载。在 Diana10.1 软件中，如前文所述，当需要施加的点荷载、线荷载或者面荷载并不在已经生成的建模点上时，如果用户直接在几何模型面上创立生成点、线、面会影响材料属性赋予以及模型划分，因此这里采用 Diana 软件印刻投影（Imprint projection）特色功能来施加荷载，这样已经生成的点就不会影响后续的材料属性赋予以及网格划分。

首先在结构模型的上方创立点坐标(0,1.3,0.4)，为了方便沿着 Z 方向直接投影印刻，这里选择创立点的坐标高于壳单元高度的坐标。点击快捷键菜单栏选择投影（Project）方式来进行印刻，如图 5.2-19 所示。在 Operation 选择 Edge，选中被投影对象 Edge，在下方的 tool selection 中选中要投影的点，方向选择整体坐标系下的 Z 轴方向，勾选 Imprint，即可完成印刻，如图 5.2-20、图 5.2-21 所示。

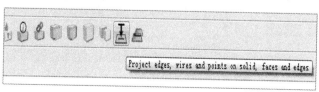

图 5.2-19　投影印刻在快捷菜单栏图标位置

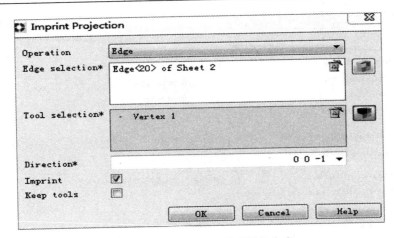

图 5.2-20 投影印刻界面操作方式

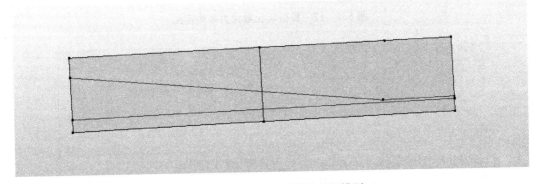

图 5.2-21 投影印刻后的几何模型

接下来施加荷载。点击 Load 模块下的 Add a new load case,点击 Rename 命名为 load,右击选择 Attach load,选择点荷载集中力 Force,施加沿着 Z 轴负向 10 kN 大小的集中力。如图 5.2-22 所示。

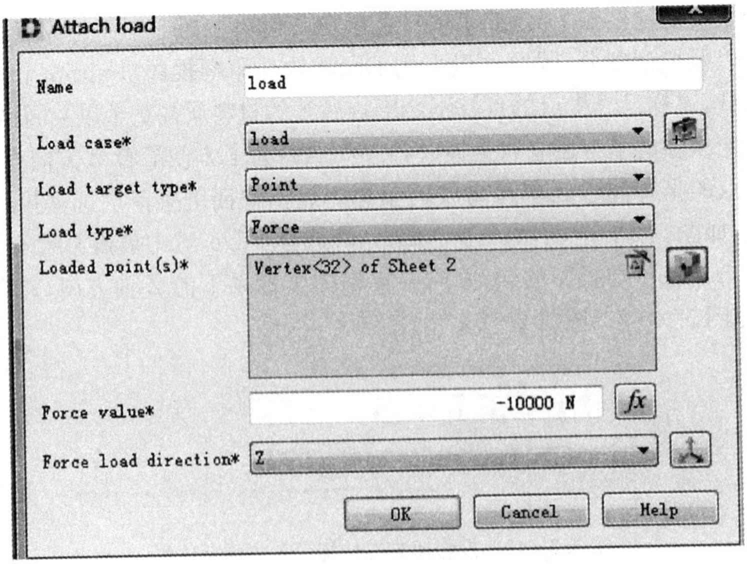

图 5.2-22 外荷载施加方式

接下来添加预应力折线筋荷载工况。创建荷载工况 2,命名为 tenin,加载目标类型选择实体类型(Solid),荷载类型选择预应力荷载(Post tensioning load)施加预应力筋荷载。与之前例题方式不同,本例题采用一端锚固(One end)的方式,其中张拉力(Nodal anchor force)为 180 kN,锚固回缩长度(Anchor retention length)为 0.01 m,库仑摩擦系数(Coulomb friction coefficient)为 0.22,握裹系数(Wobble factor)为 0.01/m。各项设置如图 5.2-23 所示。再用同样方式得到 teninphase 直线筋荷载,各参数设置均与 tenin 相同,如图 5.2-24 所示。

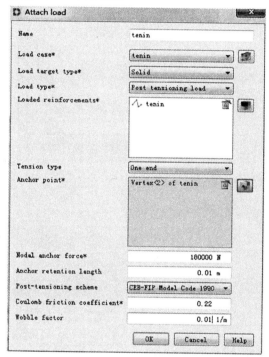

图 5.2-23 施加 tenin 折线组预应力筋

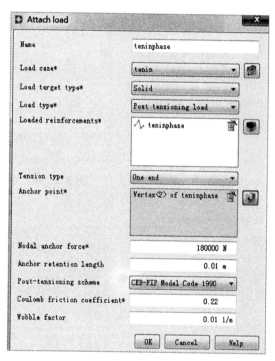

图 5.2-24 施加 teninphase 直线组预应力筋

再点击 Define a global load,命名为 gravity,添加重力荷载工况,如图 5.2-25 所示。

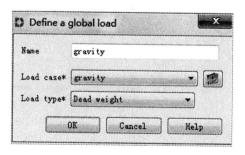

图 5.2-25 添加重力荷载

设置荷载工况组合。为了方便以后张拉备用束,特地将 tenin 折线筋和重力荷载设置为一组作为初始荷载工况添加,集中荷载外荷载设定为一组,直线筋设定为一组。分别命名为荷载组合 1,2,3,如图 5.2-26 所示。由于本模型需要考虑收缩徐变效应,因此采用之前介绍的 Time dependency 时变设置功能将所有的荷载组合赋予时变特性。以一年作为收缩徐变考察期限(其中 Time dependency 关键点为 0 s,2419200 s,15768000 s,31536000 s)。这里设置在初始时候开始施加除直线预应力筋荷载外的所有荷载。时变性能设置如图 5.2-27 所示。

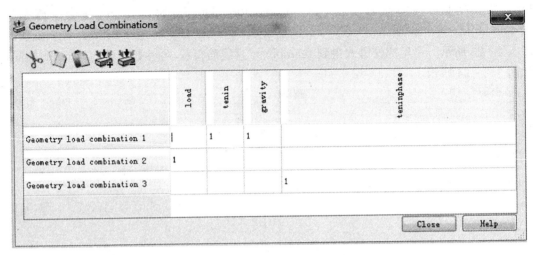

图 5.2-26 添加荷载工况组合

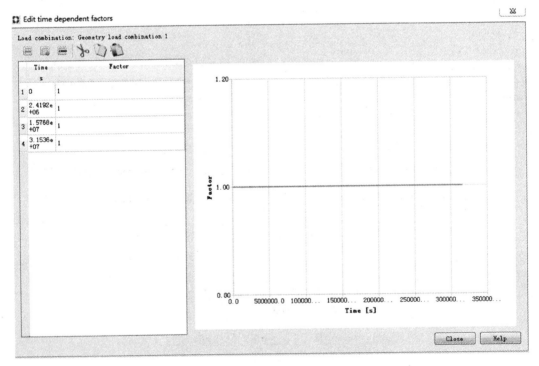

图 5.2-27 荷载系数—时间曲线设置

支座约束条件设置成简支,在几何模型的边(Edge)施加正对称半结构约束。左边约束 X,Z 方向平动位移,右边约束 Y 方向平动位移和 X,Z 方向的转动位移以防止平面向外转动,如图 5.2-28 所示。点击菜单栏 Analysis→Attach Support 设置约束,其中约束条件及信息如图 5.2-29、图 5.2-30 所示。

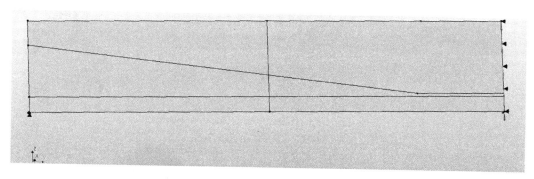

图 5.2‑28　添加约束后的几何模型图

图 5.2‑29　半结构左边部分的约束信息

图 5.2‑30　正对称半结构位置处约束信息

进入网格划分流程。本例中采用单元尺寸(Element size)的方式进行划分。点击选中模型→右击选择 Set mesh properties 进行网格划分。其中曲壳单元的单元尺寸为 0.1 m，网格类别为六面体/矩形方式。单元中点处位置采用线性插值(Linear interpolation)。如图 5.2‑31 所示。采用同样的单元尺寸方式对界面单元进行网格划分，操作类型为对边(Edge)进行操作，划分长度同样选择为 0.1 m，如图 5.2‑32 所示。

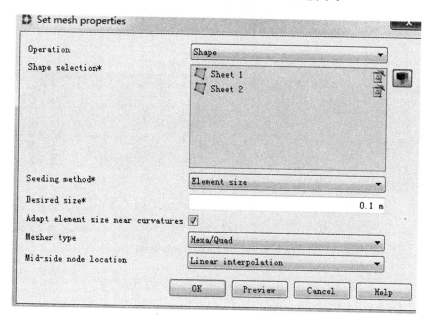

图 5.2‑31　曲壳单元网格划分

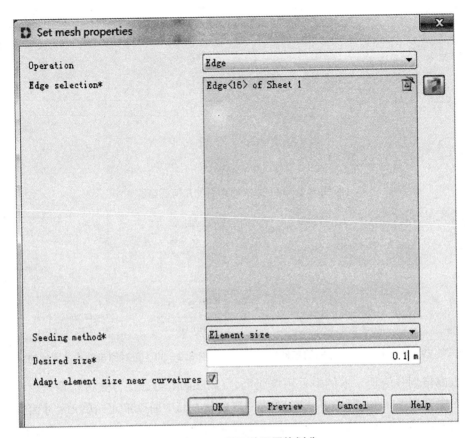

图 5.2-32 界面单元网格划分

点击快捷键图标 Generate mesh of a shape 按钮生成网格,如图 5.2-33 所示。查看 Meshing 模块下网格单元类型信息,确定生成了需要的曲壳单元类型和界面单元信息,如图 5.2-34所示。

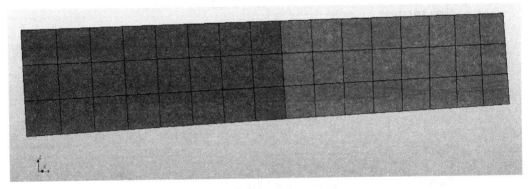

图 5.2-33 网格划分后的模型

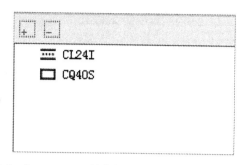

图 5.2-34 Meshing 信息栏生成的曲壳单元和界面单元

接下来进入分析模块设置。在阶段性分析中点击生成分析,右击选中 Add command→phased 设置初始阶段,在第一阶段的 phase 阶段取消勾选直线筋组的 teninphase 钢筋组,将 teninphase 直线筋组作为第二阶段添加,如图 5.2-35 所示。

右击 Analysis1 → Add command → Structural nonlinear 进行非线性结构分析,在 Evaluate 模块的选择界面单元中对配筋进行评估(Evaluate reinforcements in Interface elements),并且右击非线性分析模块,添加荷载组合 1 为初始荷载(Start steps),命名为 tenin。将预应力折线筋 tenin 设置为初始荷载,由于预应力筋是有粘结的,因此添加物理非线性模块选择预应力筋单元完全粘结(Fully bonded reinforcements),如图 5.2-36 所示。在荷载步中设置预应力筋荷载 tenin,并且将荷载组选择为第一组。

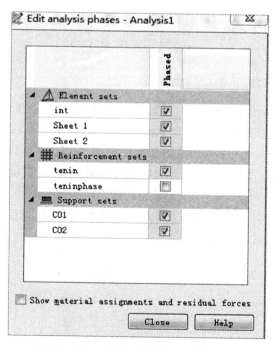

图 5.2-35 阶段性分析第一阶段添加项

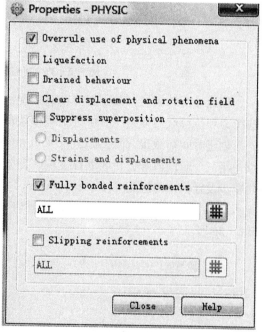

图 5.2-36 有粘结预应力筋物理非线性设置

折线预应力筋载工况设置如图 5.2-37 所示。

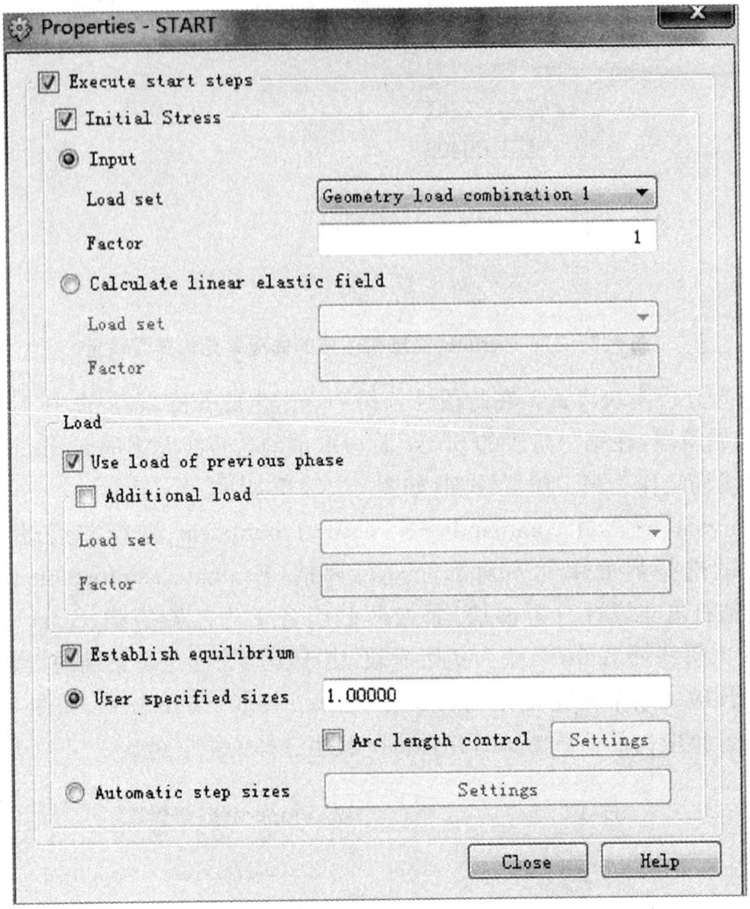

图 5.2-37　折线预应力筋荷载工况设置

创建外荷载模块。分别选择荷载工况组合对应的第 2 组,设置最大迭代步数为 20 步,选择力和位移收敛准则。其中外荷载模块荷载步数设置为 1 个荷载步,步长因子设置为 1。

创建时间荷载步,命名为 creep and shrinkage。这里选择与之前相同的 28 天、半年以及一年三个时间段来分析收缩徐变的时变效应。由于时间步长对应的叠加效应,因此时间荷载步设置依次对应为 2.419 20e+06 s,1.334 80e+07 s,1.576 80e+07 s,如图 5.2-38 所示。收缩徐变荷载最大迭代步数设置为 50 步。收敛准则选择力和位移收敛准则,如图 5.2-39 所示。

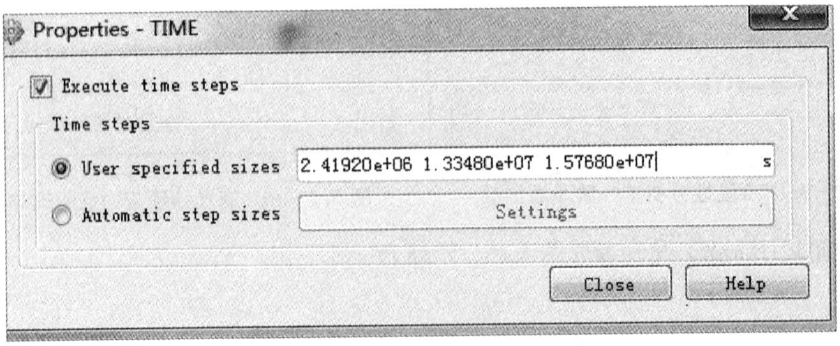

图 5.2-38　时间荷载步的设定

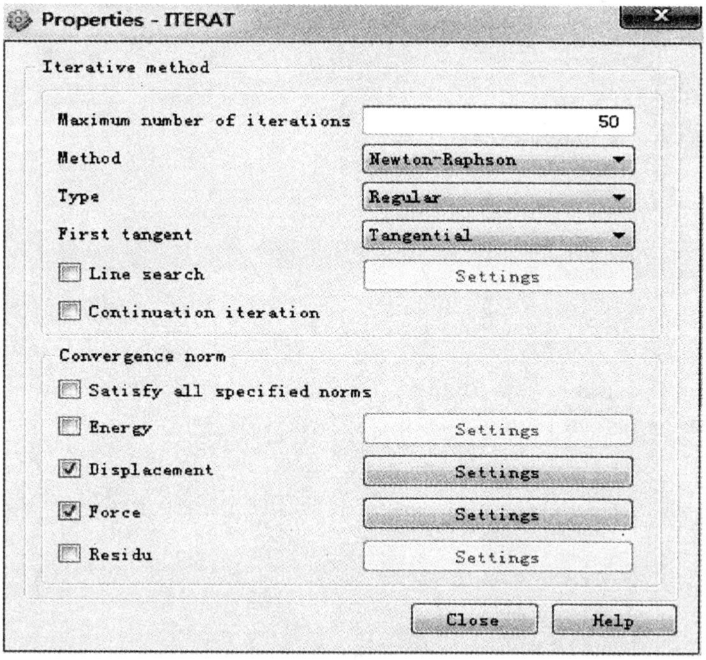

图 5.2-39　收缩徐变工况收敛准则的设定

同样的方式创建第二阶段分析。在该分析中，选中 teninphase 组，考虑到第二阶段施加 teninphase 预应力筋荷载时，先前的所有单元荷载及收缩徐变仍然存在，因此这里选择所有单元全部勾选，如图 5.2-40 所示。

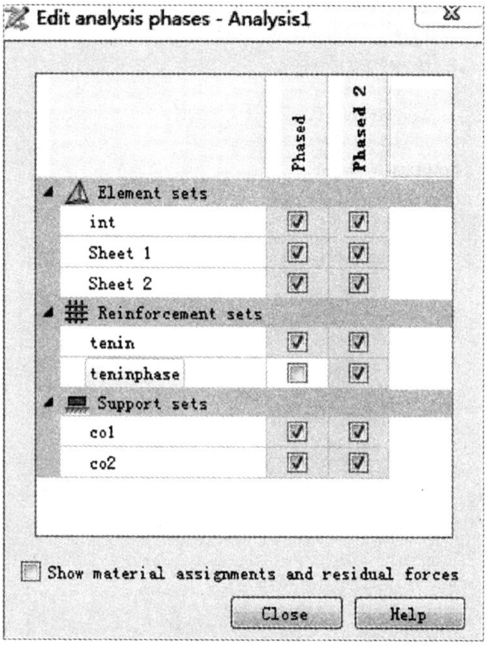

图 5.2-40　阶段性分析第二阶段添加项

用上述同样的方法创建结构非线性分析模块 Structural nonlinear 2，建立初始荷载，荷载组合号为 3 号，如图 5.2-41 所示。与之前操作一样，物理属性设置为完全粘结。为了有效地模拟二次加固特征，勾选选项如图 5.2-42 所示。

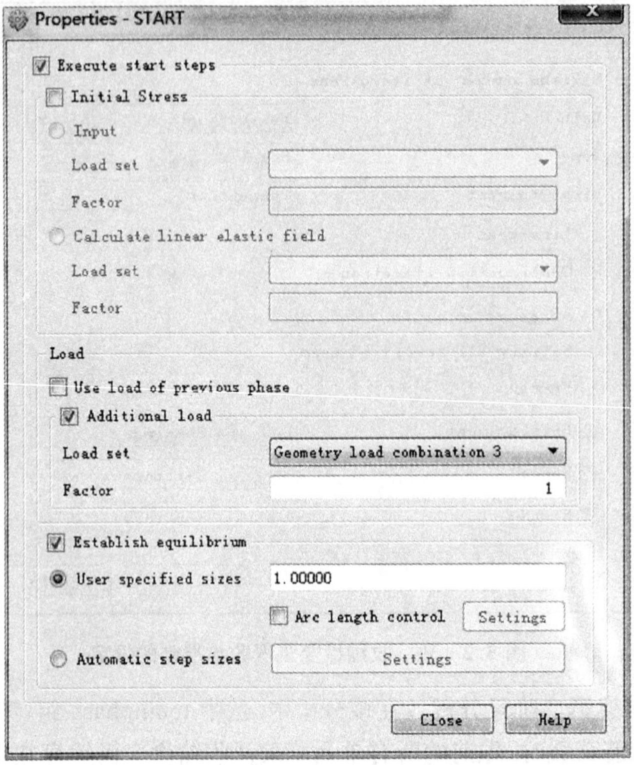

图 5.2-41　第二阶段直线筋初始荷载添加

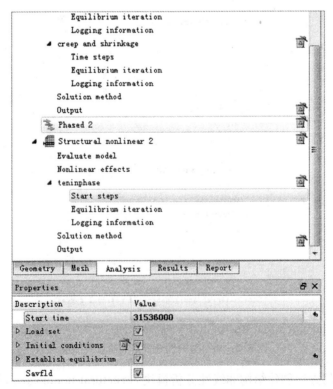

图 5.2-42　第二阶段直线筋开始作用时间

计算模块设置完毕之后,右击 OUTPUT,进行输出结果选项设置。选择整体坐标系下的位移、支座反力、开裂指数以及总应变作为输出变量,点击 Run analysis 按钮,开始计算。如图 5.2-43 所示。

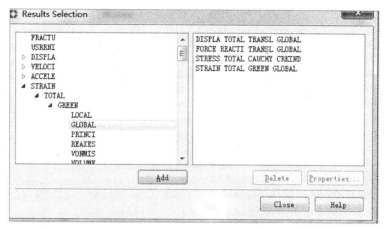

图 5.2-43 OUTPUT 输出项设置

计算完毕,一年时间内收缩徐变的作用效果见各阶段位移云图。

由于预应力筋作为初始荷载添加,因此一开始会有上挠反拱的效果,初始位移图如图 5.2-44 所示。

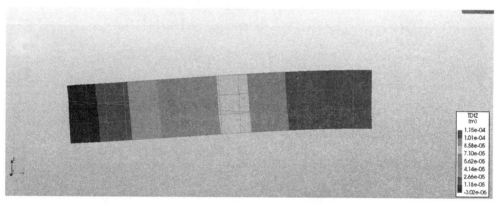

图 5.2-44 预应力筋作用下初始阶段出现反拱

重力荷载、长期荷载和收缩徐变共同作用 28 天后的位移云图如图 5.2-45 所示。

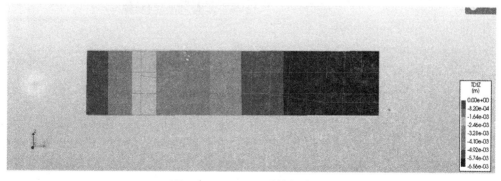

图 5.2-45 28 天后的位移云图

长期荷载和收缩徐变共同作用半年后的位移云图如图 5.2-46 所示。

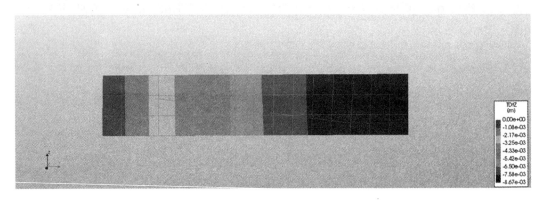

图 5.2-46 半年后的位移云图

长期荷载和收缩徐变共同作用一年后的位移云图如图 5.2-47 所示。

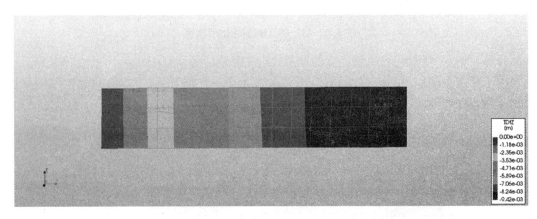

图 5.2-47 一年以后的位移云图

1 年后，第二阶段预应力筋开始作用，位移云图如图 5.2-48 所示。

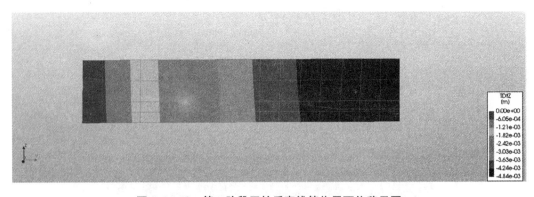

图 5.2-48 第二阶段开始后直线筋作用下位移云图

选择菜单栏 Viewer，点击节点选择（Node Selection）生成节点，这时框选跨中节点，右击选择显示节点号，显示 102，点击位移结果选项，右击显示表格，就会生成 102 号节点对应的位移值。用同样的方式选择阶段二位移云图状态下同样位置处的节点。显示结果分别如图 5.2-49 和图 5.2-50 所示。

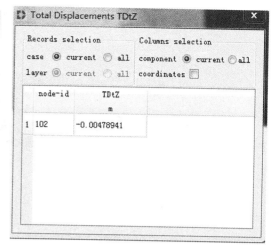

图 5.2‑49　一年后跨中 102 号节点位移值　　图 5.2‑50　添加直线筋后跨中 102 号节点位移值

查看开裂指数。点击 Element results→Crack Indices→Icr,查看结构的开裂指数分布云图,如图 5.2‑51—图 5.2‑55 所示,从而预判整个结构的裂缝开展程度。

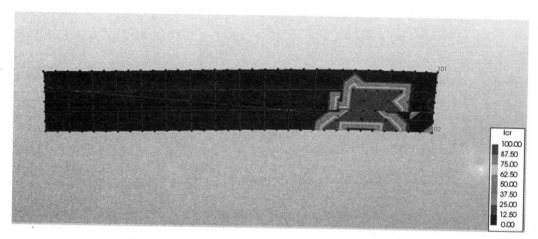

图 5.2‑51　初始预应力筋张拉时的开裂指数分布云图

图 5.2‑52　28 天后的开裂指数分布云图

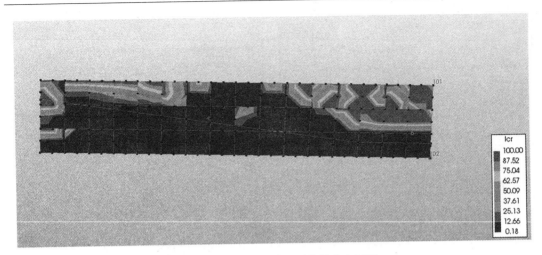

图 5.2-53 半年后的开裂指数分布云图

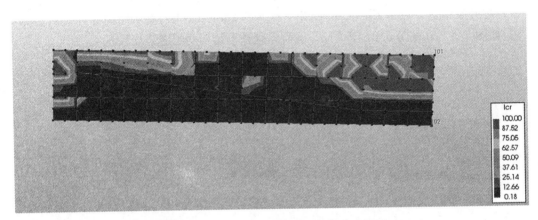

图 5.2-54 一年左右时间开裂指数分布云图

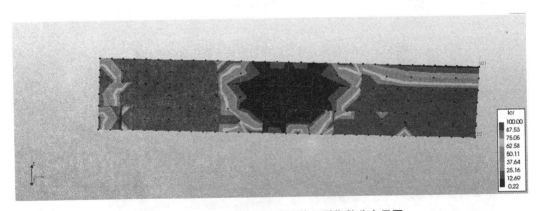

图 5.2-55 第二阶段张拉直线筋开裂指数分布云图

由开裂指数分布云图可见,在长期荷载作用下,两段梁之间部分和跨中部分因为受拉率先出现裂缝的可能性非常大,当开裂指数小于1时,就会有裂缝产生。当直线筋在第二阶段启动张拉时,受拉区混凝土开裂指数增加,开裂的可能性进一步降低。

接下来考虑预应力筋回缩长度不同的情况对非线性计算结果的影响。在 Output 中设置 All Primaries 以便在输出结果中产生所有基本项,点击 Run Analysis 开始计算。在 Result→Reinforcements→Result→Reinforcement Cross-Section Force 中查看折线预应力筋的应力云图,发现在初始预应力筋为 180 kN 的情况下,时间荷载步为一年的非线性分析结束后预应力筋数值在 25.2 kN 以下,结果严重失真,如图 5.2-56 所示。这是因为在之前的数值模拟中预应力钢绞线的回缩长度(Rentention length)采用默认长度 0.01 m,回缩长度设置过大,导致结果失真,不符合实际情况,这时应考虑按照回缩与模型中钢绞线长度在实际情况下比值的数量级设置合理的回缩长度。此外,由于设定的网格尺寸单元过大(0.1 m),计算结果往往不够精确而失真。而直接减少回缩长度,往往会造成计算结果不收敛,因此需要设置更小预应力筋的回缩长度和单元网格尺寸。

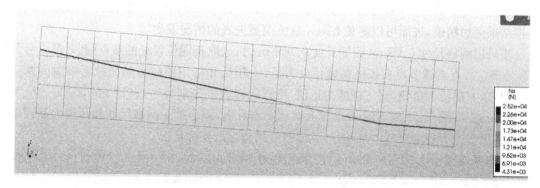

图 5.2-56　回缩长度采用默认值(0.01 m)时折线筋一年后的拉力云图

在 load 中找到 tenin 荷载组,将预应力筋的回缩长度设定为 0.001 m,其他参数不变,在网格划分设定中将单元尺寸 Element size 改为 0.05 m,点击 Generate mesh of a shape 重新划分网格,然后点击 Analysis 模块下的 Run Analysis 开始计算。计算完毕后,查看模型在同样施加 180 kN 初始预应力值时的位移云图和一年后的折线钢绞线应力云图,如图 5.2-57 所示。其中单元网格尺寸为 0.05 m,回缩长度为 0.001 m 位移。

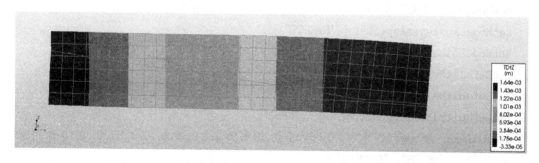

图 5.2-57　单元网格尺寸 0.05 m、回缩长度 0.001 m 时位移云图
（注意跨中位置处的反拱值）

在 Result→Reinforcements→Result→Reinforcement Cross-Section Force 中查看折线预应力筋的拉力云图、时间荷载步为一年的非线性分析结束后折线筋应力云图如图 5.2-58 所示。这时候可以看出预应力筋的拉力云图在 107～158 kN 范围内波动,更贴近真实情况。

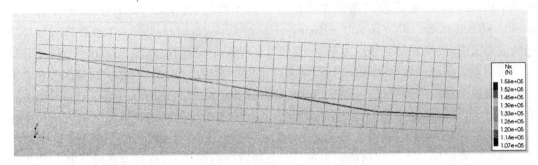

图 5.2-58　单元网格尺寸 0.05 m、回缩长度 0.001 m 时折线筋一年后的拉力云图

通过对比不难看出,采用较小的回缩长度和较小的单元网格尺寸会使得有限元非线性模拟结果更加精确,进而可以避免有限元数值模型失真的情况发生。

当采用网格尺寸 0.05 m,回缩长度 0.001 m 时,选取相同位置处的节点(此时节点号为 343 号节点),查看该节点在张拉初始预应力后,一年后和备用预应力筋张拉后一瞬间跨中挠度值,以上拱值为负值,以下挠值为正值,可得三者数值分别为 −1.628 mm,6.70 mm 和 1.503 mm。将图 5.2-56 和图 5.2-58 对比不难看出,缩小单元网格尺寸和预应力筋回缩长度,备用预应力筋的恢复效果越好,且模拟效果越接近实际情况。

模型界面操作生成的 Python 语言命令流(供参考)。

```
#################################################################
# DianaIE 10.1 update 2017-04-25 13:38:53
# Python 3.3.4
# Session recorded at 2017-06-09 22:22:17
#################################################################
    newProject( "PSC", 10 )
    setModelAnalysisAspects( [ "STRUCT" ] )
    setModelDimension( "3D" )
    setDefaultMeshOrder( "QUADRATIC" )
    setDefaultMesherType( "HEXQUAD" )
    setDefaultMidSideNodeLocation( "ONSHAP" )
    createSheet( "Sheet 1", [[ 0, 0, 0 ],[ 0, 0.8, 0 ],[ 0, 0.8, 0.3 ],[ 0, 0, 0.3 ]] )
    setUnit( "TEMPER", "CELSIU" )
    saveProject(  )
    arrayCopy( [ "Sheet 1" ], [ 0, 0.8, 0 ], [ 0, 0, 0 ], [ 0, 0, 0 ], 1 )
    saveProject(  )
    createPolyline( "tenin", [[ 0, 0, 0.22 ],[ 0, 1.3, 0.06 ],[ 0, 1.6, 0.06 ]],
False )
    saveProject(  )
    createLine( "teninphase", [ 0, 0, 0.05 ], [ 0, 1.6, 0.05 ] )
```

```
saveProject( )
addMaterial( "concrete", "CONCDC", "MC1990", [ "CREEP", "CRKIDX", "SHRINK" ] )
setParameter( "MATERIAL", "concrete", "MC90CO /GRADE", "C50" )
setParameter( "MATERIAL", "concrete", "MC90CO /RH", 69 )
setParameter( "MATERIAL", "concrete", "MC90CO /RH", 69 )
setParameter( "MATERIAL", "concrete", "CONCDI /YOUNG", 3.8629e + 10 )
setParameter( "MATERIAL", "concrete", "CONCDI /YOUN28", 3.45e + 10 )
setParameter( "MATERIAL", "concrete", "CONCDI /POISON", 0.15 )
setParameter( "MATERIAL", "concrete", "CONCDI /THERMX", 1e - 05 )
setParameter( "MATERIAL", "concrete", "CONCDI /DENSIT", 2500 )
setParameter( "MATERIAL", "concrete", "CONCDI /FCK28", 50000000 )
setParameter( "MATERIAL", "concrete", "CONCDI /FCM28", 58000000 )
setParameter( "MATERIAL", "concrete", "CONCSH /CURAGE", 86400 )
addGeometry( "Element geometry 1", "SHEET", "CURSHL", [ ] )
rename( "GEOMET", "Element geometry 1", "concrete" )
setParameter( "GEOMET", "concrete", "THICK", 0.09 )
setParameter( "GEOMET", "concrete", "LOCAXS", True )
setParameter( "GEOMET", "concrete", "LOCAXS /XAXIS", [ 1, 0, 0 ] )
setParameter( "GEOMET", "concrete", "LOCAXS /XAXIS", [ 0, 1, 0 ] )
setParameter( "GEOMET", "concrete", "LOCAXS /XAXIS", [ 0, 1, 0 ] )
clearReinforcementAspects( [ "Sheet 1", "Sheet 2" ] )
setElementClassType( "SHAPE", [ "Sheet 1", "Sheet 2" ], "CURSHL" )
assignMaterial( "concrete", "SHAPE", [ "Sheet 1", "Sheet 2" ] )
assignGeometry( "concrete", "SHAPE", [ "Sheet 1", "Sheet 2" ] )
resetElementData( "SHAPE", [ "Sheet 1", "Sheet 2" ] )
saveProject( )
addMaterial( "tenin", "REINFO", "VMISES", [ ] )
setParameter( "MATERIAL", "tenin", "LINEAR /ELASTI /YOUNG", 1.95e + 11 )
setParameter( "MATERIAL", "tenin", "PLASTI /HARDI1 /YLDSTR", 1.86e + 09 )
addGeometry( "Element geometry 2", "RELINE", "REBAR", [ ] )
rename( "GEOMET", "Element geometry 2", "tenin" )
setParameter( "GEOMET", "tenin", "REIEMB /CROSSE", 0.000139 )
setReinforcementAspects( [ "tenin" ] )
assignMaterial( "tenin", "SHAPE", [ "tenin" ] )
assignGeometry( "tenin", "SHAPE", [ "tenin" ] )
resetElementData( "SHAPE", [ "tenin" ] )
setReinforcementDiscretization( [ "tenin" ], "SECTION" )
saveProject( )
addMaterial( "teninphase", "REINFO", "VMISES", [ ] )
```

```
setParameter( "MATERIAL", "teninphase", "LINEAR /ELASTI /YOUNG", 1.95e+11 )
setParameter( "MATERIAL", "teninphase", "PLASTI /HARDI1 /YLDSTR", 1.86e+09 )
addGeometry( "Element geometry 3", "RELINE", "REBAR", [ ] )
rename( "GEOMET", "Element geometry 3", "teninphase" )
setParameter( "GEOMET", "teninphase", "REIEMB /CROSSE", 0.000139 )
setReinforcementAspects( [ "teninphase" ] )
assignMaterial( "teninphase", "SHAPE", [ "teninphase" ] )
assignGeometry( "teninphase", "SHAPE", [ "teninphase" ] )
resetElementData( "SHAPE", [ "teninphase" ] )
setReinforcementDiscretization( [ "teninphase" ], "SECTION" )
saveProject(  )
addMaterial( "int", "INTERF", "NONLIF", [ ] )
setParameter( "MATERIAL", "int", "LINEAR /IFTYP", "LIN3D" )
setParameter( "MATERIAL", "int", "LINEAR /ELAS4 /DSNY", 1e+16 )
setParameter( "MATERIAL", "int", "LINEAR /ELAS4 /DSSX", 10000 )
setParameter( "MATERIAL", "int", "LINEAR /ELAS4 /DSSZ", 1e+12 )
setParameter( "MATERIAL", "int", "NONLIN /IFNOTE", "NOTENS" )
addGeometry( "Element geometry 4", "LINE", "SHLLIF", [ ] )
rename( "GEOMET", "Element geometry 4", "int" )
setParameter( "GEOMET", "int", "THICK", 0.09 )
setParameter( "GEOMET", "int", "THICK", 0.09 )
setParameter( "GEOMET", "int", "THKDIR", "PARALL" )
setParameter( "GEOMET", "int", "YAXIS", [ 1, 0, 0 ] )
setParameter( "GEOMET", "int", "YAXIS", [ 0, 1, 0 ] )
saveProject(  )
setParameter( "GEOMET", "int", "YAXIS", [ 0, -1, 0 ] )
setParameter( "GEOMET", "int", "YAXIS", [ 0, -1, 0 ] )
createLineConnection( "int" )
setParameter( "GEOMETRYCONNECTION", "int", "CONTYP", "INTER" )
setParameter( "GEOMETRYCONNECTION", "int", "MODE", "AUTO" )
attachTo( "GEOMETRYCONNECTION", "int", "SOURCE", "Sheet 1", [[ 0, 0.8, 0.15 ]] )
setElementClassType( "GEOMETRYCONNECTION", "int", "SHLLIF" )
assignMaterial( "int", "GEOMETRYCONNECTION", "int" )
assignGeometry( "int", "GEOMETRYCONNECTION", "int" )
resetElementData( "GEOMETRYCONNECTION", "int" )
saveProject(  )
createVertex( "Vertex 1", [ 0, 1.32, 0.4 ] )
saveProject(  )
projection( "SHAPEEDGE", "Sheet 2", [[ 0, 1.2, 0.3 ]], [ "Vertex 1" ], [ 0, 0,
```

```
    -1 ], True )
        removeShape( [ "Vertex 1" ] )
        saveProject(   )
        addSet( "GEOMETRYLOADSET", "Geometry load case 1" )
        rename( "GEOMETRYLOADSET", "Geometry load case 1", "Geometry load case 1" )
        rename( "GEOMETRYLOADSET", "Geometry load case 1", "load" )
        createPointLoad( "load", "load" )
        setParameter( "GEOMETRYLOAD", "load", "FORCE /VALUE", -10000 )
        setParameter( "GEOMETRYLOAD", "load", "FORCE /DIRECT", 3 )
        attach( "GEOMETRYLOAD", "load", "Sheet 2", [[ 0, 1.32, 0.3 ]] )
        saveProject(   )
        addSet( "GEOMETRYLOADSET", "Geometry load case 2" )
        rename( "GEOMETRYLOADSET", "Geometry load case 2", "tenin" )
        createBodyLoad( "tenin", "tenin" )
        setParameter( "GEOMETRYLOAD", "tenin", "LODTYP", "POSTEN" )
        setParameter( "GEOMETRYLOAD", "tenin", "POSTEN /TENTYP", "ONEEND" )
        setParameter( "GEOMETRYLOAD", "tenin", "POSTEN /ONEEND /FORCE1", 180000 )
        setParameter( "GEOMETRYLOAD", "tenin", "POSTEN /ONEEND /RETLE1", 0.01 )
        setParameter( "GEOMETRYLOAD", "tenin", "POSTEN /SHEAR", 0.22 )
        setParameter( "GEOMETRYLOAD", "tenin", "POSTEN /WOBBLE", 0.01 )
        attach( "GEOMETRYLOAD", "tenin", [ "tenin" ] )
        attachTo( "GEOMETRYLOAD", "tenin", "POSTEN /ONEEND /PNTS1", "tenin", [[ 0, 0, 0.22 ]] )
        saveProject(   )
        addSet( "GEOMETRYLOADSET", "Geometry load case 3" )
        rename( "GEOMETRYLOADSET", "Geometry load case 3", "teninphase" )
        createBodyLoad( "teninphase", "teninphase" )
        setParameter( "GEOMETRYLOAD", "teninphase", "LODTYP", "POSTEN" )
        setParameter( "GEOMETRYLOAD", "teninphase", "POSTEN /TENTYP", "ONEEND" )
        setParameter( "GEOMETRYLOAD", "teninphase", "POSTEN /ONEEND /FORCE1", 180000 )
        setParameter( "GEOMETRYLOAD", "teninphase", "POSTEN /ONEEND /RETLE1", 0.01 )
        setParameter( "GEOMETRYLOAD", "teninphase", "POSTEN /SHEAR", 0.22 )
        setParameter( "GEOMETRYLOAD", "teninphase", "POSTEN /WOBBLE", 0.01 )
        attach( "GEOMETRYLOAD", "teninphase", [ "teninphase" ] )
        attachTo ( " GEOMETRYLOAD ", " teninphase ", " POSTEN /ONEEND /PNTS1 ", "teninphase", [[ 0, 0, 0.05 ]] )
        saveProject(   )
        addSet( "GEOMETRYSUPPORTSET", "C01" )
        createPointSupport( "C01", "C01" )
```

```
setParameter( "GEOMETRYSUPPORT", "C01", "AXES", [ 1, 2 ] )
setParameter( "GEOMETRYSUPPORT", "C01", "TRANSL", [ 1, 0, 1 ] )
setParameter( "GEOMETRYSUPPORT", "C01", "ROTATI", [ 0, 0, 0 ] )
attach( "GEOMETRYSUPPORT", "C01", "Sheet 1", [[ 0, 0, 0 ]] )
saveProject( )
addSet( "GEOMETRYSUPPORTSET", "Geometry support set 2" )
rename( "GEOMETRYSUPPORTSET", "Geometry support set 2", "C02" )
createPointSupport( "C02", "C02" )
setParameter( "GEOMETRYSUPPORT", "C02", "AXES", [ 1, 2 ] )
setParameter( "GEOMETRYSUPPORT", "C02", "TRANSL", [ 0, 1, 0 ] )
setParameter( "GEOMETRYSUPPORT", "C02", "ROTATI", [ 1, 0, 1 ] )
attach( "GEOMETRYSUPPORT", "C02", "Sheet 2", [[ 0, 1.6, 0 ]] )
remove( "GEOMETRYSUPPORT", "C02" )
createLineSupport( "C02", "C02" )
setParameter( "GEOMETRYSUPPORT", "C02", "AXES", [ 1, 2 ] )
setParameter( "GEOMETRYSUPPORT", "C02", "TRANSL", [ 0, 1, 0 ] )
setParameter( "GEOMETRYSUPPORT", "C02", "ROTATI", [ 1, 0, 1 ] )
attach( "GEOMETRYSUPPORT", "C02", "Sheet 2", [[ 0, 1.6, 0.15 ]] )
saveProject( )
setElementSize( [ "Sheet 1", "Sheet 2" ], 0.1, -1, True )
setMesherType( [ "Sheet 1", "Sheet 2" ], "HEXQUAD" )
setMidSideNodeLocation( [ "Sheet 1", "Sheet 2" ], "LINEAR" )
saveProject( )
setElementSize( "Sheet 1", 1, [[ 0, 0.8, 0.15 ]], 0.1, 0, True )
saveProject( )
generateMesh( [] )
hideView( "GEOM" )
showView( "MESH" )
addAnalysis( "Analysis1" )
remove( "REINFORCEMENTSET", [ "teninphase", "tenin" ] )
remove( "ELEMENTSET", [ "Sheet 2", "Sheet 1", "int" ] )
showView( "GEOM" )
setDefaultGeometryLoadCombinations( )
setGeometryLoadCombinationFactor( "Geometry load combination 1", "load", 1 )
addSet( "GEOMETRYLOADSET", "Geometry load case 4" )
undo( 1 )
addSet( "GEOMETRYLOADSET", "GRAVITY" )
createModelLoad( "GRAVITY", "GRAVITY" )
saveProject( )
```

```
setGeometryLoadCombinationFactor( "Geometry load combination 1", "load", 1 )
remove( "GEOMETRYLOADCOMBINATION", "Geometry load combination 1" )
remove( "GEOMETRYLOADCOMBINATION", "Geometry load combination 2" )
setGeometryLoadCombinationFactor( "Geometry load combination 3", "GRAVITY", 1 )
addGeometryLoadCombination( "" )
setGeometryLoadCombinationFactor( "Geometry load combination 2", "load", 1 )
setGeometryLoadCombinationFactor( "Geometry load combination 3", "tenin", 1 )
setGeometryLoadCombinationFactor ( " Geometry load combination 3 ", "teninphase", 1 )
setGeometryLoadCombinationFactor( "Geometry load combination 2", "load", 1 )
remove( "GEOMETRYLOADCOMBINATION", "Geometry load combination 2" )
remove( "GEOMETRYLOADCOMBINATION", "Geometry load combination 3" )
addGeometryLoadCombination( "" )
setGeometryLoadCombinationFactor( "Geometry load combination 1", "tenin", 1 )
setGeometryLoadCombinationFactor( "Geometry load combination 1", "GRAVITY", 1 )
addGeometryLoadCombination( "" )
setGeometryLoadCombinationFactor( "Geometry load combination 2", "load", 1 )
addGeometryLoadCombination( "" )
saveProject( )
setGeometryLoadCombinationFactor ( " Geometry load combination 3 ", "teninphase", 1 )
saveProject( )
setTimeDependentLoadFactors ( " GEOMETRYLOADCOMBINATION ", " Geometry load combination 1", [ 0, 2419200, 15768000, 31536000 ], [ 1, 1, 1, 1 ] )
setTimeDependentLoadFactors ( " GEOMETRYLOADCOMBINATION ", " Geometry load combination 2", [ 0, 2419200, 15768000, 31536000 ], [ 1, 1, 1, 1 ] )
saveProject( )
saveProject( )
setTimeDependentLoadFactors ( " GEOMETRYLOADCOMBINATION ", " Geometry load combination 3", [ 0, 2419200, 15768000, 31536000, 63072000 ], [ 1, 1, 1, 1, 1 ] )
generateMesh( [] )
hideView( "GEOM" )
showView( "MESH" )
renameAnalysis( "Analysis1", "Analysis1" )
renameAnalysis( "Analysis1", "Analysis1" )
renameAnalysis( "Analysis1", "Analysis1" )
removeAnalysis( "Analysis1" )
addAnalysis( "Analysis1" )
addAnalysisCommand( "Analysis1", "PHASE", "Phased" )
```

```
    renameAnalysis( "Analysis1", "Analysis1" )
    setActivePhase( "Analysis1", "Phased" )
    setActiveInPhase ( " Analysis1", " REINFORCEMENTSET", [ " teninphase" ],
[ "Phased" ], False )
    addAnalysisCommand( "Analysis1", "NONLIN", "Structural nonlinear" )
    addAnalysisCommandDetail ( " Analysis1", " Structural nonlinear", " MODEL /
EVALUA /REINFO /INTERF" )
    setAnalysisCommandDetail ( " Analysis1", " Structural nonlinear", " MODEL /
EVALUA /REINFO /INTERF", True )
    removeAnalysisCommandDetail( "Analysis1", "Structural nonlinear", " EXECUT
(1)" )
    setActivePhase( "Analysis1", "Phased" )
    setAnalysisCommandDetail ( " Analysis1", " Structural nonlinear", " EXECUT /
EXETYP", "START" )
    renameAnalysisCommandDetail( " Analysis1", " Structural nonlinear", " EXECUT
(1)", "new execute block 2" )
    renameAnalysisCommandDetail( " Analysis1", " Structural nonlinear", " EXECUT
(1)", "tenin" )
    addAnalysisCommandDetail( "Analysis1", "Structural nonlinear", "EXECUT(1) /
START /INITIA /STRESS" )
    setAnalysisCommandDetail( "Analysis1", "Structural nonlinear", "EXECUT(1) /
START /INITIA /STRESS", True )
    setAnalysisCommandDetail( "Analysis1", "Structural nonlinear", "EXECUT(1) /
ITERAT /MAXITE", 20 )
    setAnalysisCommandDetail( "Analysis1", "Structural nonlinear", "EXECUT(1) /
ITERAT /CONVER /DISPLA /NOCONV", "CONTIN" )
    setAnalysisCommandDetail( "Analysis1", "Structural nonlinear", "EXECUT(1) /
ITERAT /CONVER /FORCE /NOCONV", "CONTIN" )
    saveProject( )
    setActivePhase( "Analysis1", "Phased" )
    setAnalysisCommandDetail ( " Analysis1", " Structural nonlinear", " EXECUT /
EXETYP", "LOAD" )
    renameAnalysisCommandDetail( " Analysis1", " Structural nonlinear", " EXECUT
(2)", "load" )
    setAnalysisCommandDetail( "Analysis1", "Structural nonlinear", "EXECUT(2) /
LOAD /LOADNR", 2 )
    setActivePhase( "Analysis1", "Phased" )
    setAnalysisCommandDetail ( " Analysis1", " Structural nonlinear", " EXECUT /
EXETYP", "TIME" )
```

    renameAnalysisCommandDetail("Analysis1", "Structural nonlinear", " EXECUT (3)", "creep and shrinkage")
    saveProject( )
    setAnalysisCommandDetail("Analysis1", "Structural nonlinear", "EXECUT(3) / TIME /STEPS /EXPLIC /SIZES", "2419200 13348000 15768000")
    setAnalysisCommandDetail("Analysis1", "Structural nonlinear", "EXECUT(3) / TIME /STEPS /EXPLIC /SIZES", "2.41920e+06 1.33480e+07 1.57680e+07")
    setAnalysisCommandDetail("Analysis1", "Structural nonlinear", "EXECUT(3) / TIME /STEPS /EXPLIC /SIZES", "2.41920e+06 1.33480e+07 1.57680e+07")
    setAnalysisCommandDetail("Analysis1", "Structural nonlinear", "EXECUT(3) / TIME /STEPS /EXPLIC /SIZES", "2.41920e+06 1.33480e+07 1.57680e+07")
    setAnalysisCommandDetail("Analysis1", "Structural nonlinear", "EXECUT(3) / TIME /STEPS /EXPLIC /SIZES", "2.41920e+06 1.33480e+07 1.57680e+07")
    setAnalysisCommandDetail("Analysis1", "Structural nonlinear", "EXECUT(3) / ITERAT /MAXITE", 50)
    setAnalysisCommandDetail("Analysis1", "Structural nonlinear", "EXECUT(3) / ITERAT /CONVER /DISPLA /NOCONV", "CONTIN")
    setAnalysisCommandDetail("Analysis1", "Structural nonlinear", "EXECUT(3) / ITERAT /CONVER /FORCE /NOCONV", "CONTIN")
    saveProject( )
    addAnalysisCommand("Analysis1", "PHASE", "Phased 1")
    setActivePhase("Analysis1", "Phased 1")
    setActivePhase("Analysis1", "Phased 1")
    setActivePhase("Analysis1", "Phased 1")
    setActivePhase("Analysis1", "Phased 1")
    setActivePhase("Analysis1", "Phased 1")
    setActivePhase("Analysis1", "Phased 1")
    setActivePhase("Analysis1", "Phased 1")
    setActivePhase("Analysis1", "Phased 1")
    renameAnalysisCommand("Analysis1", "Phased 1", "Phased 1")
    setActivePhase("Analysis1", "Phased 1")
    setActivePhase("Analysis1", "Phased 1")
    setActivePhase("Analysis1", "Phased 1")
    renameAnalysisCommand("Analysis1", "Phased 1", "Phased 1")
    setActivePhase("Analysis1", "Phased 1")
    setActivePhase("Analysis1", "Phased 1")
    setActivePhase("Analysis1", "Phased 1")
    renameAnalysisCommand("Analysis1", "Phased 1", "Phased 2")
    addAnalysisCommand("Analysis1", "NONLIN", "Structural nonlinear 1")

```
    setActivePhase( "Analysis1", "Phased 2" )
    renameAnalysisCommand( "Analysis1", "Structural nonlinear 1", "Structural nonlinear 2" )
    addAnalysisCommandDetail( "Analysis1", "Structural nonlinear 2", "MODEL /EVALUA /REINFO /INTERF" )
    setAnalysisCommandDetail( "Analysis1", "Structural nonlinear 2", "MODEL /EVALUA /REINFO /INTERF", True )
    addAnalysisCommandDetail( "Analysis1", "Structural nonlinear", "EXECUT(1) /PHYSIC" )
    setAnalysisCommandDetail( "Analysis1", "Structural nonlinear", "EXECUT(1) /PHYSIC /BOND", True )
    setAnalysisCommandDetail( "Analysis1", "Structural nonlinear", "EXECUT(1) /PHYSIC /LIQUEF", False )
    saveProject( )
    removeAnalysisCommandDetail( "Analysis1", "Structural nonlinear 2", "EXECUT(1)" )
    setActivePhase( "Analysis1", "Phased 2" )
    setAnalysisCommandDetail( "Analysis1", "Structural nonlinear 2", "EXECUT /EXETYP", "START" )
    renameAnalysisCommandDetail( "Analysis1", "Structural nonlinear 2", "EXECUT(1)", "teninphase" )
    addAnalysisCommandDetail( "Analysis1", "Structural nonlinear 2", "EXECUT(1) /START /INITIA /STRESS" )
    setAnalysisCommandDetail( "Analysis1", "Structural nonlinear 2", "EXECUT(1) /START /INITIA /STRESS", True )
    setAnalysisCommandDetail( "Analysis1", "Structural nonlinear 2", "EXECUT(1) /START /INITIA /STRESS /INPUT /LOAD", 3 )
    setAnalysisCommandDetail( "Analysis1", "Structural nonlinear 2", "EXECUT(1) /ITERAT /MAXITE", 20 )
    setAnalysisCommandDetail( "Analysis1", "Structural nonlinear 2", "EXECUT(1) /ITERAT /CONVER /DISPLA /NOCONV", "CONTIN" )
    setAnalysisCommandDetail( "Analysis1", "Structural nonlinear 2", "EXECUT(1) /ITERAT /CONVER /FORCE /NOCONV", "CONTIN" )
    saveProject( )
    runSolver( "Analysis1" )
    showView( "RESULT" )
    setResultPlot( "contours", "Total Displacements /node", "TDtZ" )
    setAnalysisCommandDetail( "Analysis1", "Structural nonlinear 2", "EXECUT(1) /START /TIME", 31536000 )
```

```
runSolver( "Analysis1" )
showView( "RESULT" )
setResultPlot( "contours", "Total Displacements /node", "TDtZ" )
setResultCase( [ "Analysis1", "Output", "Phased, Load – step 2, Load – factor 1.0000" ] )
setResultCase( [ "Analysis1", "Output", "Phased, Time – step 3, Time 0.24192E + 07" ] )
setResultCase( [ "Analysis1", "Output", "Phased, Time – step 4, Time 0.15767E + 08" ] )
setResultCase( [ "Analysis1", "Output", "Phased, Time – step 3, Time 0.24192E + 07" ] )
setResultCase( [ "Analysis1", "Output", "Phased, Load – step 2, Load – factor 1.0000" ] )
setResultCase( [ "Analysis1", "Output", "Phased, Start – step 1, Load – factor 1.0000" ] )
setResultCase( [ "Analysis1", "Output", "Phased, Load – step 2, Load – factor 1.0000" ] )
setResultCase( [ "Analysis1", "Output", "Phased, Time – step 3, Time 0.24192E + 07" ] )
setResultCase( [ "Analysis1", "Output", "Phased, Time – step 4, Time 0.15767E + 08" ] )
setResultCase( [ "Analysis1", "Output", "Phased, Time – step 5, Time 0.31535E + 08" ] )
setResultCase( [ "Analysis1", "Output", "Phased 2, Start – step 1, Load – factor 1.0000" ] )
setResultCase( [ "Analysis1", "Output", "Phased, Time – step 5, Time 0.31535E + 08" ] )
setResultCase( [ "Analysis1", "Output", "Phased, Time – step 4, Time 0.15767E + 08" ] )
setResultCase( [ "Analysis1", "Output", "Phased, Time – step 3, Time 0.24192E + 07" ] )
setResultCase( [ "Analysis1", "Output", "Phased, Load – step 2, Load – factor 1.0000" ] )
setResultCase( [ "Analysis1", "Output", "Phased, Start – step 1, Load – factor 1.0000" ] )
setResultCase( [ "Analysis1", "Output", "Phased, Load – step 2, Load – factor 1.0000" ] )
setResultCase( [ "Analysis1", "Output", "Phased, Time – step 3, Time 0.24192E + 07" ] )
```

```
    setResultCase( [ "Analysis1", "Output", "Phased, Time – step 4, Time 0.15767E +
08" ] )
    setResultCase( [ "Analysis1", "Output", "Phased, Time – step 5, Time 0.31535E +
08" ] )
    setResultCase( [ "Analysis1", "Output", "Phased 2, Start – step 1, Load – factor
1.0000" ] )
    setResultCase( [ "Analysis1", "Output", "Phased, Time – step 5, Time 0.31535E +
08" ] )
    setResultCase( [ "Analysis1", "Output", "Phased, Time – step 4, Time 0.15767E +
08" ] )
    setResultCase( [ "Analysis1", "Output", "Phased, Time – step 3, Time 0.24192E +
07" ] )
    setResultCase( [ "Analysis1", "Output", "Phased, Load – step 2, Load – factor
1.0000" ] )
    setResultCase( [ "Analysis1", "Output", "Phased, Start – step 1, Load – factor
1.0000" ] )
    setAnalysisCommandDetail( "Analysis1", "Structural nonlinear 2", "EXECUT(1) /
START /LOAD /PREVIO", False )
    runSolver( "Analysis1" )
    showView( "RESULT" )
    setResultPlot( "contours", "Total Displacements /node", "TDtZ" )
    setResultCase( [ "Analysis1", "Output", "Phased, Load – step 2, Load – factor
1.0000" ] )
    setResultCase( [ "Analysis1", "Output", "Phased, Time – step 3, Time 0.24192E +
07" ] )
    setResultCase( [ "Analysis1", "Output", "Phased, Load – step 2, Load – factor
1.0000" ] )
    setResultCase( [ "Analysis1", "Output", "Phased, Time – step 3, Time 0.24192E +
07" ] )
    setResultCase( [ "Analysis1", "Output", "Phased, Time – step 4, Time 0.15767E +
08" ] )
    setResultCase( [ "Analysis1", "Output", "Phased, Time – step 5, Time 0.31535E +
08" ] )
    setResultCase( [ "Analysis1", "Output", "Phased 2, Start – step 1, Load – factor
1.0000" ] )
    showIds( "NODE", [ 102 ] )
    showIds( "NODE", [ 102, 101 ] )
    showIds( "NODE", [ 102 ] )
    setResultCase( [ "Analysis1", "Output", "Phased, Time – step 5, Time 0.31535E +
```

08"])

setResultCase(["Analysis1", "Output", "Phased, Time-step 4, Time 0.15767E+08"])

setResultCase(["Analysis1", "Output", "Phased, Time-step 3, Time 0.24192E+07"])

setResultCase(["Analysis1", "Output", "Phased, Load-step 2, Load-factor 1.0000"])

setResultCase(["Analysis1", "Output", "Phased, Start-step 1, Load-factor 1.0000"])

setResultCase(["Analysis1", "Output", "Phased, Load-step 2, Load-factor 1.0000"])

setResultCase(["Analysis1", "Output", "Phased, Time-step 3, Time 0.24192E+07"])

setResultCase(["Analysis1", "Output", "Phased, Time-step 4, Time 0.15767E+08"])

setResultCase(["Analysis1", "Output", "Phased, Time-step 5, Time 0.31535E+08"])

setResultCase(["Analysis1", "Output", "Phased 2, Start-step 1, Load-factor 1.0000"])

setResultCase(["Analysis1", "Output", "Phased, Time-step 5, Time 0.31535E+08"])

setResultCase(["Analysis1", "Output", "Phased 2, Start-step 1, Load-factor 1.0000"])

setAnalysisCommandDetail("Analysis1", "Structural nonlinear", "OUTPUT(1)/SELTYP", "USER")

addAnalysisCommandDetail("Analysis1", "Structural nonlinear", "OUTPUT(1)/USER")

addAnalysisCommandDetail("Analysis1", "Structural nonlinear", "OUTPUT(1)/USER/DISPLA(1)/TOTAL/TRANSL/GLOBAL")

addAnalysisCommandDetail("Analysis1", "Structural nonlinear", "OUTPUT(1)/USER/FORCE(1)/REACTI/TRANSL/GLOBAL")

addAnalysisCommandDetail("Analysis1", "Structural nonlinear", "OUTPUT(1)/USER/STRESS(1)/TOTAL/CAUCHY/GLOBAL")

removeAnalysisCommandDetail("Analysis1", "Structural nonlinear", "OUTPUT(1)/USER/STRESS(1)")

addAnalysisCommandDetail("Analysis1", "Structural nonlinear", "OUTPUT(1)/USER/STRESS(1)/TOTAL/CAUCHY/CRKIND")

addAnalysisCommandDetail("Analysis1", "Structural nonlinear", "OUTPUT(1)/USER/STRAIN(1)/TOTAL/GREEN/GLOBAL")

saveProject( )
setAnalysisCommandDetail("Analysis1", "Structural nonlinear 2", "OUTPUT(1) / SELTYP", "USER")
addAnalysisCommandDetail("Analysis1", "Structural nonlinear 2", "OUTPUT(1) / USER")
addAnalysisCommandDetail("Analysis1", "Structural nonlinear 2", "OUTPUT(1) / USER /DISPLA(1) /TOTAL /TRANSL /GLOBAL")
addAnalysisCommandDetail("Analysis1", "Structural nonlinear 2", "OUTPUT(1) / USER /FORCE(1) /REACTI /TRANSL /GLOBAL")
addAnalysisCommandDetail("Analysis1", "Structural nonlinear 2", "OUTPUT(1) / USER /STRESS(1) /TOTAL /CAUCHY /CRKIND")
addAnalysisCommandDetail("Analysis1", "Structural nonlinear 2", "OUTPUT(1) / USER /STRAIN(1) /TOTAL /GREEN /GLOBAL")
saveProject( )
runSolver("Analysis1")
showView("RESULT")
setResultPlot("contours", "Crack Indices /node", "Icr")
setResultLayer(2)
setResultLayer(3)
setResultLayer(1)
setResultCase(["Analysis1", "Output", "Phased, Load-step 2, Load-factor 1.0000"])
setResultCase(["Analysis1", "Output", "Phased, Time-step 3, Time 0.24192E+07"])
setResultCase(["Analysis1", "Output", "Phased, Time-step 4, Time 0.15767E+08"])
setResultCase(["Analysis1", "Output", "Phased, Time-step 5, Time 0.31535E+08"])
setResultCase(["Analysis1", "Output", "Phased 2, Start-step 1, Load-factor 1.0000"])
saveProject( )

回缩长度 0.001 m、单元网格尺寸 0.05 m 时，Python 语言命令流供参考（仅贴出该部分操作之后的命令流，其余部分与上述相同）。

setParameter("GEOMETRYLOAD", "tenin", "POSTEN /ONEEND /RETLE1", 0.001)
generateMesh([])
hideView("GEOM")
showView("MESH")
runSolver("Analysis1")
showView("RESULT")

```
showView( "GEOM" )
hideView( "GEOM" )
showView( "GEOM" )
show( "SHAPE", [ "Sheet 2" ] )
show( "SHAPE", [ "tenin" ] )
show( "SHAPE", [ "teninphase" ] )
show( "SHAPE", [ "Sheet 1" ] )
show( "GEOMETRYCONNECTION", [ "int" ] )
setElementSize( [ "Sheet 1", "Sheet 2" ], 0.05, -1, True )
setMesherType( [ "Sheet 1", "Sheet 2" ], "HEXQUAD" )
setMidSideNodeLocation( [ "Sheet 1", "Sheet 2" ], "LINEAR" )
setElementSize( "Sheet 1", 1, [[ 0, 0.8, 0.15 ]], 0.05, 0, True )
saveProject( )
generateMesh( [] )
hideView( "GEOM" )
showView( "MESH" )
runSolver( "Analysis1" )
showView( "RESULT" )
setResultPlot( "contours", "Total Displacements /node", "TDtZ" )
setResultCase( [ "Analysis1", "Output", "Phased, Load-step 2, Load-factor 1.0000" ] )
setResultCase( [ "Analysis1", "Output", "Phased, Time-step 3, Time 0.24192E+07" ] )
setResultCase( [ "Analysis1", "Output", "Phased, Time-step 4, Time 0.15767E+08" ] )
setResultCase( [ "Analysis1", "Output", "Phased, Time-step 5, Time 0.31535E+08" ] )
setResultCase( [ "Analysis1", "Output", "Phased 2, Start-step 1, Load-factor 1.0000" ] )
setResultCase( [ "Analysis1", "Output", "Phased, Time-step 5, Time 0.31535E+08" ] )
setResultCase( [ "Analysis1", "Output", "Phased, Time-step 4, Time 0.15767E+08" ] )
setResultCase( [ "Analysis1", "Output", "Phased, Time-step 3, Time 0.24192E+07" ] )
setResultCase( [ "Analysis1", "Output", "Phased, Load-step 2, Load-factor 1.0000" ] )
setResultCase( [ "Analysis1", "Output", "Phased, Start-step 1, Load-factor 1.0000" ] )
```

    setResultCase(["Analysis1", "Output", "Phased, Load‐step 2, Load‐factor 1.0000"])
    setResultCase(["Analysis1", "Output", "Phased, Start‐step 1, Load‐factor 1.0000"])
    setResultPlot("contours", "Reinforcement Cross‐section Forces /node", "Nx")
    setResultCase(["Analysis1", "Output", "Phased, Load‐step 2, Load‐factor 1.0000"])
    setResultCase(["Analysis1", "Output", "Phased, Time‐step 3, Time 0.24192E+07"])
    setResultCase(["Analysis1", "Output", "Phased, Time‐step 4, Time 0.15767E+08"])
    setResultCase(["Analysis1", "Output", "Phased, Time‐step 5, Time 0.31535E+08"])
    setResultCase(["Analysis1", "Output", "Phased 2, Start‐step 1, Load‐factor 1.0000"])
    setResultCase(["Analysis1", "Output", "Phased, Time‐step 5, Time 0.31535E+08"])

5.3 案例三：预制节段拼装梁随机场分析案例

使用 Diana10.1 软件不仅可以针对钢筋混凝土结构的确定状态进行传统的土木工程中受弯、受剪、长期性能、水化热和施工阶段分析等非线性分析，同时还可以运用随机场理论针对结构不确定性状态进行一定程度的研究。而在实际情况中，结构往往由于材料参数（例如混凝土抗压强度、混凝土弹性模量）、构件加工尺寸和荷载大小等方面离散性而存在着不确定性。这些不确定性对结构在承载力极限状态下的极限承载力、正常使用极限状态下的结构挠度和结构可靠度等方面均具有较大的影响。Diana10.1 软件仅适用于基于混凝土材料本身不确定性状态下的随机场分析，而对于外界因素的不确定性尚无法充分考虑。目前 Diana10.1 软件中可用的随机场模型为 JCSS 概率模型。本节以预制节段拼装梁承受短期集中荷载作为载体，对 Diana10.1 软件中基于随机场 JCSS 概率模型在预测跨中挠度、裂缝位置、裂缝宽度的应用进行简单的介绍。并且采用嵌入式、桁架杆弹性粘结滑移、梁单元杆件弹性粘结滑移三种方式显示上述结果在 JCSS 概率模型下的随机场预测结果。

案例介绍：本案例为全长 12.8 m 的两键齿简支预制节段拼装梁，由两个节段组成，单个节段长度为 6.4 m，节段拼装梁高度为 2.4 m，键齿突出部分长度为 0.24 m，采用 3D 普通曲壳单元建模（Regular Curved Shell elements），混凝土强度等级为 C45，采用软件自带的基于 JCSS 概率模型建立随机场有限元数值模型。在整个梁三等分点处承受对称集中荷载，初始加载状态下施加的基准荷载大小为 100kN，在后续的软件荷载步中设置逐级加载直至出现明显裂缝时停止。预应力筋为一根 1×7 直径 15.24 mm 的对称折线筋，端部最大高度距底部 1.5 m，最低高度 0.6 m，弹性模量为 $1.95×10^{11} N/m^2$，屈服强度 1860 MPa。预应力筋分别采用嵌入式 Bar 钢筋单元、桁架杆件粘结滑移和梁单元杆件粘结滑移类型进行 JCSS 概率模型下随机场的数值模拟。初始加载状态下具体尺寸如图 5.3-1 所示。

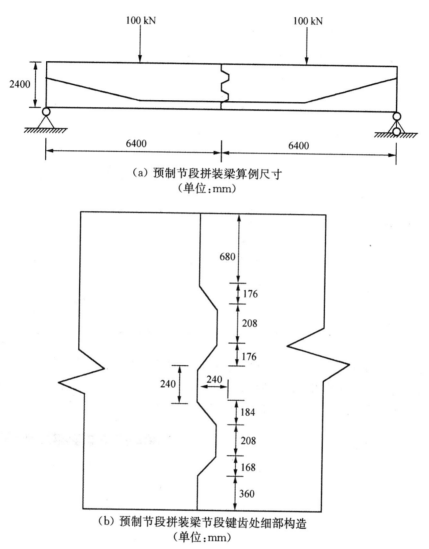

(a) 预制节段拼装梁算例尺寸
(单位：mm)

(b) 预制节段拼装梁节段键齿处细部构造
(单位：mm)

图 5.3-1

注：作为 **Diana10.1 算例例题**，本书例题分析条件均为假定值，关于分析结果妥当性另当别论，望谅解。

学习要义：

(1) 学习采用曲壳单元建立预制节段拼装梁两键齿几何模型。

(2) 学习基于 JCSS 概率模型建立预应力筋为嵌入式和粘结滑移等不同状态下的预制节段拼装梁随机场数值模型，并且根据随机场计算结果预测裂缝分布和裂缝宽度。

(3) 复习建立界面单元中库仑摩擦的本构类型。

(4) 学习预应力筋采用梁单元杆件下的粘结滑移本构设置方式。

首先，打开 DianaIE，弹出 New project 对话框，在原先 G 盘中创立的名称为例题的文件夹内创立 PSB-Random 文件，选择分析类型为结构，3D 建模，模型最大尺寸为 1000 m，网格单元形状为六面体/四边形形状,默认阶数为二阶,单元中间节点确定方式为线性插值。如图 5.3-2 所示。

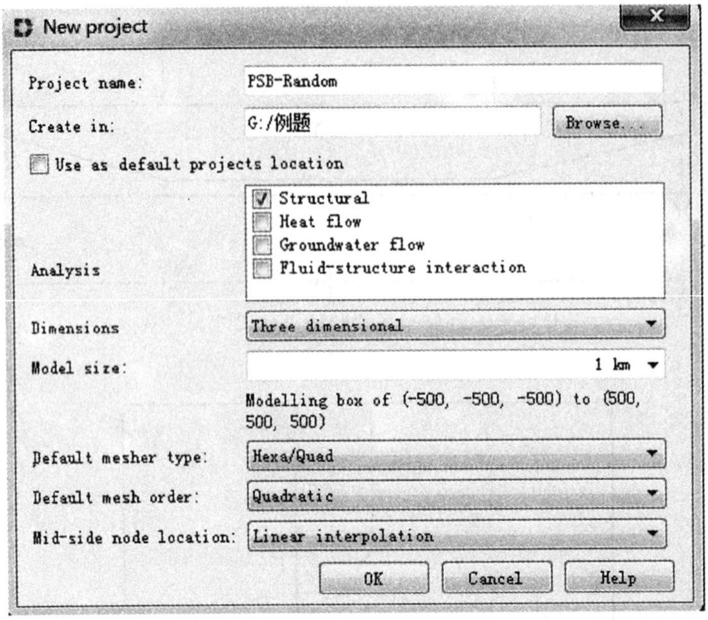

图 5.3-2 New Project 界面

点击 Add a sheet 创立左边部分两键齿节段的几何模型,依次输入各点坐标。点击 OK,生成左边节段的两键齿几何模型 Sheet 1,采用同样的操作方式依次输入右边节段的两键齿几何模型 Sheet 2,分别如图 5.3-3 和图 5.3-4 所示。

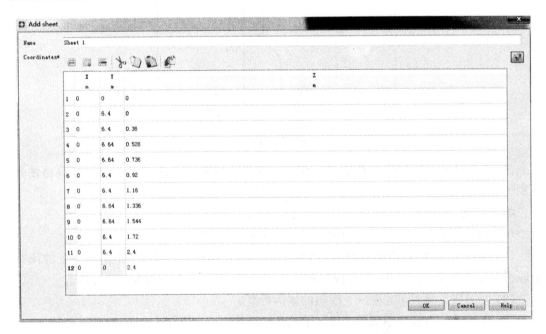

图 5.3-3 左边部分两键齿节段各点几何坐标

	X m	Y m	Z m
1	0	6.4	0
2	0	12.8	0
3	0	12.8	2.4
4	0	6.4	2.4
5	0	6.4	1.72
6	0	6.64	1.544
7	0	6.64	1.336
8	0	6.4	1.16
9	0	6.4	0.92
10	0	6.64	0.736
11	0	6.64	0.528
12	0	6.4	0.36

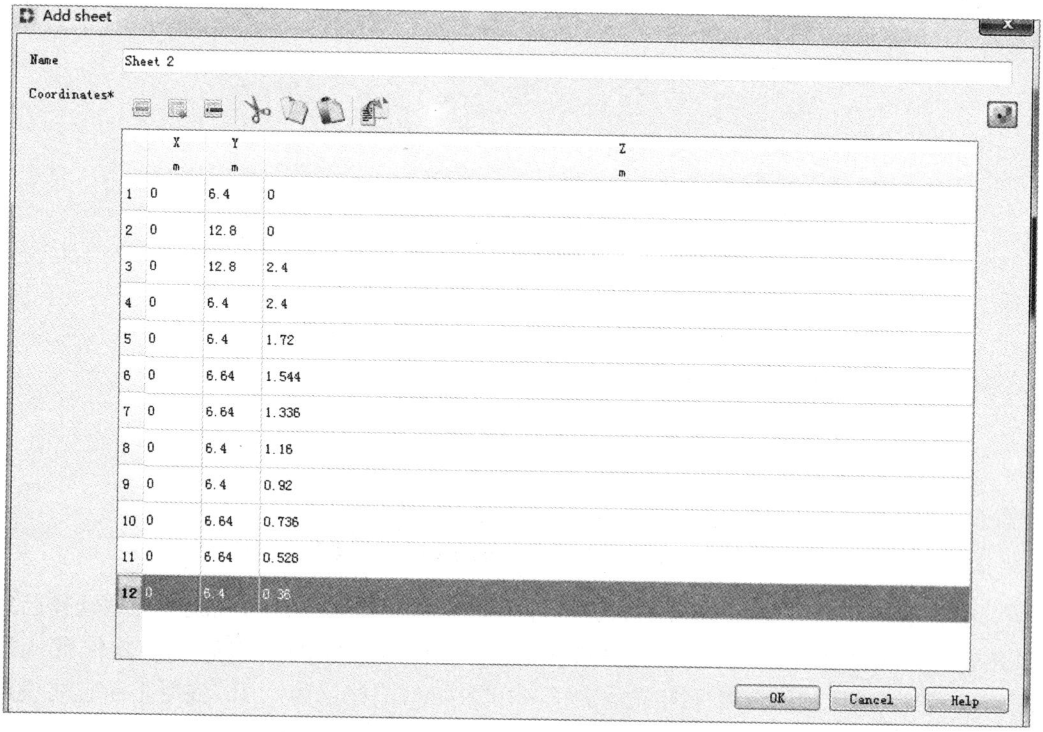

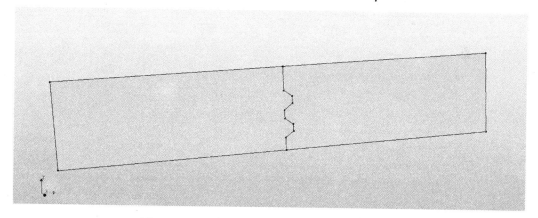

图 5.3－4　右边两键齿节段各点几何坐标

点击 OK，生成如图 5.3－5 所示的几何模型。

图 5.3－5　两节段两键齿预制节段拼装梁几何模型

点击 Add a polyline 快捷图标，创立折线预应力筋，分别输入坐标(0,0,1.5)，(0,4.3,0.6)，(0,8.6,0.6)，(0,12.8,1.5)，将折线命名为 tenin，点击 OK，创立折线预应力筋的几何模型。

定义随机场，鼠标拾取 Sheet 1 和 Sheet 2，右击选择 Property assignments 选项，选择混凝土材料单元类型为常规的曲壳单元(Regular Curved Shell elements)，点击下方添加材料属性图标，弹出材料属性对话框 Add new material，命名为 concrete，材料类型选择为混凝土设计规范(Concrete design codes)，材料模型为 JCSS 概率模型规范(JCSS

Probabilistic model code),在 Aspects to include 中选择 JCSS 随机场(JCSS Random field)。如图 5.3-6 所示。

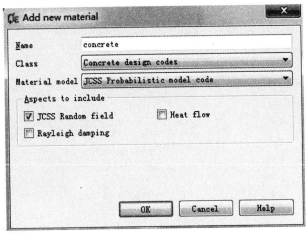

图 5.3-6 JCSS 随机场概率模型界面

点击 OK,弹出 Edit material 材料编辑对话框,如图 5.3-7 所示,混凝土类型选择已掺和(Ready mixed),混凝土强度等级选择 C45,平均抗压强度的平均值选择软件默认的 $5.0376 \times 10^7 \text{N/m}^2$,抗压强度标准差为 $5.17 \times 10^6 \text{N/m}^2$,泊松比为 0.15,混凝土断裂能设定为 150 N/m,密度为 2500 kg/m³。

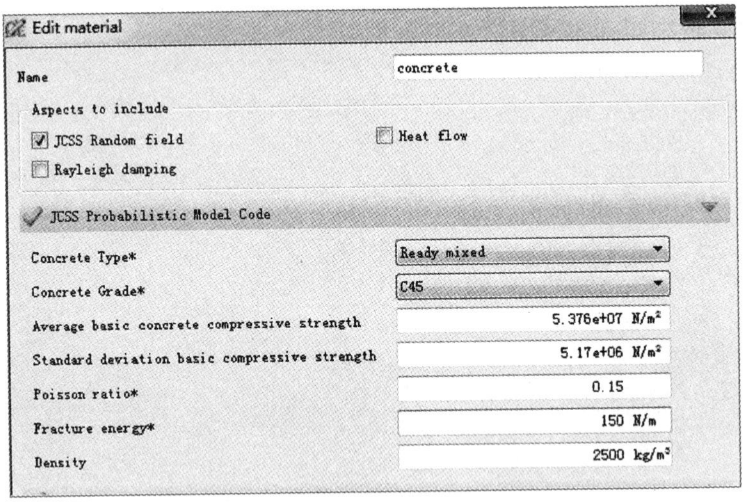

图 5.3-7 JCSS 概率模型材料参数设定

定义 JCSS 随机场,在随机场生成器(Random field generator)一栏中选择协方差矩阵分解法(Covariance matrix decomposition method)。由于模型的长度方向为整体坐标系下的 Y 方向,厚度方向为整体坐标系下的 X 方向。因此设定整体坐标系 X 方向的步数(Number of steps in global X-direction)为 1,Y 方向的步数为 10,Z 方向的步数为 1。矩阵分解方法为 Cholesky modified by Fenton 方法,协方差函数类型(Covariance type)为平方指数(Squared exponential)函数,概率分布为对数正态分布(Log-normal),临界值(Threshold value)选择软件默认值 0.5,相关长度为 5 m。如图 5.3-8 所示。

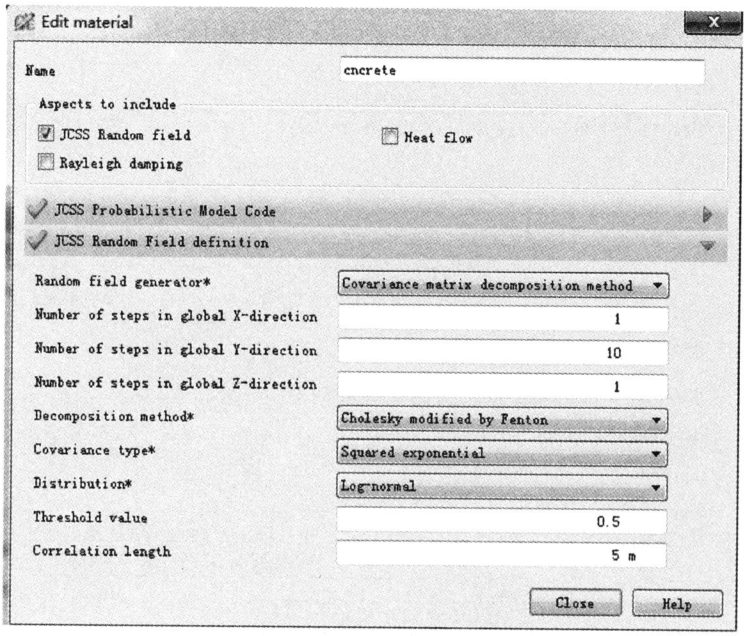

图 5.3-8 JSCC 随机场定义

定义混凝土材料的壳单元几何属性,点击图标 ⇌ ,添加截面几何属性。命名为 concrete,厚度设定为 0.8 m。局部坐标系下的 x 轴对应整体坐标系的 Y 轴方向,因此 Element x aixs 选择 0,1,0。如图 5.3-9 所示。截面几何特性赋予完毕后,点击 OK,完成混凝土材料基于 JCSS 概率模型下的随机场定义。

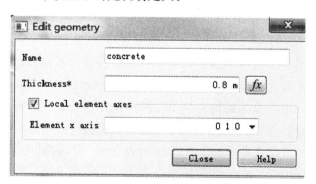

图 5.3-9 壳单元几何属性定义

定义折线预应力筋材料和几何属性,与之前章节部分操作类型相同,选择配筋和桩基类型(Reinforcements and pile foundations)下的线弹性本构类型,赋予预应力钢绞线材料属性,输入弹性模量为 $1.95 \times 10^{11} \mathrm{~N/m^2}$,屈服强度为 1860 MPa,截面面积为 $1.39 \times 10^{-4} \mathrm{~m^2}$,这里不再赘述。

创立点 Vertex 1 和 Vertex 2,坐标分别为 (0,4.3,3) 和 (0,8.6,3)。先选中 Vertex 1,点击投影印刻快捷图标 ,选中 Sheet 1 上方所在的边,将点 Vertex 1 投影印刻在这条边上,采用同样操作方式将 Vertex 2 投影印刻到右边节段对应边上。Vertex 2 投影印刻操作如图 5.3-10所示。

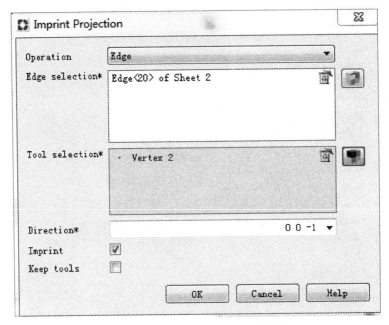

图 5.3-10　Vertex 2 投影印刻操作

点击 Load 栏下方图标 Define a global load，命名为 gravity，添加重力荷载。再点击 Load 栏下方图标 Add a new load case，创立荷载工况 2，命名为 load，施加对称作用的集中荷载，将集中荷载 100 kN 施加到 Vertex 1 和 Vertex 2 投影印刻的点上。再施加预应力荷载，命名为 postte，预应力荷载采用两端张拉（Both ends）的方式，荷载大小为 120 kN，两端的预应力筋回缩长度均设定为 1×10^{-4} m，库仑摩擦系数为 0.22，握裹系数 0.01/m。施加预应力荷载的界面操作过程与之前相同，这里不再赘述。

创立界面单元，选择节段之间键齿接缝部分作为界面单元，点击快捷图标，弹出界面单元编辑对话框，命名为 int，连接类型选择界面单元（Interface elements），界面单元形状选择边（Edge），界面单元类型为结构壳单元的界面单元（Structural Shell Interface）。

为了较为精确地模拟预制节段拼装梁节段之间键齿接缝面在短期集中荷载作用下的力学行为特性，界面材料本构选择库仑摩擦（Coulomb friction），如图 5.3-11 所示。

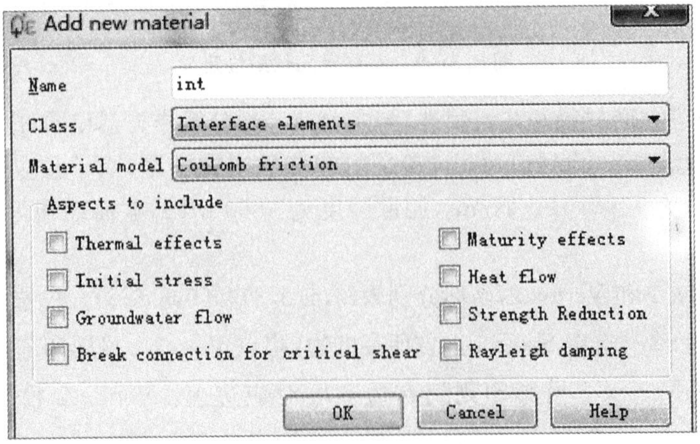

图 5.3-11　库仑摩擦本构类型定义界面

点击 OK,弹出编辑材料属性对话框。在对话框线弹性材料属性(Linear material properties)一栏选择界面单元类型为 3D line interface between shells,定义法向压缩刚度 (Normal stiffness modulus - y)为 3.65e16 N/m³,两个切向的抗剪刚度均为 3.65e12 N/m³, 3.65e12 N/m³。如图 5.3-12 所示。

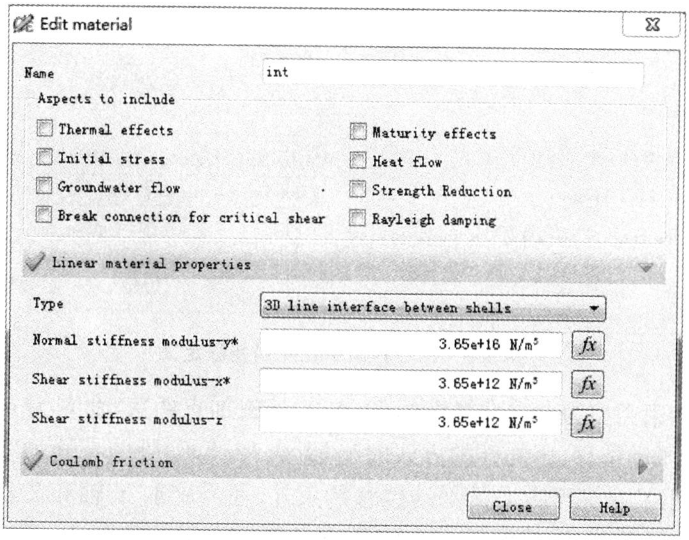

图 5.3-12 界面单元库仑摩擦各向材料刚度定义

修改参考系 Refercence System 栏下方 Units 单位制,将温度单位改为摄氏度 (celsius),将角度单位改为度(degree),粘滞系数设定为 0,摩擦角设定为 30°,膨胀角为 25°, 界面单元张开模式为不张开(No opening)。如图 5.3-13 所示。

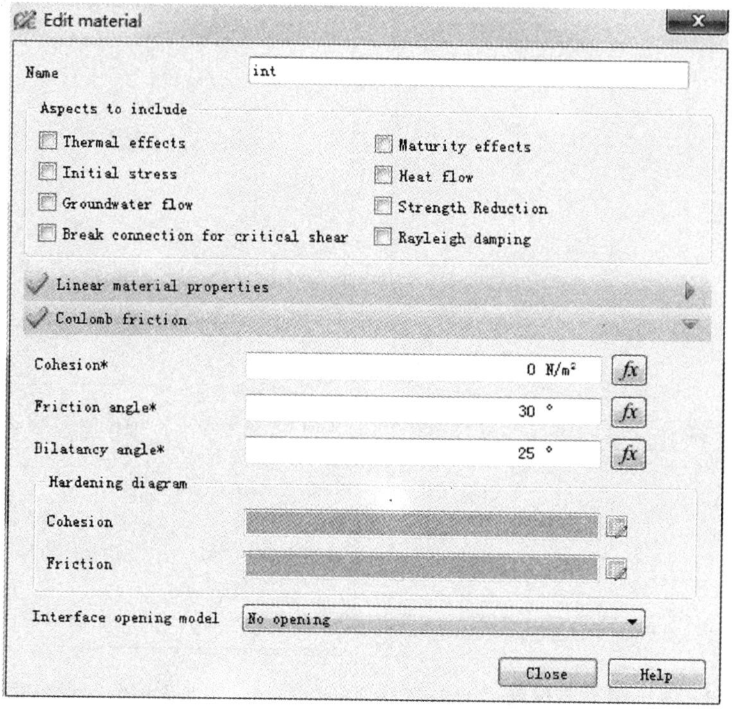

图 5.3-13 库仑摩擦角和膨胀角的定义

定义界面单元的截面几何特性,厚度为 0.8 m,界面单元的单元方向为平行于壳平面方向(Parallel to shell plane),平行于壳平面的单元方向向量对应整体坐标系下的 Y 轴负方向(0,-1,0)。如图 5.3-14 所示。

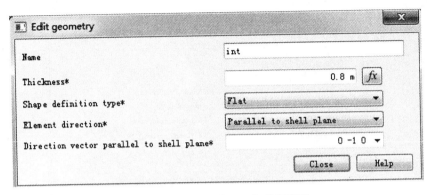

图 5.3-14 界面单元几何特征定义

施加约束,约束类型为点。选择坐标(0,0,0)处施加点约束,命名为 co1,选择约束 X,Y,Z 三个方向的平动位移,用同样的方式对坐标(0,12.8,0)处的点施加约束,约束名称为 co2。除了约束 X,Y,Z 三个方向的平动位移约束外,还需施加 Y 轴和 Z 轴的转动约束 R2 和 R3,co1 和 co2 的约束信息如图 5.3-15 和图 5.3-16 所示。

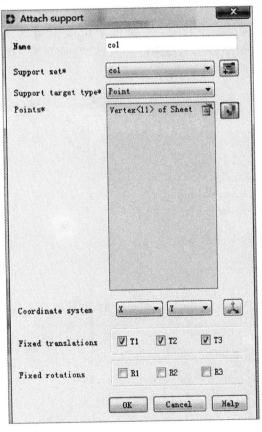 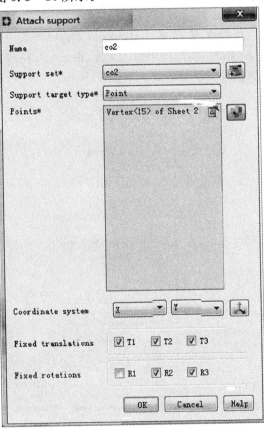

图 5.3-15 co1 约束信息　　　　　图 5.3-16 co2 约束信息

设立荷载组合。将重力荷载 gravity 和预应力荷载 postte 添加到同一组荷载工况 1,作为后面的初始应力添加。设定集中荷载 load 为荷载工况 2。如图 5.3 - 17 所示。

图 5.3 - 17　荷载工况分组

点击快捷键图标 Set mesh properties of a shape ◉ 按钮,选择 Sheet 1 和 Sheet 2,设定混凝土曲壳单元网格尺寸为 0.4 m,网格类型为六面体/四边形。重复之前的选择,其中曲壳网格单元中点位置确定方式为线性插值。线对线界面单元尺寸为 0.4 m,网格划分分别如图 5.3 - 18 和图 5.3 - 19 所示。

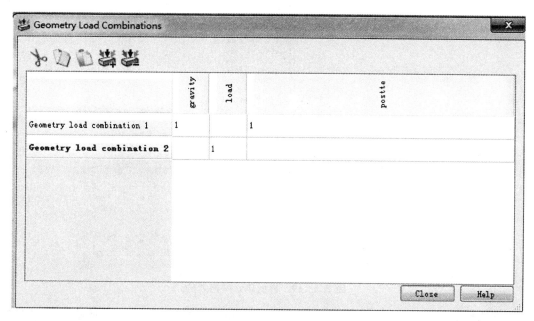

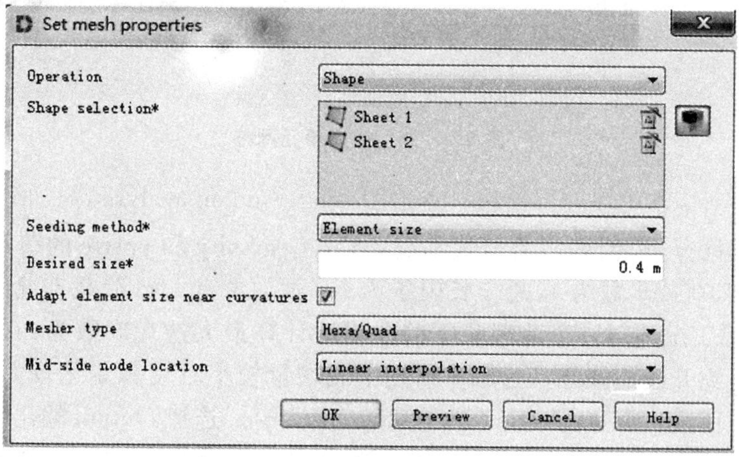

图 5.3 - 18　混凝土曲壳单元网格划分

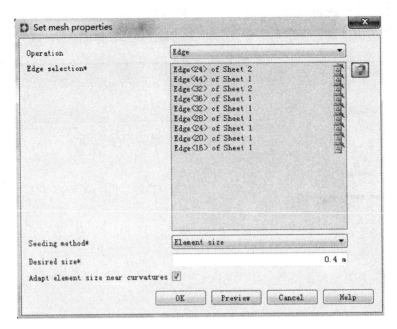

图 5.3-19　界面单元网格划分

点击快捷键图标 Generate mesh of a shape 按钮,生成如图 5.3-20 所示的单元网格。其中节段间键齿处红线部分为界面单元。

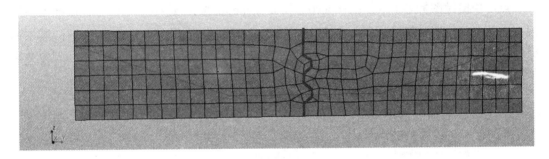

图 5.3-20　生成单元网格

建立 Analysis 非线性分析模块。点击图标 Add an analysis,添加分析工况,右击选择结构非线性分析,创立初始荷载工况,将含有 gravity 和 postte 的荷载组合 1 作为初始荷载工况,荷载步数为 1,总步长因子为 1。添加含有 load 荷载工况所在的荷载组合 2,所有荷载工况的每一个荷载步下的非线性计算最大迭代次数为 50 次,考虑到在荷载作用下直到开裂的非线性过程中后期荷载增量较小时位移变化较大,因此设定荷载工况 2 中的荷载步设置为 14.0000 0.200000(10)。选择 Output 输出结果中的整体坐标系下的各向位移、开裂信息、整体坐标系下的裂缝宽度和主应力方向裂缝应变等信息作为 Result 中显示结果。各项输出结果如图 5.3-21 所示。

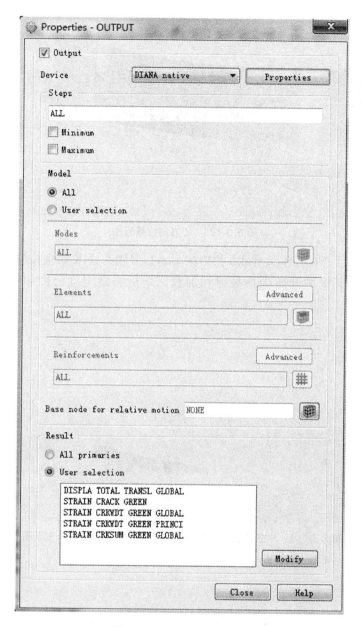

图 5.3-21 输出结果选择

点击 Run analysis,进行非线性计算。待计算结果完成后,点击 Result 选项,查看随机场模型计算完毕后预测的各方向跨中挠度。其中加载完毕后 Z 方向的随机场预测位移云图如图 5.3-22 所示。

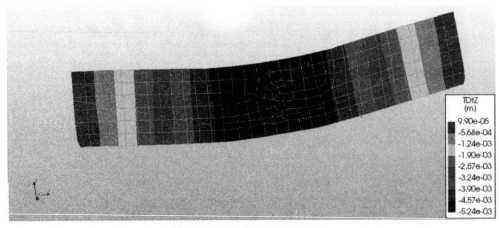

图 5.3-22　Z 方向位移云图

点击 Results 下方的 Crack-width 栏中的 Ecw1，分别查看 X 向主拉应力方向预测裂缝分布和宽度云图，其中刚开始出现裂缝和加载完毕后裂缝宽度云图如图 5.3-23 和图 5.3-24 所示。

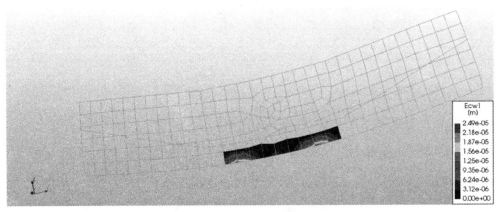

图 5.3-23　X 主拉应力方向裂缝刚出现时位置和宽度分布云图

图 5.3-24　加载完毕时 x 主拉应力方向裂缝位置和宽度分布云图

点击 Element results 下方 Crack-width 栏中 EcwYY 和 EcwZZ，查看加载完毕后采用 JCSS 概率模型预测的 Y 方向和 Z 方向的裂缝宽度分布云图如图 5.3-25 和图 5.3-26 所示。以 Y 向裂缝宽度为例,可以看出 JCSS 概率模型预测右边节段部分受拉区最下方 Y 向裂缝宽度最大。

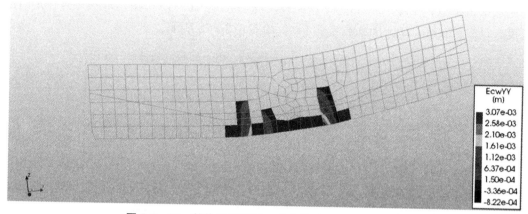

图 5.3-25　整体坐标系下 Y 方向的裂缝宽度分布云图

图 5.3-26　Z 方向的裂缝宽度分布云图

点击 Element results 下方 Crack Strains 一栏中 Eknn,查看拉伸方向的裂缝分布位置预测云图,如图 5.3-27 所示。由图中可以看出,基于 JCSS 随机场概率模型预测结果表明,在节段拼装梁梁体跨中受拉区开裂比较明显,并且开裂较为明显的位置距离键齿根部有一段距离。

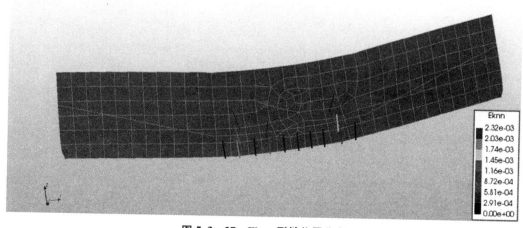

图 5.3-27　Eknn 裂缝位置分布

将数值模型中的预应力筋的本构模型改为桁架杆粘结滑移,法向刚度为 $1\times10^{16}\,\text{N/m}^2$,切向刚度设置为 $1\times10^{9}\,\text{N/m}^2$,在钢筋粘结滑移界面单元破坏模型中选择 Doerr 立方粘结滑移函数(Cubic bond-slip function by Doerr),滑移参数为 $20\,\text{N/m}^2$,剪切滑移开始稳定时的长度(Shear slip at start plateau)为 0.1 m。如图 5.3-28 所示。

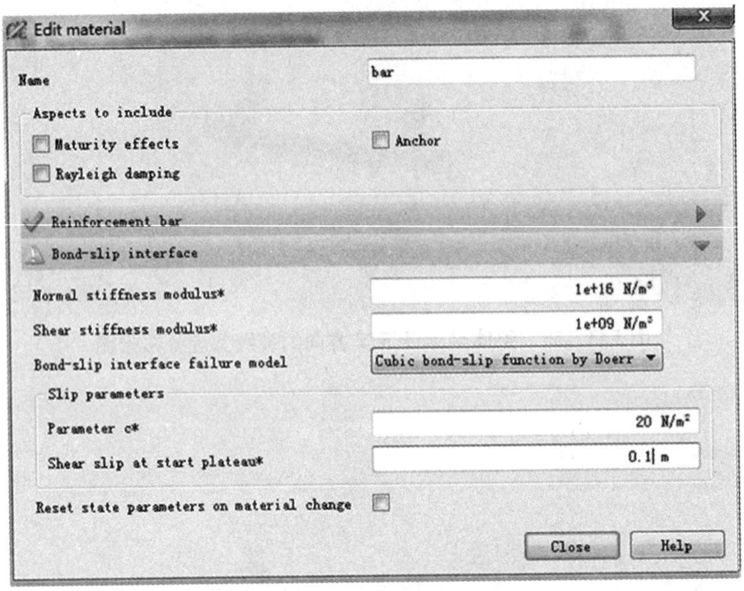

图 5.3-28 预应力筋选用粘结滑移材料模型

在几何特性参数中选择 Truss bondslip,截面面积为 139 mm²,周长 95.76 mm。其他部分参数不变,点击 Generate mesh of a shape 快捷按钮重新划分网格,将最大迭代步数设置为 50 次,重新开始计算,查看 Result 下方的 EcwYY 显示 Y 方向刚开裂时和加载完毕后的裂缝宽度云图,分别如图 5.3-29 和图 5.3-30 所示。

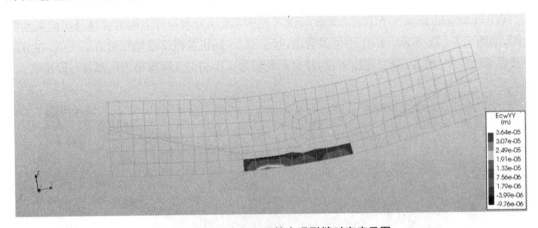

图 5.3-29 Y 方向一开始出现裂缝时宽度云图

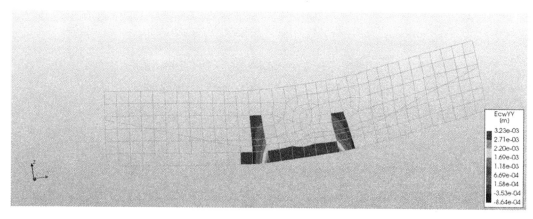

图 5.3-30　加载完毕后 Y 方向裂缝宽度云图

再查看预制节段拼装梁在初始状态竖向 Z 方向的位移反拱云图和加载完毕后竖向位移云图,分别如图 5.3-31 和图 5.3-32 所示。

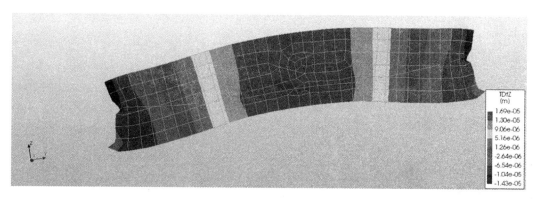

图 5.3-31　Truss bondslip 粘结滑移模型下初始反拱状态下 Z 向位移云图

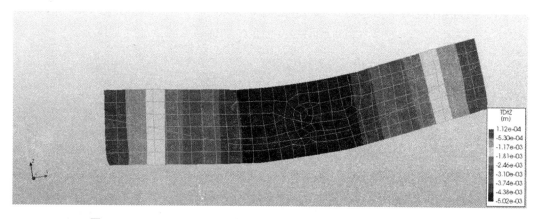

图 5.3-32　Truss bondslip 粘结滑移模型下加载完毕后的 Z 向位移云图

点击 Output 栏下方的 Reinforcement Result 中 Reinforcement Cross-section Force,查看钢筋拉力云图,如图 5.3-33 和图 5.3-34 所示。由拉力云图可得,当加载完毕后在键齿两侧对称位置处出现最大预应力值。且最大预应力值的位置不断靠近键齿部分。

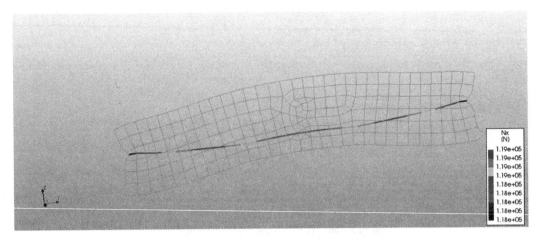

图 5.3-33 Truss bondslip 粘结滑移模型初始状态下拉力云图

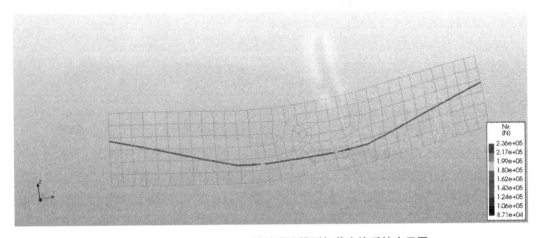

图 5.3-34 Truss bondslip 粘结滑移模型加载完毕后拉力云图

在几何截面特性中将 Truss bondslip 改为梁单元杆件 Circular beam bondslip，其中横截面直径为 15.24 mm，梁单元的 z 轴对应的是整体坐标系下的 X 轴。如图 5.3-35 所示。

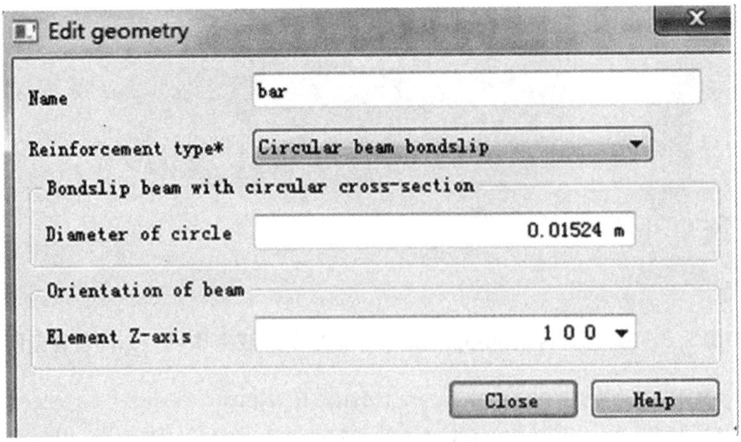

图 5.3-35 Circular beam bondslip 截面几何参数定义

重新划分网格并且计算完毕后,点击 Output 栏下方的 TDtZ 查看在梁单元杆件 Circular beam bondslip 粘结滑移模型下的预制节段拼装梁初始状态和加载完毕后竖向位移云图,分别如图 5.3-36 和图 5.3-37 所示。

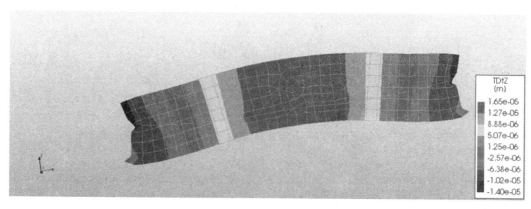

图 5.3-36　Circular beam bondslip 粘结滑移模型下竖向初始反拱云图

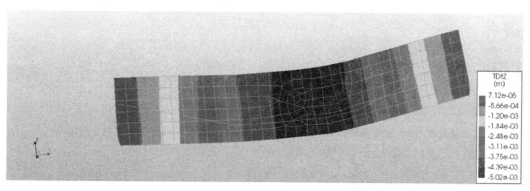

图 5.3-37　Circular beam bondslip 粘结滑移模型下加载完毕竖向位移云图

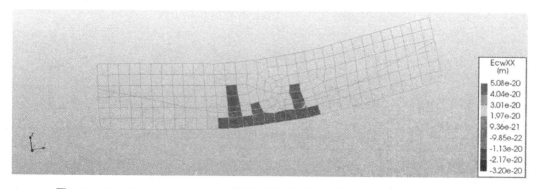

图 5.3-38　Circular beam bondslip 粘结滑移模型下加载完毕后 X 向的裂缝宽度云图

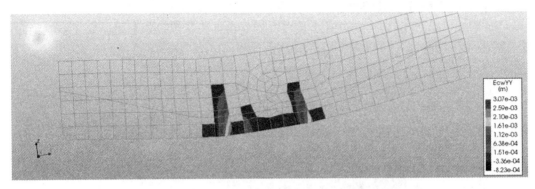

图 5.3-39　Circular beam bondslip 粘结滑移模型下 Y 方向的裂缝宽度云图

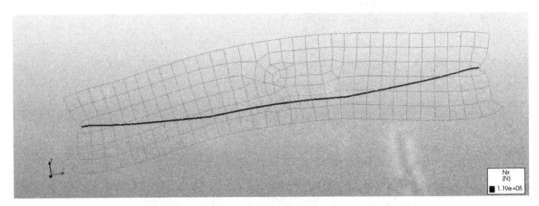

图 5.3-40　梁单元杆件 Circular beam bondslip 粘结滑移模型下初始拉力云图

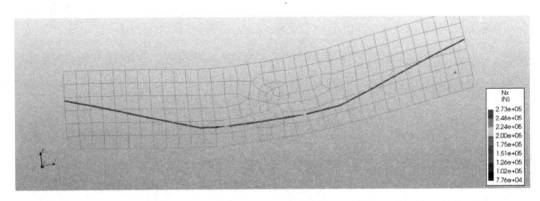

图 5.3-41　梁单元杆件 Circular beam bondslip 粘结滑移模型下加载完毕后拉力云图

经验总结：

(1) 由上述对比可以发现，经过非线性计算后，当预应力筋分别采用嵌入式塑性模型、Truss bondslip 粘结滑移桁架杆模型和 Circular beam bondslip 粘结滑移梁单元杆件状态下的竖向位移值差别不大，但采用梁单元杆件进行粘结滑移模拟时，初始状态下沿着折线预应力筋各处拉力值相同，随着荷载的不断加大，在左右节段均有预应力数值最大点出现，且对称分布。这与图 5.3-34 中显示桁架杆粘结滑移模式下最大预应力值仅在右边节段内出现有所差别。三者 Y 向的裂缝宽度和裂缝位置分布也稍有差别。

(2) 经过作者的多次重复计算发现,由于材料本身具有的离散型和随机场本身就是基于不确定性概率模型计算,这种概率模型的计算结果可能与整体坐标系下各坐标轴在随机场生成器中的步数大小(Number of steps)有关,因此在 Diana10.1 软件中每一次计算结果同上一次比较都会有一定的偏差。以裂缝出现的位置而言,既有可能在左侧节段中出现,也有可能在右侧节段中出现,还可能在左右节段同时出现。此外,以跨中竖向位移(Z方向)而言,每次随机场模型位移云图结果也有些许的差别。而本案例仅限于展示如何建立 Diana10.1 中的随机场模型,鉴于案例中计算次数过少,对于使用该软件进行相关领域研究的用户而言,则需要在进行一定数量的抽样和计算后综合汇总出较为精确的数值模拟预测结果。

Python 语言命令流如下(以嵌入式为例)。

```
newProject( "PSB-Random", 1000 )
setModelAnalysisAspects( [ "STRUCT" ] )
setModelDimension( "3D" )
setDefaultMeshOrder( "QUADRATIC" )
setDefaultMesherType( "HEXQUAD" )
setDefaultMidSideNodeLocation( "LINEAR" )
saveProject(  )
createSheet( "Sheet 1", [[ 0, 0, 0 ],[ 0, 6.4, 0 ],[ 0, 6.4, 0.36 ],[ 0, 6.64, 0.528 ],[ 0, 6.64, 0.736 ],[ 0, 6.4, 0.92 ],[ 0, 6.4, 1.16 ],[ 0, 6.64, 1.336 ],[ 0, 6.64, 1.544 ],[ 0, 6.4, 1.72 ],[ 0, 6.4, 2.4 ],[ 0, 0, 2.4 ]] )
saveProject(  )
createSheet( "Sheet 2", [[ 0, 6.4, 0 ],[ 0, 12.8, 0 ],[ 0, 12.8, 2.4 ],[ 0, 6.4, 2.4 ],[ 0, 6.4, 1.72 ],[ 0, 6.64, 1.544 ],[ 0, 6.64, 1.336 ],[ 0, 6.4, 1.16 ],[ 0, 6.4, 0.92 ],[ 0, 6.64, 0.736 ],[ 0, 6.64, 0.528 ],[ 0, 6.4, 0.36 ]] )
createPolyline( "tenin", [[ 0, 0, 1.5 ],[ 0, 4.3, 0.6 ],[ 0, 8.6, 0.6 ],[ 0, 12.8, 1.5 ]], False )
saveProject(  )
addMaterial( "cncrete", "CONCDC", "JCSSPR", [ "JCSSRF" ] )
setParameter( "MATERIAL", "cncrete", "JCSSMC/JCSSG1/JCSSGR", "C45" )
setParameter( "MATERIAL", "cncrete", "JCSSMC/DENSIT", 2500 )
setParameter( "MATERIAL", "cncrete", "JCSSMC/GF1", 150 )
setParameter( "MATERIAL", "cncrete", "JCSSMC/GF1", 150 )
setParameter( "MATERIAL", "cncrete", "JCSSMC/GF1", 150 )
removeParameter( "MATERIAL", "cncrete", "JCSSMC/DENSIT" )
setParameter( "MATERIAL", "cncrete", "JCSSRF/COVARI/NX", 1 )
setParameter( "MATERIAL", "cncrete", "JCSSRF/COVARI/NY", 10 )
setParameter( "MATERIAL", "cncrete", "JCSSRF/COVARI/DECMTH", "FENTON" )
setParameter( "MATERIAL", "cncrete", "JCSSMC/DENSIT", 2500 )
addGeometry( "Element geometry 1", "SHEET", "CURSHL", [] )
```

```
rename( "GEOMET", "Element geometry 1", "concrete" )
setParameter( "GEOMET", "concrete", "THICK", 0.8 )
setParameter( "GEOMET", "concrete", "LOCAXS", True )
setParameter( "GEOMET", "concrete", "LOCAXS /XAXIS", [ 1, 0, 0 ] )
setParameter( "GEOMET", "concrete", "LOCAXS /XAXIS", [ 0, 1, 0 ] )
setParameter( "GEOMET", "concrete", "LOCAXS /XAXIS", [ 0, 1, 0 ] )
clearReinforcementAspects( [ "Sheet 1", "Sheet 2" ] )
setElementClassType( "SHAPE", [ "Sheet 1", "Sheet 2" ], "CURSHL" )
assignMaterial( "cncrete", "SHAPE", [ "Sheet 1", "Sheet 2" ] )
assignGeometry( "concrete", "SHAPE", [ "Sheet 1", "Sheet 2" ] )
resetElementData( "SHAPE", [ "Sheet 1", "Sheet 2" ] )
saveProject( )
addMaterial( "bar", "REINFO", "LINEAR", [] )
setParameter( "MATERIAL", "bar", "LINEAR /ELASTI /YOUNG", 1.95e + 11 )
addGeometry( "Element geometry 2", "RELINE", "REBAR", [] )
rename( "GEOMET", "Element geometry 2", "bar" )
setParameter( "GEOMET", "bar", "REIEMB /CROSSE", 0.000139 )
setReinforcementAspects( [ "tenin" ] )
assignMaterial( "bar", "SHAPE", [ "tenin" ] )
assignGeometry( "bar", "SHAPE", [ "tenin" ] )
resetElementData( "SHAPE", [ "tenin" ] )
setReinforcementDiscretization( [ "tenin" ], "SECTION" )
saveProject( )
createVertex( "Vertex 1", [ 0, 4.3, 3 ] )
saveProject( )
createVertex( "Vertex 2", [ 0, 8.6, 0 ] )
removeShape( [ "Vertex 2" ] )
createVertex( "Vertex 2", [ 0, 8.6, 3 ] )
saveProject( )
projection( SHAPEEDGE, "Sheet 1", [[ 0, 3.2, 2.4 ]], [ "Vertex 1" ], [ 0, 0, - 1 ], True )
removeShape( [ "Vertex 1" ] )
projection( SHAPEEDGE, "Sheet 2", [[ 0, 9.6, 2.4 ]], [ "Vertex 2" ], [ 0, 0, - 1 ], True )
removeShape( [ "Vertex 2" ] )
saveProject( )
addSet( "GEOMETRYLOADSET", "gravity" )
createModelLoad( "gravity", "gravity" )
addSet( "GEOMETRYLOADSET", "Geometry load case 2" )
```

```
rename( "GEOMETRYLOADSET", "Geometry load case 2", "load" )
createPointLoad( "load", "load" )
setParameter( "GEOMETRYLOAD", "load", "FORCE /VALUE", -100000 )
setParameter( "GEOMETRYLOAD", "load", "FORCE /DIRECT", 3 )
attach( "GEOMETRYLOAD", "load", "Sheet 1", [[ 0, 4.3, 2.4 ]] )
attach( "GEOMETRYLOAD", "load", "Sheet 2", [[ 0, 8.6, 2.4 ]] )
saveProject( )
addSet( "GEOMETRYLOADSET", "Geometry load case 3" )
rename( "GEOMETRYLOADSET", "Geometry load case 3", "postte" )
createBodyLoad( "postte", "postte" )
setParameter( "GEOMETRYLOAD", "postte", "LODTYP", "POSTEN" )
setParameter( "GEOMETRYLOAD", "postte", "POSTEN /BOTHEN /FORCE1", 120000 )
setParameter( "GEOMETRYLOAD", "postte", "POSTEN /BOTHEN /FORCE2", 120000 )
setParameter( "GEOMETRYLOAD", "postte", "POSTEN /BOTHEN /RETLE1", 0.0001 )
setParameter( "GEOMETRYLOAD", "postte", "POSTEN /BOTHEN /RETLE2", 0.0001 )
setParameter( "GEOMETRYLOAD", "postte", "POSTEN /SHEAR", 0.22 )
setParameter( "GEOMETRYLOAD", "postte", "POSTEN /WOBBLE", 0.01 )
attach( "GEOMETRYLOAD", "postte", [ "tenin" ] )
attachTo( "GEOMETRYLOAD", "postte", "POSTEN /BOTHEN /PNTS1", "tenin", [[ 0, 0, 1.5 ]] )
attachTo( "GEOMETRYLOAD", "postte", "POSTEN /BOTHEN /PNTS2", "tenin", [[ 0, 12.8, 1.5 ]] )
saveProject( )
addMaterial( "int", "INTERF", "FRICTI", [] )
setParameter( "MATERIAL", "int", "LINEAR /IFTYP", "LIN3D" )
setParameter( "MATERIAL", "int", "LINEAR /ELAS4 /DSNY", 3.65e+16 )
setParameter( "MATERIAL", "int", "LINEAR /ELAS4 /DSSX", 3.65e+12 )
setParameter( "MATERIAL", "int", "LINEAR /ELAS4 /DSSZ", 3.65e+12 )
setParameter( "MATERIAL", "int", "COULOM /COHESI", 0 )
setUnit( "ANGLE", "DEGREE" )
setUnit( "TEMPER", "CELSIU" )
setParameter( "MATERIAL", "int", "COULOM /COHESI", 0 )
setParameter( "MATERIAL", "int", "COULOM /PHI", 30 )
setParameter( "MATERIAL", "int", "COULOM /PSI", 25 )
setParameter( "MATERIAL", "int", "COULOM /PSI", 25 )
addGeometry( "Element geometry 3", "LINE", "SHLLIF", [] )
rename( "GEOMET", "Element geometry 3", "int" )
setParameter( "GEOMET", "int", "THICK", 0.8 )
setParameter( "GEOMET", "int", "THKDIR", "PARALL" )
```

```
setParameter( "GEOMET", "int", "YAXIS", [ 0, -1, 0 ] )
createLineConnection( "int" )
setParameter( "GEOMETRYCONNECTION", "int", "CONTYP", "INTER" )
setParameter( "GEOMETRYCONNECTION", "int", "MODE", "AUTO" )
attachTo( "GEOMETRYCONNECTION", "int", "SOURCE", "Sheet 1", [[ 0, 6.52,
1.632 ],[ 0, 6.64, 1.44 ],[ 0, 6.52, 1.248 ],[ 0, 6.4, 1.04 ],[ 0, 6.52, 0.828 ],[ 0,
6.64, 0.632 ],[ 0, 6.4, 0.18 ]] )
attachTo( "GEOMETRYCONNECTION", "int", "SOURCE", "Sheet 2", [[ 0, 6.4, 2.06 ],
[ 0, 6.52, 0.444 ]] )
setElementClassType( "GEOMETRYCONNECTION", "int", "SHLLIF" )
assignMaterial( "int", "GEOMETRYCONNECTION", "int" )
assignGeometry( "int", "GEOMETRYCONNECTION", "int" )
resetElementData( "GEOMETRYCONNECTION", "int" )
saveProject(  )
addSet( "GEOMETRYSUPPORTSET", "co1" )
createPointSupport( "co1", "co1" )
setParameter( "GEOMETRYSUPPORT", "co1", "AXES", [ 1, 2 ] )
setParameter( "GEOMETRYSUPPORT", "co1", "TRANSL", [ 1, 1, 1 ] )
setParameter( "GEOMETRYSUPPORT", "co1", "ROTATI", [ 0, 0, 0 ] )
attach( "GEOMETRYSUPPORT", "co1", "Sheet 1", [[ 0, 0, 0 ]] )
addSet( "GEOMETRYSUPPORTSET", "Geometry support set 2" )
rename( "GEOMETRYSUPPORTSET", "Geometry support set 2", "co2" )
createPointSupport( "co2", "co2" )
setParameter( "GEOMETRYSUPPORT", "co2", "AXES", [ 1, 2 ] )
setParameter( "GEOMETRYSUPPORT", "co2", "TRANSL", [ 1, 1, 1 ] )
setParameter( "GEOMETRYSUPPORT", "co2", "ROTATI", [ 0, 1, 1 ] )
attach( "GEOMETRYSUPPORT", "co2", "Sheet 2", [[ 0, 12.8, 0 ]] )
setDefaultGeometryLoadCombinations(  )
setGeometryLoadCombinationFactor( "Geometry load combination 1", "gravity",
1 )
remove( "GEOMETRYLOADCOMBINATION", "Geometry load combination 1" )
remove( "GEOMETRYLOADCOMBINATION", "Geometry load combination 2" )
remove( "GEOMETRYLOADCOMBINATION", "Geometry load combination 3" )
addGeometryLoadCombination( "" )
setGeometryLoadCombinationFactor( "Geometry load combination 1", "gravity", 1 )
setGeometryLoadCombinationFactor( "Geometry load combination 1", "postte", 1 )
addGeometryLoadCombination( "" )
setGeometryLoadCombinationFactor( "Geometry load combination 2", "load", 1 )
saveProject(  )
```

```
setElementSize( [ "Sheet 1", "Sheet 2" ], 0.4, -1, True )
setMesherType( [ "Sheet 1", "Sheet 2" ], "HEXQUAD" )
setMidSideNodeLocation( [ "Sheet 1", "Sheet 2" ], "LINEAR" )
saveProject( )
setElementSize( "Sheet 1", 1, [[ 0, 6.52, 1.632 ],[ 0, 6.52, 1.248 ],[ 0, 6.4,
1.04 ],[ 0, 6.52, 0.828 ],[ 0, 6.64, 0.632 ],[ 0, 6.52, 0.444 ],[ 0, 6.4, 0.18 ]],
0.4, 0, True )
setElementSize( "Sheet 2", 1, [[ 0, 6.4, 2.06 ],[ 0, 6.64, 1.44 ]], 0.4, 0, True )
saveProject( )
generateMesh( [] )
hideView( "GEOM" )
showView( "MESH" )
addAnalysis( "Analysis5" )
addAnalysisCommand( "Analysis5", "NONLIN", "Structural nonlinear" )
renameAnalysis( "Analysis5", "Analysis5" )
removeAnalysisCommandDetail( "Analysis5", "Structural nonlinear", "EXECUT
(1)" )
setAnalysisCommandDetail ( "Analysis5", "Structural nonlinear", "EXECUT/
EXETYP", "START" )
renameAnalysisCommand ( "Analysis5", "Structural nonlinear", "Structural
nonlinear" )
renameAnalysisCommandDetail( "Analysis5", "Structural nonlinear", "EXECUT
(1)", "tenin" )
addAnalysisCommandDetail( "Analysis5", "Structural nonlinear", "EXECUT(1)/
START/INITIA/STRESS" )
setAnalysisCommandDetail( "Analysis5", "Structural nonlinear", "EXECUT(1)/
START/INITIA/STRESS", True )
setAnalysisCommandDetail( "Analysis5", "Structural nonlinear", "EXECUT(1)/
START/LOAD/PREVIO", False )
setAnalysisCommandDetail ( "Analysis5", "Structural nonlinear", "EXECUT/
EXETYP", "LOAD" )
renameAnalysisCommandDetail( "Analysis5", "Structural nonlinear", "EXECUT
(2)", "load" )
setAnalysisCommandDetail( "Analysis5", "Structural nonlinear", "EXECUT(2)/
LOAD", False )
setAnalysisCommandDetail( "Analysis5", "Structural nonlinear", "EXECUT(2)/
LOAD", True )
setAnalysisCommandDetail( "Analysis5", "Structural nonlinear", "EXECUT(2)/
LOAD/LOADNR", 2 )
```

```
    setAnalysisCommandDetail( "Analysis5", "Structural nonlinear", "EXECUT(2) /
LOAD /STEPS /EXPLIC /SIZES", "14.0000 0.200000(10)" )
    setAnalysisCommandDetail( "Analysis5", "Structural nonlinear", "EXECUT(2) /
ITERAT /MAXITE", 50 )
    setAnalysisCommandDetail( "Analysis5", "Structural nonlinear", "EXECUT(2) /
ITERAT /CONVER /DISPLA /NOCONV", "CONTIN" )
    setAnalysisCommandDetail( "Analysis5", "Structural nonlinear", "EXECUT(2) /
ITERAT /CONVER /FORCE /NOCONV", "CONTIN" )
    setAnalysisCommandDetail( "Analysis5", "Structural nonlinear", "EXECUT(1) /
ITERAT /MAXITE", 50 )
    setAnalysisCommandDetail( "Analysis5", "Structural nonlinear", "EXECUT(1) /
ITERAT /CONVER /DISPLA /NOCONV", "CONTIN" )
    setAnalysisCommandDetail( "Analysis5", "Structural nonlinear", "EXECUT(1) /
ITERAT /CONVER /FORCE /NOCONV", "CONTIN" )
    setAnalysisCommandDetail( "Analysis5", "Structural nonlinear", "OUTPUT(1) /
SELTYP", "USER" )
    addAnalysisCommandDetail( "Analysis5", "Structural nonlinear", "OUTPUT(1) /
USER" )
    addAnalysisCommandDetail( "Analysis5", "Structural nonlinear", "OUTPUT(1) /
USER /DISPLA(1) /TOTAL /TRANSL /GLOBAL" )
    addAnalysisCommandDetail( "Analysis5", "Structural nonlinear", "OUTPUT(1) /
USER /STRAIN(1) /CRACK /GREEN" )
    addAnalysisCommandDetail( "Analysis5", "Structural nonlinear", "OUTPUT(1) /
USER /STRAIN(2) /CRKWDT /GREEN /GLOBAL" )
    addAnalysisCommandDetail( "Analysis5", "Structural nonlinear", "OUTPUT(1) /
USER /STRAIN(3) /CRKWDT /GREEN /PRINCI" )
    addAnalysisCommandDetail( "Analysis5", "Structural nonlinear", "OUTPUT(1) /
USER /STRAIN(4) /CRKSUM /GREEN /GLOBAL" )
    runSolver( "Analysis5" )
    showView( "RESULT" )
```

第六章　Diana 不足之处和对后续版本建议

作者近两年的软件使用中，深刻体会到无论是 9.4 版还是 10.1 版，Diana 软件均是针对混凝土结构非线性分析一款性能优异的软件。然而，在使用的过程中，也发现了以下几个尚需改进的小瑕疵，在这里列举出来供用户和开发者参考。也希望读者在使用过程中真正了解该软件的利弊，扬长避短。

（1）在混凝土长期性能中宜加入抗冻性和抗渗性方面的研究。并且可在后续升级版中增加纤维增强复合材料(FRP)，根据 FRP 合成材料类型不同力学特性设置时用户可选择诸如 CFRP，AFRP 和 GFRP 等材料本构选项。

（2）欧洲"CEB-FIP 1990"规范、JSCE 规范以及欧洲 2010 规范中混凝土本构材料采用"规范＋收缩徐变＋开裂模型"计算时非线性计算情况比较复杂，收敛性往往较差甚至不收敛，希望 Diana 高版本的升级研发中可以得到改进。

（3）裂缝开展部分，如能在有限元后处理界面中显示哪几个点裂缝宽度最大，哪几个点裂缝宽度最小，那么试验结果与数值模拟的对比就会一目了然。

（4）相比 Diana9.4 版本，10.0 版本后处理部分没有设置较多的 GUI 图形界面操作，查看后处理结果的设置放在了左下角不起眼的部位，增大了用户调整后处理查看结果的难度，尤其是混凝土结构或者预应力筋的应力云图范围跨越过大，且最大云图数量显示只有 20 个。与 9.4 版本相比，10.0 版本虽然在背景颜色上做了一些改进，但没有像 9.4 版本那样在生成应力云图的界面中自动显示单元某个结果的荷载步、最大值和最小值的具体数值，并且图像显示设置(Plot settings)在操作界面和变形图等方面有一些不完善之处。

（5）在前处理的 GUI 界面操作中，菜单栏下方只有 Generate mesh of a shape 这种生成网格划分的快捷键按钮，而在实际建模中，由于计算精度的要求或是非线性收敛性的要求，网格划分的网格数量往往会非常多，划分成功后一旦需要调试或者是修改模型只能手动删除数量庞大的网格，不仅耗时而且非常麻烦，同时还无法显示删除网格后的模型图形，需要重新退出后再打开才可以。建议可以设置一个类似于 Generate mesh of a shape 这种快速取消网格划分的快捷键，以满足复杂模型计算下的用户需求。与此同时，软件在形成网格的过程中有时会出现运行卡顿、迟缓、时间耗费较长、CPU 内存占用过高以及"吃内存"的现象。

（6）Diana10.1 版本中尚没有类似收缩徐变和钢筋松弛的钢筋锈蚀的物理模型，模拟钢筋锈蚀的总体思路依然是减少钢筋面积来等效，与实际情况往往差别较大。建议在后续版本中可以像软件中的松弛模型一样，考虑增加钢筋锈蚀的物理模型或本构模型，以便能够更加真实地模拟钢筋锈蚀造成的结构病害影响。

（7）目前绝大部分有限元软件只能输入一个固定不变的值来模拟温度和湿度的影响，

而对长期监测或者需要对每天不断变化的温度和湿度对混凝土结构长期性能的影响的模拟并不是很精确。建议后续的升级版本中可进一步通过对各规范(如 CEB-FIP)本构中的温湿度与时间关系,设置类似于时间—荷载曲线的方式,添加温度—时间或相对环境湿度—时间曲线完善各类规范下温度和湿度等动态变化环境对结构长期性能影响的模拟。

(8) Diana 软件记忆性有待增强,在反复操作和长时间运行的情况下多次出现了性能不够可靠的现象,也经常会出现关于软件系统自身问题的红色错误提示,尤其是在长时间运行后导入以 .dat 为后缀的模型数据库文件和以 .py 为后缀的命令流文件时不如其他有限元软件顺畅,需要重复打开的 dpf 文件的反应也比较慢甚至会停止运行。这种记忆性的缺陷尤其体现在网格划分后会对划分前已经添加的约束和荷载产生记忆性遗漏,比如计算往往会提示约束没有显示在网格划分结束后的属性参数中,这也是该版本软件设计尚待完善之处。

(9) 使用过程中往往需要调试模型,调试模型的过程中又需要经常修改模型。由于 Python 语言格式具有随时性和随意性,这时每一步修改操作都会被 Python 语言记录,往往 Python 中多种重复语句过多,直接提取的 Python 语言会像涂改后的草稿一样看起来凌乱不堪。建议可以在 Diana 软件中设置一种界面操作完毕之后自动转化成 Python 语言命令流的功能,同时设置从 Python 语言命令流直接转化成 .dpf 文件的功能,完全实现不同后缀和功能的文件之间自如转化,这样就会节约很多用来精简、修改、提取 Python 命令流的时间,提高建模效率。

(10) 模型计算中的二进制中转文件 .dnb 和 .ff 往往是用户不需要关注的中转文件,但往往占用大量的空间。此外,在同一个文件夹内一旦计算生成一个 .dnb 文件,其他名称的 .dpf 文件就无法再打开,使用时不仅浪费时间,也非常不方便。建议升级版的 Diana 软件中可以考虑在应用菜单栏上设置可供软件自动删去 .ff 和 .dnb 二进制中转文件的相关选项,以节约计算时间,进一步提高软件的计算效率。

(11) 当前的 Diana10.1 软件在使用鼠标选中或是捕捉几何体的时候,由于软件操作系统的精度问题,本应为零的坐标点、线或平面往往会出现一个极小的非零值(例如本书 3.5 节命令流中本应为 0 的坐标值却在软件中显示出一个极小的非零值 $6.4817536e-34$),这样极容易给读者造成不必要的误解。建议在 Diana 后续版本中解决这一问题。

作为有限元非线性分析的一款强大的软件,作者在使用这款软件的同时也期待上述的建议和不足之处可以在新版的 Diana 软件中得到进一步的完善。瑕不掩瑜。尽管该软件还存在着上述小问题需要在更高版本的软件中加以完善,但是作者相信,后续不断升级的高版本 Diana 软件会在优异的人机操作性、兼容性以及可靠性上更上一层楼,在非线性领域分析中做到独树一帜,成为越来越多用户和土木工程专家推崇的一款高级结构分析软件。

附录　Diana10.1快捷操作及默认术语

放大:滚轮键向上
缩小:滚轮键向下
旋转:按住鼠标滚轮键将鼠标左右转动
平移:同时按住 Ctrl 键和鼠标滚轮键拖动图形
保存模型及文本文件:Ctrl+S
复制文本文件:Ctrl+C
撤回上一步操作:Ctrl+Z
运行模型:F5
Windows+左方向键:模型调到电脑屏幕左边
Windows+右方向键:模型调到电脑屏幕右边
1,0,0:代表整体坐标系下的 X 方向
0,1,0:代表整体坐标系下的 Y 方向
0,0,1:代表整体坐标系下的 Z 方向

参考文献

[1] 宋守坛. 高速铁路预制拼装箱梁桥抗弯及接缝抗剪试验研究与理论分析[D]. 南京:东南大学,2015.

[2] Shamass, Rabee, X. Zhou, et. al. Finite-Element Analysis of Shear-Off Failure of Keyed Dry Joints in Precast Concrete Segmental Bridges[J]. Journal of Bridge Engineering 20.6(2014):04014084.

[3] DIANA User's Manual-Element Library, Release 9.3. (2008). TNO Building and Construction Research, Holland.

[4] DIANA User's Manual-Material Library, Release 9.3. (2008). TNO Building and Construction Research, Holland.

[5] DIANA User's Manual-Analysis Example, Release 9.3. (2008). TNO Building and Construction Research, Holland.

[6] DIANA-从入门到精通—混凝土非线性分析.

[7] 孙海林,叶列平,丁建彤. 混凝土徐变计算分析方法[C]// 学术讨论会,2004.

[8] 贾明杰. 五分钟帮你认识DIANA中的各类裂缝模型//敦橪. 网络培训会,2017.

[9] Concrete modeling and analysis Tutorials and experiences// DIANA.

[10] 刘铁. 混凝土箱梁桥基于可靠度和全寿命成本的维护策略研究[D]. 南京:东南大学,2013.

[11] 贾明杰. 基于Diana中不同裂缝模型的混凝土梁开裂模拟//敦橪. 网络培训会,2017.

[12] R 克拉夫,J 彭津. 结构动力学[M]. 高等教育出版社,2006.

[13] 贾明杰. 钢筋混凝土剪力墙的滞回性能分析//敦橪网络培训会,2018